Chemical Engineering Methods and Technology

nova science publishers

www.novapublishers.com

Chemical Engineering Methods and Technology

Chromophores with Nonlinear Optical Properties and Their Applications
Maria Marinescu, PhD (Editor)
2023. ISBN: 979-8-88697-525-3 (Softcover)
2023. ISBN: 979-8-88697-726-4 (eBook)

Properties and Uses of Calcium Silicate
Amanda G. Carlton (Editor)
2022. ISBN: 979-8-88697-128-6 (Hardcover)
2022. ISBN: 979-8-88697-184-2 (eBook)

Chemical Vapor Deposition (CVD): Methods and Technologies
Levi Karlsson (Editor)
2021. ISBN: 978-1-53619-949-9 (Hardcover)
2021. ISBN: 978-1-53619-990-1 (eBook)

Fluidized Bed Reactors: Principles and Applications
Kristian Beike (Editor)
2020. ISBN: 978-1-53617-727-5 Softcover)
2020. ISBN: 978-1-53617-733-6 (eBook)

Microfiltration: Principles, Process and Applications
Youssef El Rayess, PhD (Editor)
2019. ISBN: 978-1-53616-320-9 (Hardcover)
2019. ISBN: 978-1-53616-347-6 (eBook)

More information about this series can be found at
https://novapublishers.com/product-category/series/chemical-engineering-methods-and-technology/

Karl Roland
Editor

Triazoles and Their Uses

nova
science publishers
www.novapublishers.com

NOTICE TO THE READER

Library of Congress Cataloging-in-Publication Data

ISBN: 979-8-88697-800-1

Published by Nova Science Publishers, Inc. † New York

Contents

Preface vii

Chapter 1 **A Multicomponent-Cascade Reaction for the Synthesis of Substituted 1,2,3-Triazoles** 1
Qin-Pei Wu, Zhi-Bin Xu and Lin-Lin Zhang

Chapter 2 **Design and Synthesis of New 1,2,3-Triazole Hybrids Possessing Antitubercular and Antimicrobial Activities** 47
Pramod S. Phatak, Adinath D. Badar and Kishan P. Haval

Chapter 3 **The Distribution of Phytosanitary Products in Sediment Cores from the Bizerte Lagoon (Tunisia)** 63
Mouna Necibi and Nadia Mzoughi

Chapter 4 **The Distribution of Phytosanitary Products in Sediments from the Bizerte Lagoon, Tunisia** 83
Mouna Necibi and Nadia Mzoughi

Chapter 5 **Tautomerism in 1,2,3-Triazoles** 103
Fernanda Petzold Pauli, Adriane Antônia Pereira Amaral, Acácio Silva de Souza, Ruan Carlos Busquet Ribeiro, Luana da Silva Magalhães Forezi, Fernando de Carvalho da Silva and Vitor Francisco Ferreira

Chapter 6 **Application of 1,2,3-Triazoles in Biosurfactants**.........153
Thorsten Heidelberg

Chapter 7 **Triazoles as Potential Corrosion Inhibitors: Recent Developments and Future Perspectives**..........169
Sowmya Ramkumar, D. Nalini, D. S. Chauhan and M. A. Quraishi

Chapter 8 **Bioactive 1,2,3-Triazoles: Applications in Agrochemistry**....................................205
Poliana Aparecida Rodrigues Gazolla, Ângela Maria Almeida Lima, Alex Ramos de Aguiar, Samira Soares Santiago, Danilo Aniceto da Silva, Roberta Silva Evangelista, Adilson Vidal Costa and Róbson Ricardo Teixeira

Index 243

Preface

This book focuses on the latest research on triazoles and their applications.

Chapter 1 - Triazoles are the most important heterocyclic compounds with a wide range of applications. Substituted 1,2,3-triazoles show high dipole moments, strong hydrogen bonds, and dipole-dipole interactions. These compounds have been extensively used in the fields of various materials, biology and medicine, like spacers to conjugate with bio-macromolecules and linkers to multi-target drugs and pharmacophores. Substituted 1,2,3-triazoles include 1,4-disubstituted, 1,5-disubstituted, 4,5-disubstituted and 1,4,5-trisubstituted 1,2,3-triazoles. Well-known synthetic methods include copper-catalyzed cycloaddition of azides to alkynes that selectively forms 1,4-disubstituted 1,2,3-triazoles and a ruthenium-catalyzed process that gives rise to 1,5-disubstituted triazole rings through cycloaddition of azides to alkynes. Non-alkyne substrates have also been successfully employed in constructing functionalized triazoles. These non-alkyne substrates include carbonyl compounds like chalcones, arylacetones, deoxybenzoins, allyl ketones, and α-thioketones.

Moreover, sustainable synthetic processes are demanded in modern organic synthesis and manufacturing industry. Multicomponent one-pot reactions are green synthetic methodologies due to reducing operative steps and wastes, decreasing purification procedures, and enhancing synthetic efficiency. In view of the compatibility of sequential reactions, it is not easy to efficiently, regioselectively, and chemoselectively construct cyclo-compounds with multicomponent one-pot reaction. This chapter describes the synthetic methodologies documented for substituted 1,2,3-triazoles with a multicomponent reaction, including one-pot four-component reaction, one-pot three-component reaction; classical and oxidative synthetic processes; terminal alkynes, internal alkynes, non-alkynes, organoazides, sodium azide, and sulfur and selenium powder substrates; regioselectively-substituted, fully-substituted, cyclo, and thio-, seleno-, and amino-functionalized 1,2,3-triazole products.

Chapter 2 - This chapter deals with synthesis of various triazoles and their uses in medicinal therapeutics. The compounds with triazole moiety are one of the significant heterocycles with broad range of biological activities such as antimicrobial, antitubercular, anticancer, etc. In present chapter, the authors have described the triazole derivatives synthesized by their research group possessing potential antitubercular and antimicrobial activities. The antitubercular activity is examined against *Mycobacterium tuberculosis* H37Rv (MTB) and most of them have shown better activity with compared to standard drugs. The cytotoxicity of the active moieties were examined against HEK 293 cell and RAW 264.7 cell line by MTT assay and no toxicity was observed even at small concentration. The antimicrobial activities were examined against both antibacterial *(S. aureus, B. cerus, B. subtilis, E. aerogenes, E. coli, S. typhi, P. aeruginosa, S. boydii, S. abony*) and antifungal (*A. niger, C. albicans, S. cerevisiae*) pathogens and have displayed potent activities.

A molecular docking study was performed against *enoyl-ACP-reductase* (InhA) and *DNA gyrase* to gain an insight into the molecular mechanism. The structure-activity relationship (SAR) is also studied. The pharmacokinetic parameters of moieties were studied and displayed acceptable drug-likeness score.

Chapter 3 - The present study investigates the vertical distribution of Triazole pesticides in two sediment cores collected from the Bizerte Lagoon. The sediment core (SC1) is located at the mouth of the channel connecting the lagoon to the Mediterranean Sea and is characterized by the industrial and urban activities of zones A and B. The sediment core (SC2) is situated in zone D, known for its intensive metallurgy and naval construction activities. The Triazole pesticide concentrations ranged from 20.64 ng g^{-1} found at 15 cm depth to 0.07 ng g^{-1} at 6 cm depth in SC1, and varied between 22.41 ng g^{-1} at 9 cm depth and 2.77 ng g^{-1} at 3 cm depth in SC2. Different profiles were observed, making evident that the levels of these pollutants depend strongly on anthropological discharges from industrial, urban, and farming sources. The temporal trends of the studied compounds were influenced by their different historical usages, different properties, or different degradation conditions in the environment.

Chapter 4 - The contamination of the marine environment by organic pollutants is of great concern. The Bizerte Lagoon is the second largest lagoon in Tunisia and is known for its industrial and fishing activities. A comprehensive assessment of the spatial distribution and partitioning of fifteen triazole pesticides in surface sediments is essential due to their social and economic impact.

Total organic carbon (TOC) content in sediment ranges from 0.05% found in station S3 to 5.22% found in station S5. The triazole pesticides were investigated in surface sediment from fourteen sampling stations in the Bizerte Lagoon. Levels of these pollutants are in the range of 0.22 and 26.61ng g^{-1}.

The fluquinconazole and hexaconazole are the most dominant compounds in the different studied sites. The distribution of triazoles is different, indicating different sources of contamination. The concentrations of triazole in surface water are generally similar to the background levels from the Mediterranean Sea and pose no threat to human and aquatic living species.

Chapter 5 - Triazoles are heterocycles with a privileged structure with important chemical and biological properties. They behave as non-hydrolyzable bioisoters of amides and due to these properties are widely used by Medicinal Chemistry in the search for new substances with relevant therapeutic applications. A large number of drugs are already in the pharmaceutical marketed and among these are 1,2,3- and 1,2,4-triazoles. Additionally, these heterocyclyls are present in a wide range of polymers used as in several types of materials. This chapter will focus on the tautomerism of 1,2,3-triazoles, covering properties such as prototropic tautomerism and structural rearrangements, which are fundamental to understanding the chemistry involved in the main reactions in which these heterocycles are involved.

Chapter 6 - Non-ionic surfactants are indispensable industrial products with wide application spectrum. Environmental concerns and performance advantages are constantly driving the market towards utilization of renewable resources. Particularly carbohydrates have gained significant relevance. Immiscibility of the precursors for surfactant antipodes cause challenges for the synthesis of sugar-based surfactants. One possibility to address these is the use of triazole-linkers, which can be easily created by click chemistry coupling. The flexibility of the coupling approach enables a customized design of the surfactants for specific target applications. Of particular relevance is the tuning of the relative surface areas for the surfactant antipodes, which determines the geometry of surfactant assemblies.

Besides this generic structure-based optimization of surfactant properties, the triazole-based coupling provides unique opportunities for life-science applications, particularly for drug delivery. These cover bio-conjugation of surfactant assemblies, particularly vesicles, for targeted drug delivery, as well as stimulated changes of surfactant assembly geometry, giving rise to potential induced drug delivery. An example, which also may find application in environmental chemistry or diagnostic sensors, are macrocyclic ethers incorporating triazoles as chelating ligands for the coordination of metal cations.

Chapter 7 - Industries are continuously searching for solutions to reduce the losses caused by the massive issue of corrosion, which consumes company earnings. Prevention is the best defense against corrosion. Customized solutions using various corrosion inhibitors are one of the best methods of corrosion prevention. Organic compounds rich in heteroatoms and electron-rich cores, serve as barriers that prevent corrosion by forming an adsorbed layer. Triazoles stand out as a significant five-member nitrogen heterocycle among nitrogen-containing heterocyclic compounds with outstanding uses. Five-membered triazole come in two different structural varieties: 1,2,3-triazole and 1,2,4-triazole. Both 1,2,3- and 1,2,4-triazoles may support a wide range of substituents (electrophiles and nucleophiles) around the core structures due to their structural features, which allows for the development of numerous unique compounds. The excellent corrosion protection action of triazole derivatives can be attributed to the presence of electron-withdrawing and electron-donating substituents that allow an improved protection performance compared to that of the parent triazole molecule. Computational studies have provided evidence that the triazole- based molecules undergo donor-acceptor type interactions with metallic substrates. Triazoles are used in a variety of commercial applications, including physiologically active compounds, agrochemicals, corrosion inhibitors, dyes, and optical brighteners. This paper deals with the review on recent developments on triazole as corrosion inhibitors and their application as corrosion inhibitors and their recent trends in the field of corrosion of different metals in a variety of corrosive environments.

Chapter 8 - In modern agriculture, farmers continuously face a battle to increase quality and productivity of crops. The optimization of agriculture efficiency demands, along with other requirements, the application of chemical crop protection agents (pesticides) to control a variety of pests, such weeds, insects, and fungi. Nowadays, there are several available pesticides that help farmers to keep under control a myriad of pests. However, the use of pesticides typically results in the development of species resistance to them. Moreover, increasing public concern with regard to environmental pollution deriving from agricultural practice strictly requires that pesticides would be endowed with low recalcitrance, and thus may be rapidly mineralized by the soil microflora. Low toxicity to mammals, high specificity, low application rates, and fast soilborne microorganism degradation are thus important characteristics sought in the development of new pesticides. In the search and development of new pesticides, researchers have prepared compounds containing the 1,2,3-triazole functionality and investigated their potential as agrochemicals. This book chapter will focus on the developments that have been reported concerning the bioactivities of 1,2,3-triazole in the agrochemistry field. Compounds presenting phytotoxic, insecticide, and fungicide activities will be covered. Results regarding the fungicide activity of 1,2,3-triazole based compounds from the authors' research group will be described. In addition, the synthetic steps involved in the preparation of the bioactive 1,2,3-triazoles will be discussed.

Chapter 1

A Multicomponent-Cascade Reaction for the Synthesis of Substituted 1,2,3-Triazoles

Qin-Pei Wu[1,*], PhD
Zhi-Bin Xu[1], PhD
and Lin-Lin Zhang[2], PhD
[1]School of Chemistry and Chemical Engineering, Beijing Institute of Technology, China
[2]Pharmaron Beijing Co., Ltd. China

Abstract

Triazoles are the most important heterocyclic compounds with a wide range of applications. Substituted 1,2,3-triazoles show high dipole moments, strong hydrogen bonds, and dipole-dipole interactions. These compounds have been extensively used in the fields of various materials, biology and medicine, like spacers to conjugate with bio-macromolecules and linkers to multi-target drugs and pharmacophores. Substituted 1,2,3-triazoles include 1,4-disubstituted, 1,5-disubstituted, 4,5-disubstituted and 1,4,5-trisubstituted 1,2,3-triazoles. Well-known synthetic methods include copper-catalyzed cycloaddition of azides to alkynes that selectively forms 1,4-disubstituted 1,2,3-triazoles and a ruthenium-catalyzed process that gives rise to 1,5-disubstituted triazole rings through cycloaddition of azides to alkynes. Non-alkyne substrates have also been successfully employed in constructing functionalized triazoles. These non-alkyne substrates include carbonyl compounds like chalcones, arylacetones, deoxybenzoins, allyl ketones, and α-thioketones.

---

* Corresponding Author's Email: qpwu@bit.edu.cn.

In: Triazoles and Their Uses
Editor: Karl Roland
ISBN: 979-8-88697-800-1

Moreover, sustainable synthetic processes are demanded in modern organic synthesis and manufacturing industry. Multicomponent one-pot reactions are green synthetic methodologies due to reducing operative steps and wastes, decreasing purification procedures, and enhancing synthetic efficiency. In view of the compatibility of sequential reactions, it is not easy to efficiently, regioselectively, and chemoselectively construct cyclo-compounds with multicomponent one-pot reaction. This chapter describes the synthetic methodologies documented for substituted 1,2,3-triazoles with a multicomponent reaction, including one-pot four-component reaction, one-pot three-component reaction; classical and oxidative synthetic processes; terminal alkynes, internal alkynes, non-alkynes, organoazides, sodium azide, and sulfur and selenium powder substrates; regioselectively-substituted, fully-substituted, cyclo, and thio-, seleno-, and amino-functionalized 1,2,3-triazole products.

Keywords: triazole, azide, cycloaddition, cascade reaction, multicomponent reaction

Introduction

Readily accessible 1,2,3-triazoles are ubiquitous structural motifs, showing a wide range of applications in the fields of organic synthesis (Akter, Rupa et al. 2022), medicinal chemistry, and material science (Qin, Lam et al. 2010, Juricek, Kouwer et al. 2011, Thirumurugan, Matosiuk et al. 2013). Basing on an unsaturated, five-membered aromatic heterocycle with a 6π-delocalized-electron ring, this heterocycle structure features good biocompatibility and marked stability under hydrolytic, oxidative, and reductive conditions. Substituted 1,2,3-triazoles further show high dipole moments, strong hydrogen bonds and dipole-dipole interactions (Lauria, Delisi et al. 2014, Beutick, Vermeeren et al. 2022). Thus, 1,2,3-triazoles appear to be versatile (Gadzikwa, Farha et al. 2009, Golas and Matyjaszewski 2010, Akter, Rupa et al. 2022), like spacers to conjugate with biomacromolecules (Liebert, Hansch et al. 2006), and linkers to multitarget drugs and pharmacophores (Devender, Gunjan et al. 2016, Jiang, Hao et al. 2019). Broad bioactivities shown by triazoles include anti-cancer, anti-HIV, anti-tuberculosis, antifungus, antibacteria, antivirus, and anti-inflammation (Bock, Speijer et al. 2007, Bozorov, Zhao et al. 2019, Rani, Singh et al. 2020).

For example, 1,2,3-triazole is used as an isostere of carboxylic acid, amide, and ester in many drugs. Compared to that of natural *cyclo*-[Pro-Tyr-

Pro-Val] (IC50 = 1.5 mM), cyclotetrapeptide analogue **I** (Figure 1) shows a 3-fold increase in antityrosinase (IC_{50} = 0.5 mM) (Bock, Speijer et al. 2007). Macrotriazoles (analogous to Migrastatin) **II** are active against the MDA-MB-361 cell line (Gabba, Robakiewicz et al. 2017). Tazobactam **III** containing β-lactam with a 1,2,3-triazole is also antibiotic with MIC values of 1.56, 0.1, and 3.13 mg/L against *Escherichia coli*, *Staphylococcus aureus*, and *Klebsiella pneumoniae*, respectively (Bonandi, Christodoulou et al. 2017). Rufinamide (1-[2,6-difluorobenzyl]-1*H*-1,2,3-triazole-4-carboxamide) **IV**, antiepileptic drug (trade name Inovelon and Xilep), was developed by Novartis Pharmaceutical (Hakimian, Cheng-Hakimian et al. 2007). 1,2,3-Triazole thioether **V** is active toward HIV-1 strain (IC_{50}: 10 nM) (Feng, Zheng et al. 2021). 1,2,3-Triazole-tethered sulfonamide–berberine hybrid compound **VI** shows ahigh antimalarial activity with IC_{50} = 0.142 μM (R = *p*-chlorophenylamino) (Batra, Rajendran et al. 2018). Cefatrizine **VII** shows β-lactam antibiotic activity against *Serratia marcescens* (MIC = 400 μg/mL) (Neu and Fu 1979, Ashok, Reddy et al. 2022) (Vala, Vala et al. 2022).

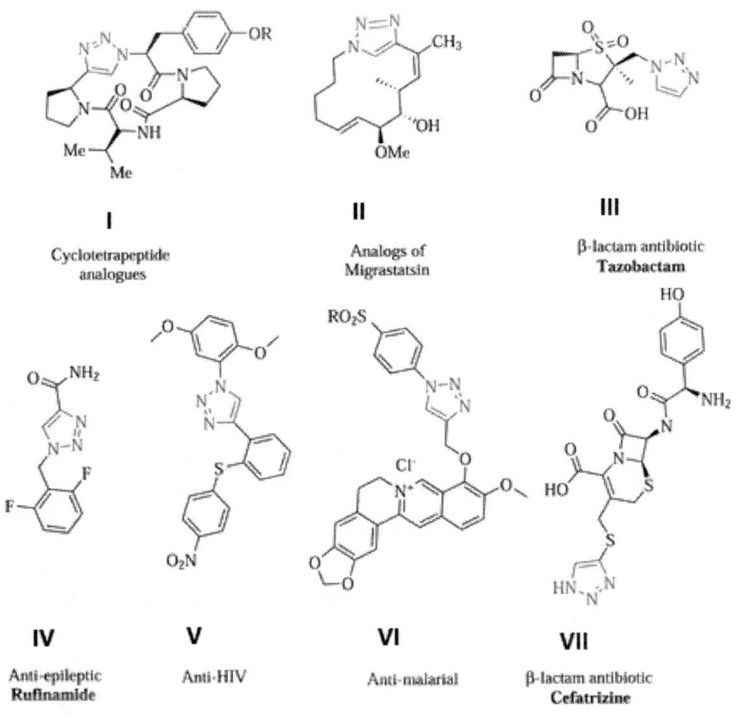

Figure 1. Drugs and bioactive compounds containing 1,2,3-triazole.

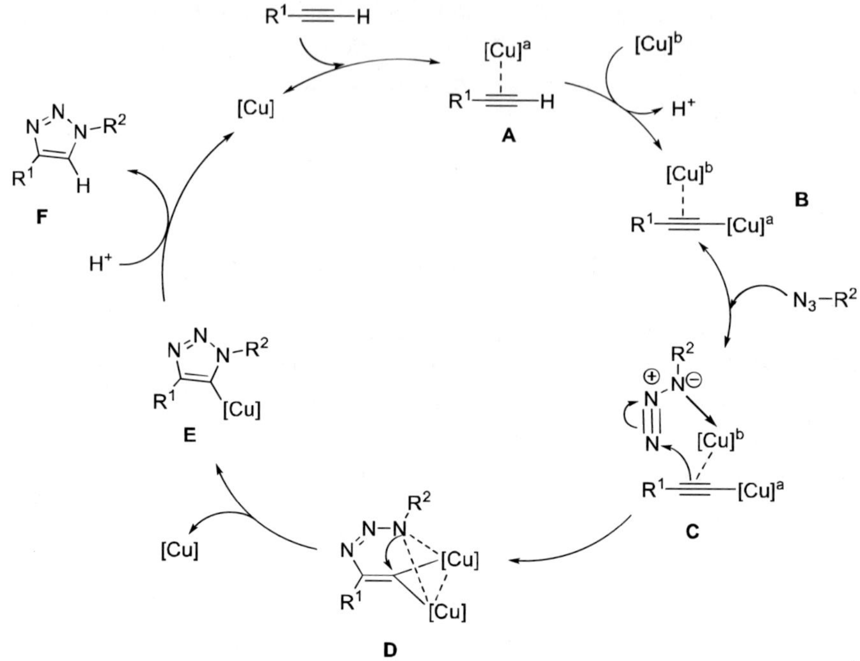

Scheme 1. Catalytic model for copper(I)-catalyzed azide-alkyne cycloaddition reaction.

In 2002, Sharpless (Rostovtsev, Green et al. 2002) and Meldal (Tornoe, Christensen et al. 2002) reported a copper(I)-catalyzed azide–alkyne cycloaddition (CuAAC) reaction forming 1,4-disubstituted 1,2,3-triazoles with high regioselectivities. In 2013, the mechanism for this CuAAC reaction was conclusively proposed by Fokin (Scheme 1) (Worrell, Malik et al. 2013). Alkynes are converted to a complex **B** of copper acetylide with two copper atoms, which coordinates an organic azide to form intermediate **D**. A thermodynamic product of copper(I) triazolide **E** is formed *via* ring closure and aromatization. The 5-copper(I) triazolide **E** is a key reactive intermediate to constructe five-membered 1,2,3-triazole ring. Thus, the interception of the copper intermediates by electrophiles and nucleophiles is extensively and intensively employed for the construction of fully substituted and functionalized 1,2,3-triazoles.

In addition, multicomponent reaction is a one-pot multistep-cascade reaction. multicomponent-cascade reactions in organic synthesis offera lot of advantages over the classical step-by-step approaches (Garbarino, Ravelli et al. 2016, Heravi and Nazari 2017). As no need for intermediate isolation and the reduction in costs and wastes, atom economy and overall yields are improved in the process of multicomponent reactions. Although significant progresses in multicomponent-cascade reactions have been achieved, one-pot multistep approaches are still challenges due to the compatibility of reaction conditions and the catalyst to perform the sequential-dandem reactions (John, Gulati et al. 2021). In this chapter, we highlight the multicomponent procedures for fabrication of the versatile 1,2,3-triazole scaffolds. Although alkyne is the general reagent, some non-alkyne species used as alternatives to alkyne are also available to construct 1,2,3-triazole rings in the multicomponent cascade process. We describe synthetic methodologies reported for 1,2,3-triazoles via multicomponent procedures in two main sections; namely, alkynes used for constructing 1,2,3-triazole rings and non-alkyne used for constructing 1,2,3-triazole rings.

Synthesis of Functionalized 1,2,3-Triazoles from Alkynes

Appukkuttan and co-workers (Appukkuttan, Dehaen et al. 2004) reported a three-component reaction for 1,4-disubstituted 1,2,3-triazoles from sodium azides, organohalides, and terminal alkynes. Under microwave irradiation and copper catalysis, organic halides first react with NaN_3 to generate the corresponding organoazides, which couples in situ with alkynes to form the expected triazole products. This procedure shows perfect regioselectivity for 1,4-disubstituted triazoles (Scheme 2).

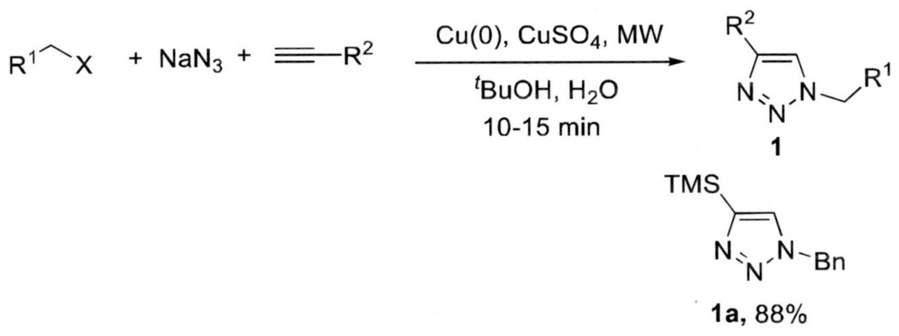

Scheme 2. Microwave-assisted preparation of 1,4-disubstituted 1,2,3-triazoles from halides, sodium azide and terminal alkynes.

In 2006, a three-component reaction of terminal ynamides, NaN_3 and alkyl bromides in the presence of $CuSO_4 \cdot 5\ H_2O$ and sodium ascorbate produces amide-substituted triazoles. In this process, alkyl bromides are in situ converted into organic azides with NaN_3 and then undergo cycloadditions with ynamides to generate the triazole products (Scheme 3) (Zhang, Li et al. 2006).

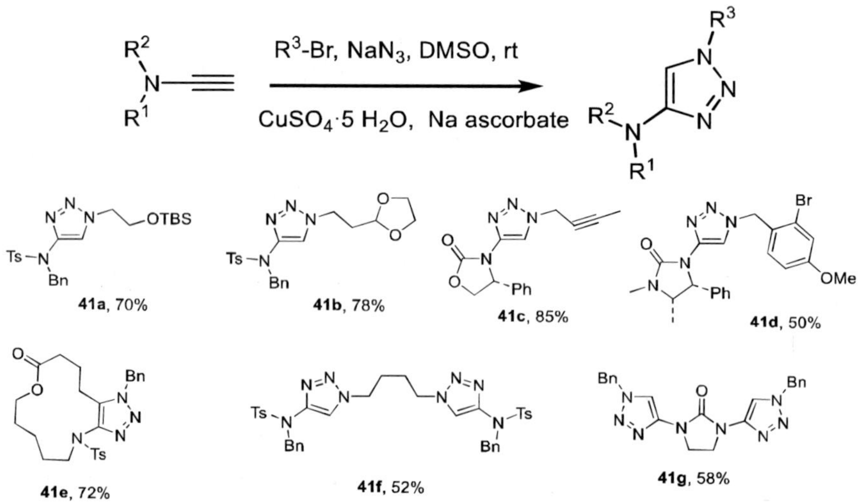

Scheme 3. One-pot synthesis of triazoles from alkyl bromides, NaN_3, and ynamides.

In 2007, Zhang reported a three-component cascade reaction, producing 5-allyl or propargyl 1,2,3-triazoles. The reagents are terminal ynamides, organoazides, and allyl or propargyl iodides, catalyzing with CuBr. Allyl iodides or propargyl iodides are excellent electrophiles suitable for trapping the key intermediate of copper (I) triazolide formed in situ to form the final 4-amino-5-allyl (or propargyl)-1,2,3-triazoles (Scheme 4) (Zhang, Hsung et al. 2007).

Another efficient three-component domino protocol catalyzed by CuO was developed for preparing 1,2-benzoisoxazole-containing 1,2,3-triazoles from commercially available aldehydes, *o*-bromoacetophenone, and sodium azide (Scheme 5) (Wang, Yu et al. 2020). This domino process involves a number of sequential reactions like aldol condensation, copper-catalyzed oxidative [3+2] cycloaddition, copper catalyzed SNAr, and denitrogenative cyclization. A reasonable reaction mechanism is described in Scheme 6. *o*-Bromoacetophenone is condensed with benzaldehyde to form 2-bromochalcone. Subsequently, a [3+2] cyclization reaction of the 2-

bromochalcone and NaN_3 forms (2-bromophenyl)(5-phenyl-4,5-dihydro-1*H*-1,2,3-triazol-4-yl)methanone (**A**), which was further oxidized in the presence of CuO or air to give (2-bromophenyl)(5-phenyl-1*H*-1,2,3-triazol-4-yl)methanone (**B**). Intermediate **B** could undergo a copper-catalyzed S_NAr with NaN_3 to afford **D** with the assistant of the intramolecular *NH*-1,2,3-triazole moiety. Finally, the target product of 3-(5-phenyl-1*H*-1,2,3-triazol-4-yl)benzo[*c*]isoxazole (**3aa**) was formed by intramolecular denitrogenative cyclization (Scheme 6).

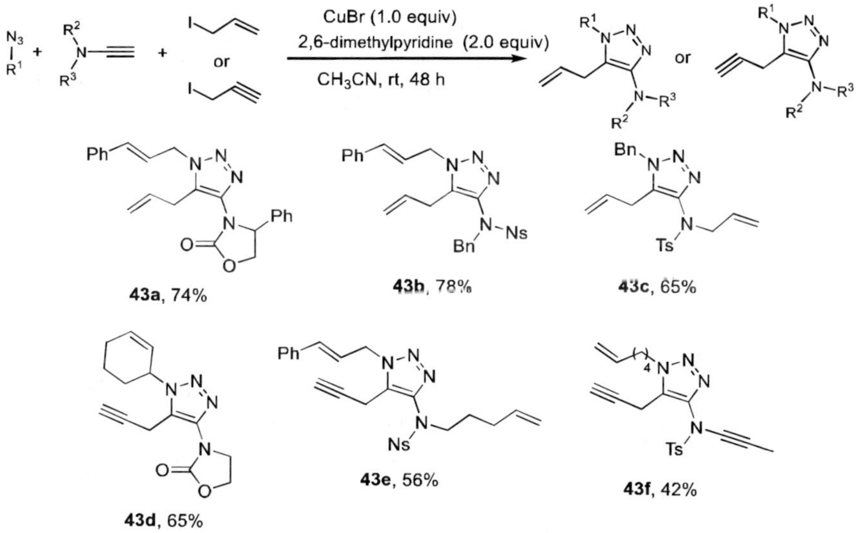

Scheme 4. Synthesis of 4-amino-5-allyl (or propargyl)-1,2,3-triazoles from ynamides, organoazides, and allyl or propargyl iodides.

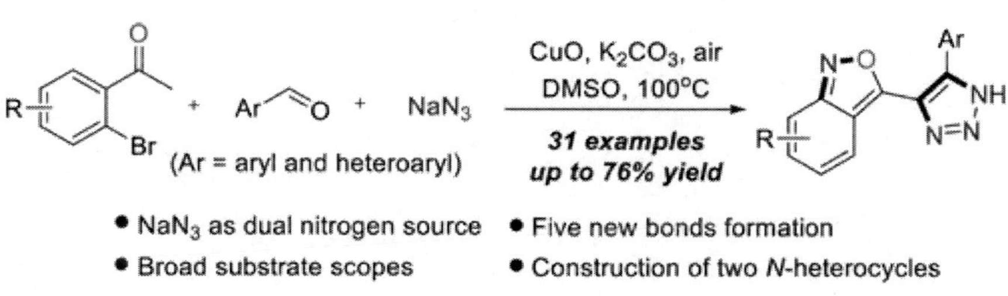

Scheme 5. Synthesis of *N*-heterocycles via copper-catalyzed domino reactions of *ortho*-functionalized aryl halides, aldehydes, and NaN_3.

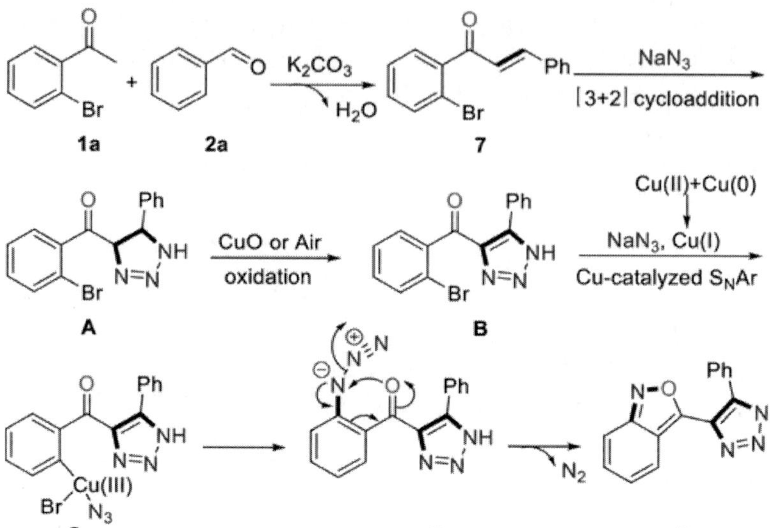

Scheme 6. A plausible reaction mechanism for copper-catalyzed domino reactions of *ortho*-functionalized aryl halides, aldehydes, and NaN_3.

Das and coworker (Das, Jena et al. 2014) reported that $AgNO_3$ and $[Ru(dppp)_2(CH_3CN)Cl][BPh_4]$ catalyze a cascade reaction of azides with terminal alkynes for tandem synthesis of 4-substituted-5-alkynyl-1,2,3-triazoles. Terminal alkynes are first converted into silver acetylide and further couples with alkyne to form a diyene. Subsequent alkyne-azide cycloaddition between the diyene and sodium azide generats 5-alkynyl 1,2,3-triazoles (Scheme 7).

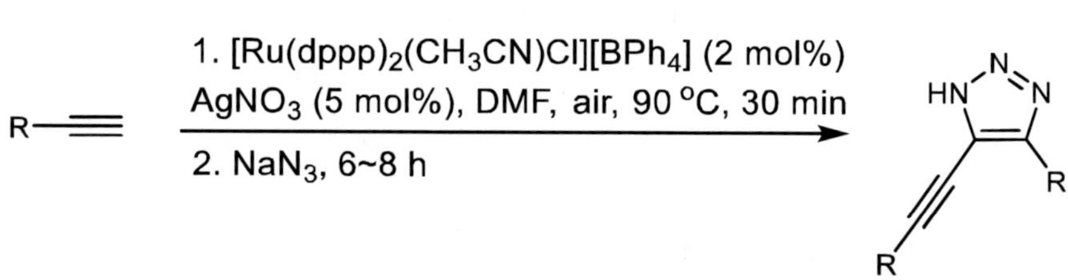

Scheme 7. Ruthenium-catalyzed tandem homocoupling-cycloaddition in the presence of silver salts.

In the presence of CuI and LiO^{t}Bu, *S*-methyl benzenesulfonothioate ($PhSO_2ER$, E = S, Se) used as an excellent electrophilic sulfenylating reagent, phenylacetylene, and the benzylazide can conduct a cascade coupling reaction to form 5-thiotriazoles at room temperature. Various 5-thio-, seleno-substituted 1,2,3-triazoles **I** were synthesized in good yields with complete regioselectivity (Scheme 8) (Wang, Peng et al. 2016). Displacement of methyl benzenesulfonothioate with $Bn_2NOCOPh$ to run this reaction, under similar conditions, a number of 5-aminotriazoles **II** are formed in excellent yields too. Using thiosulfonyl alkynes to couple with azides affords bicyclic triazoles **III** (from 5- to 14-membered rings) in moderate to good yields. A wide range of aromatic or aliphatic azides and alkynes were demonstrated to be suitable substrates, and various functional groups, such as methyl, methoxyl, halogen, cyano, thienyl, cyclopropyl, allyl, nitro, phenoxyl, cinnamyl groups, phthalimide-protected amines, and indole skeletons were all compatible with this reaction conditions (Scheme 9).

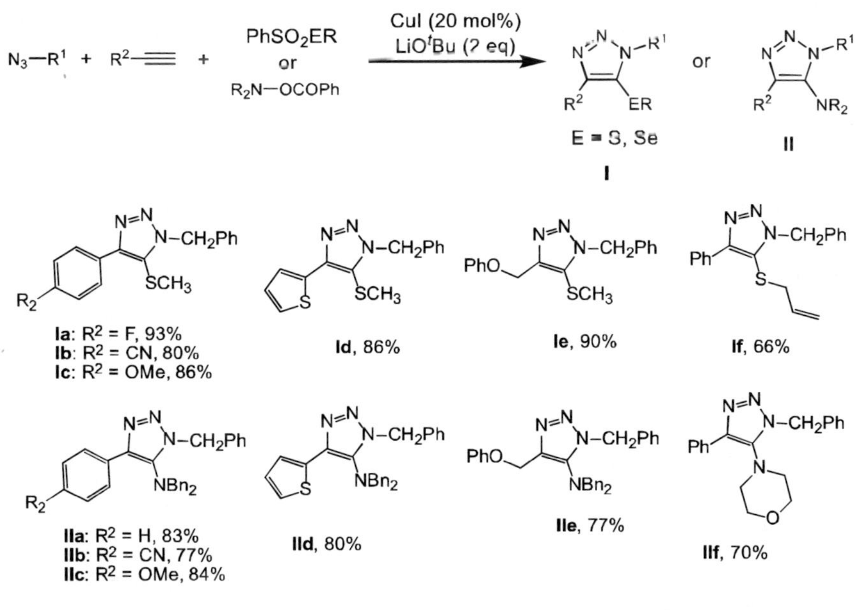

Scheme 8. Synthesis of various 5-thio-, 5-seleno-, 5-aminosubstituted 1,2,3-triazoles.

A possible reaction mechanism was illustrated in scheme 10. Under basic conditions, an alkyne is converted into copper(I) acetylide **J**. Cycloaddition of **J** to an azide produces the 5-copper(I) triazolide intermediate **K**, which is trapped by electrophilic $PhSO_2ER$ to generate the desired product (Wang, Peng et al. 2016).

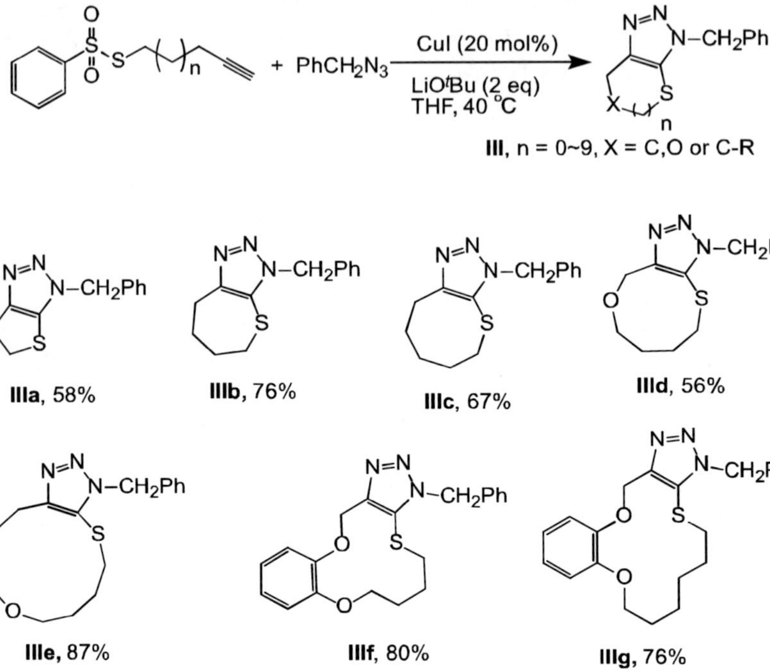

Scheme 9. Synthesis of various bicyclic 1,2,3-triazoles.

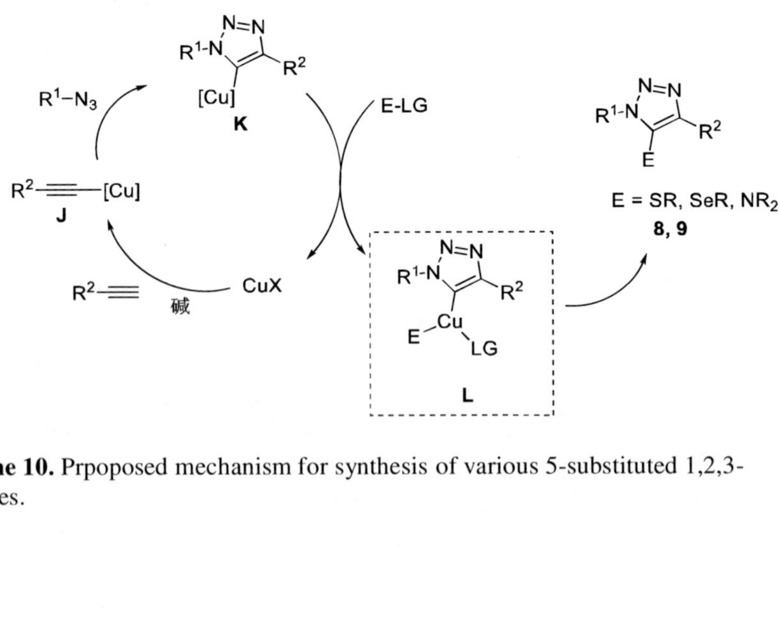

Scheme 10. Prpoposed mechanism for synthesis of various 5-substituted 1,2,3-triazoles.

Similarly, a series of triazol-5-yl disulfides are regioselectively prepared *via* treatment of copper(I) triazolides with SS-*t*-butyl *p*-toluenesulfono(dithioperoxoate), which can be finished in a three-component one-pot reaction, starting from organoazides, terminal alkynes, and SS-*t*-butyl *p*-toluenesulfono(dithioperoxoate) in the presence of CuI and LiO^{t}Bu, as described in Scheme 11 (Wang, Lin et al. 2018).

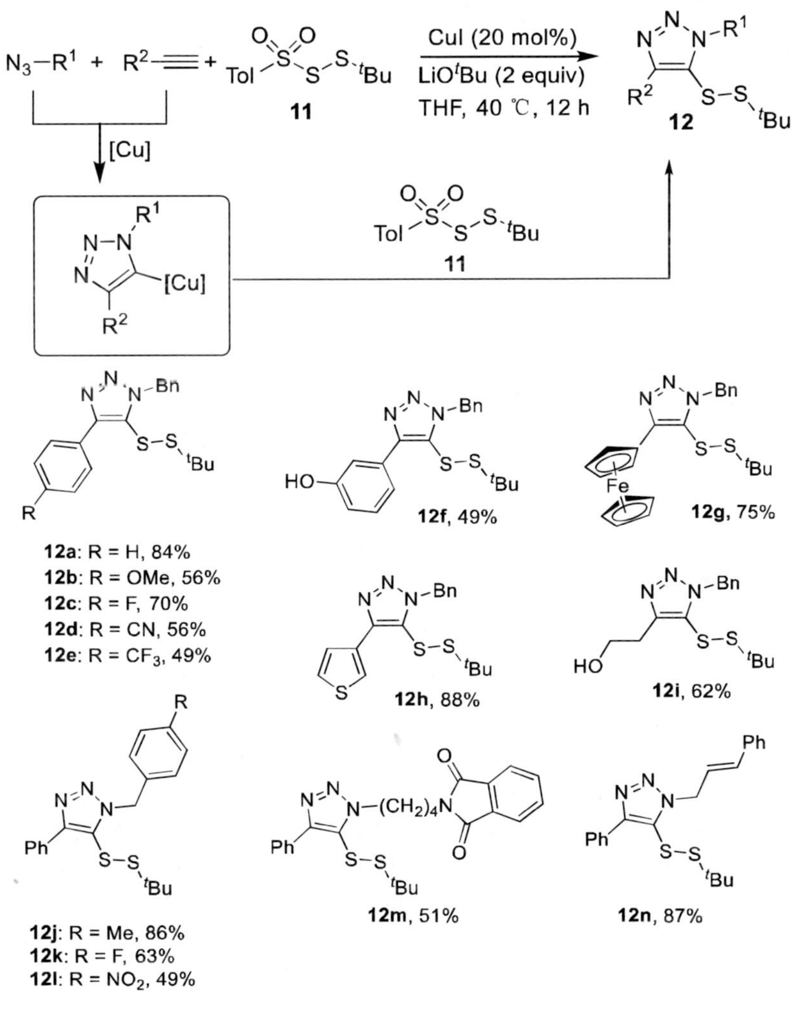

Scheme 11. Synthesis of triazol-5-yl disulfides.

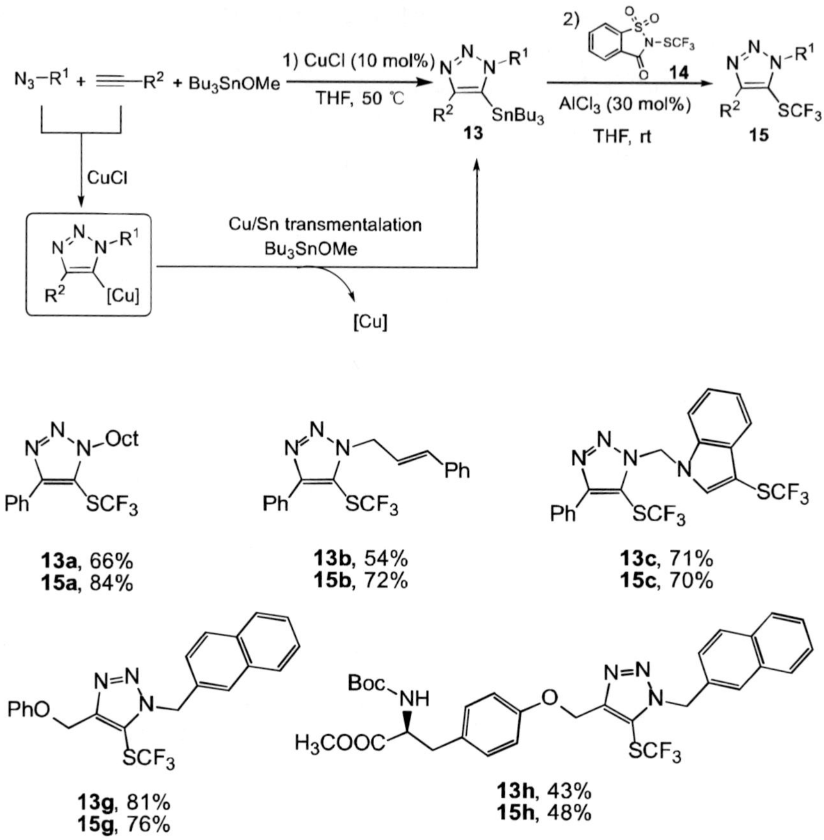

Scheme 12. Synthesis of 5-trifluoromethylthio-1,2,3-triazoles through 5-stannyl triazoles.

In addition, the 5-copper(I) triazolide can be intercepted by Bu_3SnOMe *via* a Cu/Sn transmetalation reaction to form the bench-stable 5-stannyl triazoles, which are stable under air, moisture, and light. In the presence of $AlCl_3$, the 5-stannyl triazole reacts with *N*-trifluoromethyl thiosaccharin to produce 5-trifluoromethylthio-1,2,3-triazoles in the absence of any ligand and base. A series of 5-trifluoromethylthio-triazoles were prepared with this process in good yields (Scheme 12) (Wei, Zhou et al. 2017).

We have developed a four component one-pot reaction mediated by CuI, starting from terminal alkynes, azides, sulfur or selenium powder, and alkyl halides (Scheme 13) (Zhang, et al. 2019). The expected 5-thio- or 5-seleno-triazole products were smoothly generated via interception of 5-copper(I) triazolides with sulfenylating or selenating agents, which were formed in situ from elemental S_8 or Se and alkyl halides. Besides alkyl halides, tosylates and arylboronic acids could also couple with S_8 to give sulfenylating agents in situ. Various aromatic and aliphatic alkynes demonstrate the efficiency of this procedure. Some bicyclic triazole sulfides with a five- to eight-membered ring were also successfully constructed through this sequential reaction (Scheme 14).

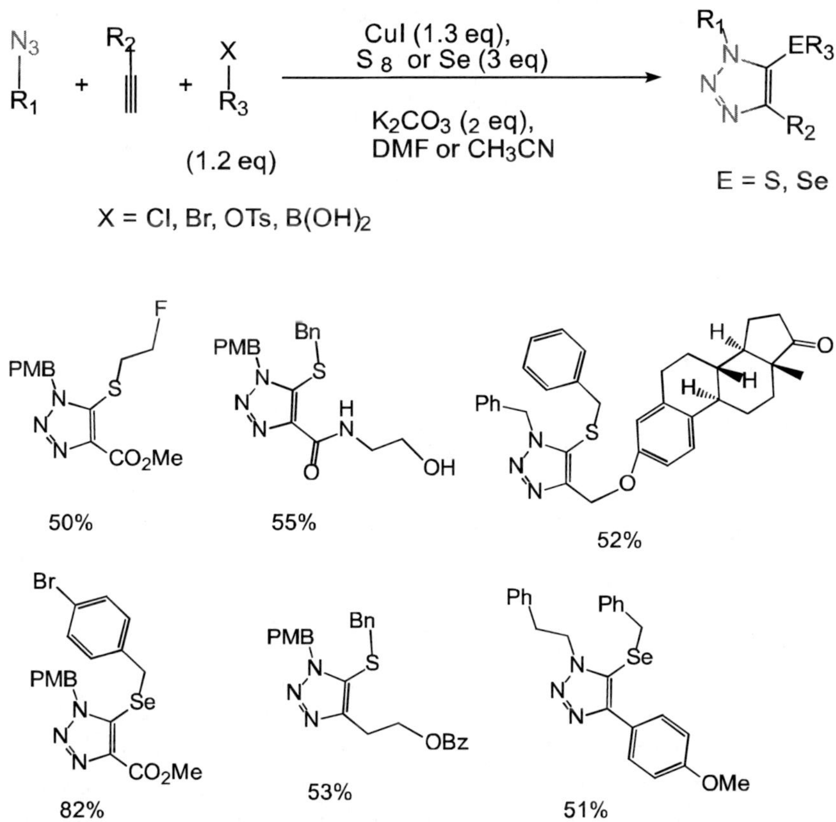

Scheme 13. Synthesis of 5-thio-, seleno-substituted 1,2,3-triazoles from alkynes, azides, alkyl halides and S_8 or Se element.

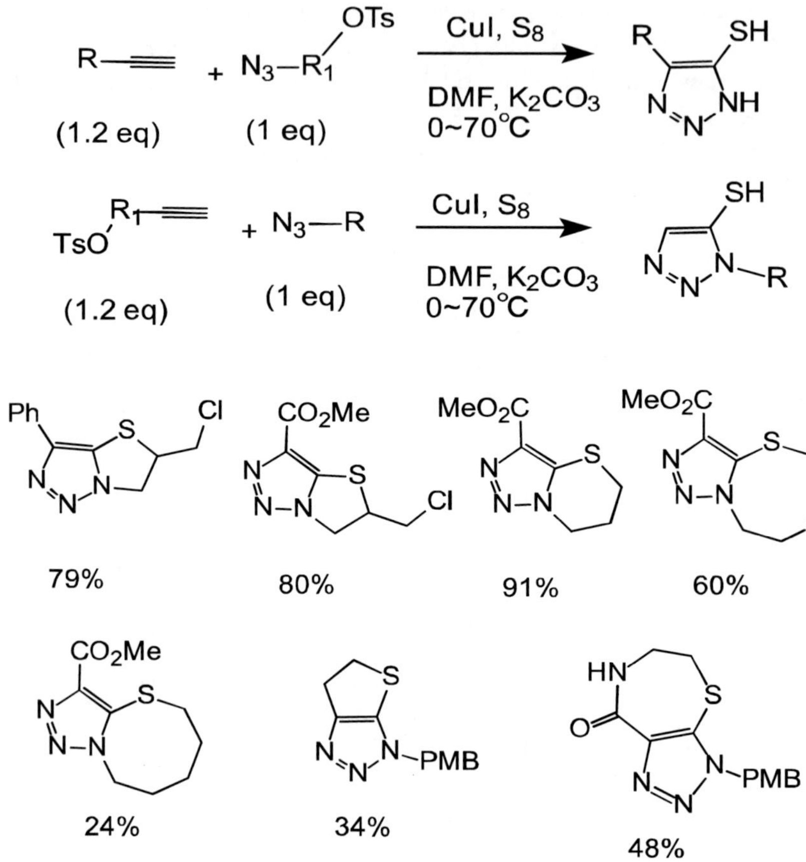

Scheme 14. Synthesis of 5-thio-, seleno-substituted 1,2,3-triazoles from alkynes, azides, and S_8 or Se element.

A possible mechanism for this four-component reaction is shown in Scheme 15. Copper sulfide **O** is generated from sulfur and CuI in basic conditions. Oxidative addition of **O** to a halide affords **P**. Reductive organo-group transfer forms **Q** and followed by oxidation reaction to form copper(II) thiolates **R**. Copper(I) acetylide **M** generated from a terminal alkyne and CuI in the presence of base, reacts with an azide to produce 5-copper(I) triazolide intermediate **N**. Attacks of **N** on **R** give key intermediate **S**. Reductive elimination of **S** affords the desired 5-thiotriazole and CuI is released (Zhang, et al. 2019).

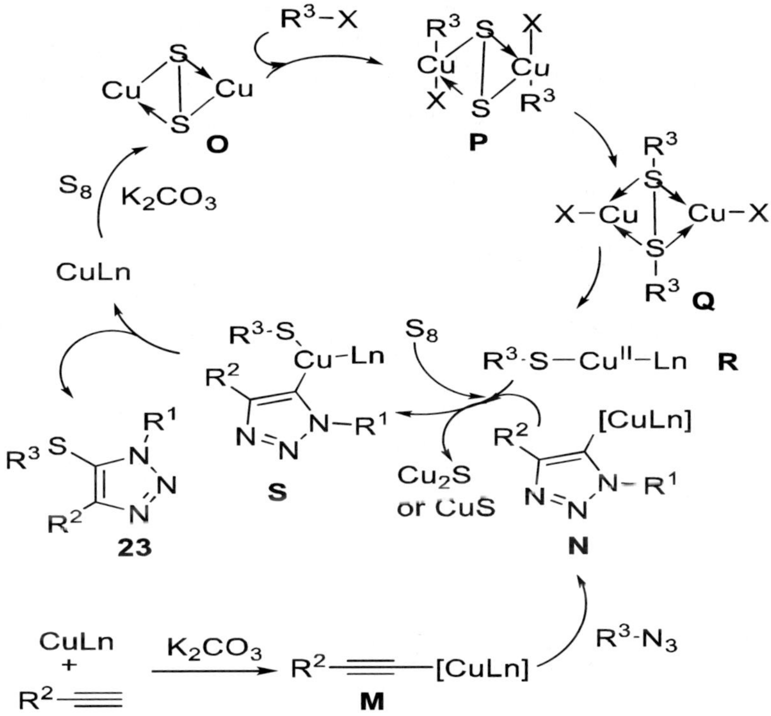

Scheme 15. Possible reaction mechanism for synthesis of 5-thio-, seleno-substituted 1,2,3-triazoles from alkynes, azides, and S_8 or Se element.

Additionally, in the presence of KF, S_8, CuI, and 1,10-phenanthroline, (trifluoromethyl) trimethylsilane ($TMSCF_3$) can be converted into a sulfenylating agent, $[(phen)Cu(SCF_3)]_2$, which is attacked by 5-copper(I) triazolides to form the final 5-F_3CS functionalized triazoles. A series of 5-trifluoromethylthio-1,2,3-triazoles were synthesized, starting from alkynes, azides, S_8 and $TMSCF_3$ (Scheme 16) (Zhang, et al. 2020).

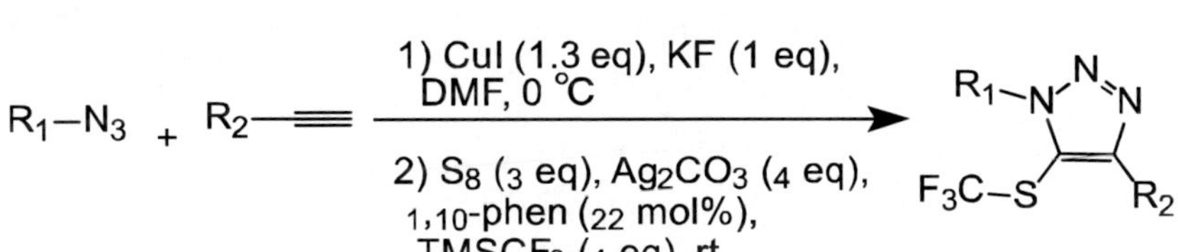

Scheme 16. Synthesis of 5-trifluoromethylthio-1,2,3-triazoles from alkynes, S_8 and $TMSCF_3$.

A zinc-mediated reaction between azido and alkynyl substrates provides a mild method for regioselective formation of 1,5-substituted 1,2,3-triazoles (Scheme 17). The initial metalation of the terminal alkynes, mediated by the amine base, forms the zinc acetylide. Coordination between azides and the zinc acetylide leads to the [3 + 2]-cycloaddition and generates 4-zinc-1,2,3-triazoles. This 4-zinc-triazole is further conveted to a number of 1,4,5-trisubstituted 1,2,3-triazoles in the presence of electrophiles. The stoichiometric quantities of $ZnEt_2$ is needed in this reaction to form the alkyl-zinc intermediate (Smith and Greaney 2013). The reaction mechanism is described in Scheme 18.

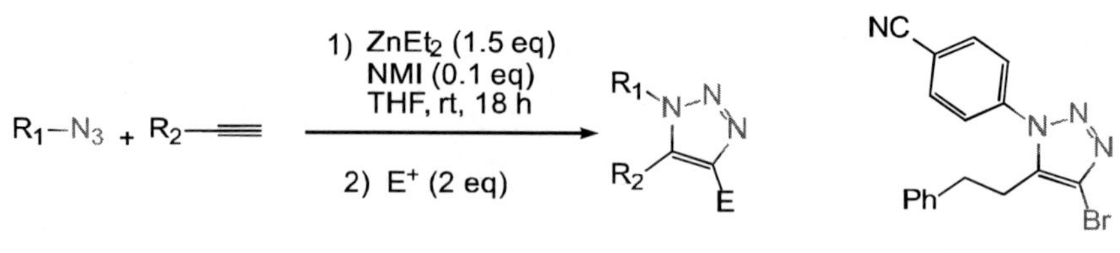

Scheme 17. Zinc-mediated synthesis of 4-functional triazoles from azides, terminal alkynes and electrophiles.

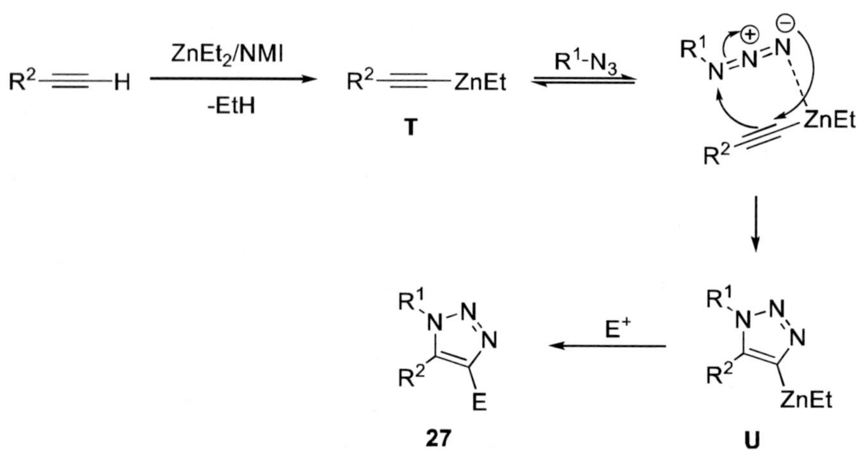

Scheme 18. Reaction mechanism of zinc-mediated synthesis of 4-functional triazoles.

On the other hand, the oxidative coupling reactions relative to the copper triazolides were also found to be feasible. Namely, incorporations of different nucleophilic groups like phosphate and fluoroalkyl onto the C-5 position in 1,2,3-triazoles (Zhu, Xing et al. 2019). In contrast to the reaction of copper triazolides with electrophilic reagents, the oxidative coupling reaction of copper triazolides with nucleophilic reagents often suffers from low efficiency and selectivity. Mild conditions are required, including optimizations of oxidants, choice of ligands, and strong basic reagents for deprotonation in some cases to achieve desirable yields (Li, Hao et al. 2013, Zhu, Xing et al. 2019).

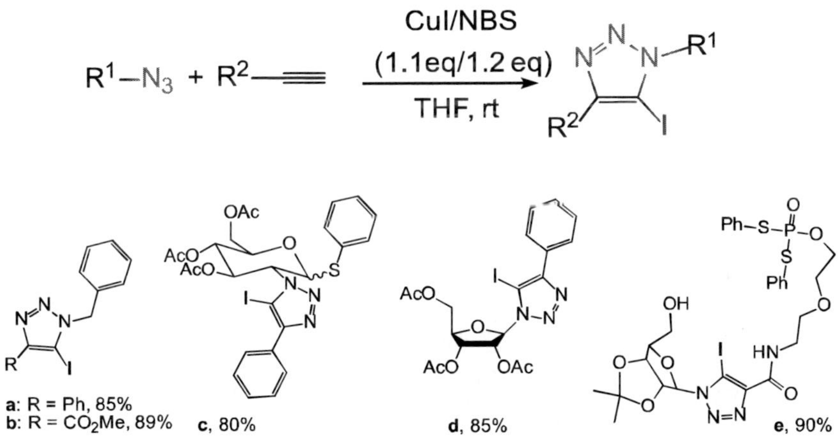

Scheme 19. One-pot procedure for preparation of 5-iodo-1,4-disubstituted-1,2,3-triazoles with CuI and NBS (The reaction was performed under room temperature in dry THF. The mole ratio of alkynes, azides, CuI, and NBS was 1:1.1:1.1:1.2, with 1 equiv of DIPEA added as the alkali).

An efficient method for preparation of 5-iodo-1,4-disubstituted-1,2,3-triazoles is achieved by a multicomponent one-pot reaction of azides with alkynes in the presence of CuI and NBS. In this process, CuI is oxidized by NBS to provide both Cu^+ and I^+. This methodology exhibits high tolerance of various sensitive groups. The in situ generated I^+ is trapped by another intermediate, copper (I) triazolides, to form 5-iodo-1,4-disubstituted-1,2,3-triazoles (Scheme 19) (Li, Zhang et al. 2008). Moreover, dispacement of CuI/NBS with CuBr/NCS, various 5-bromo-1,4-disubstituted-1,2,3-triazoles are efficiently afforded (Scheme 20) (Li, Li et al. 2011). Both system of $Cu(ClO_4)_2$-NaI and $CuCl_2$-NaI could also effectively perform similar multicomponent conversion. For these procedure, copper (II) species and sodium iodide solutions result in copper(I) species and the electrophilic triiodide ions. The former catalyzes the azide cycloaddition reaction and the latter couples to copper (I) triazolides to produce the final iodo 1,2,3-triazoles (Scheme 21) (Brotherton, Clark et al. 2012, Yan, Sander et al. 2013).

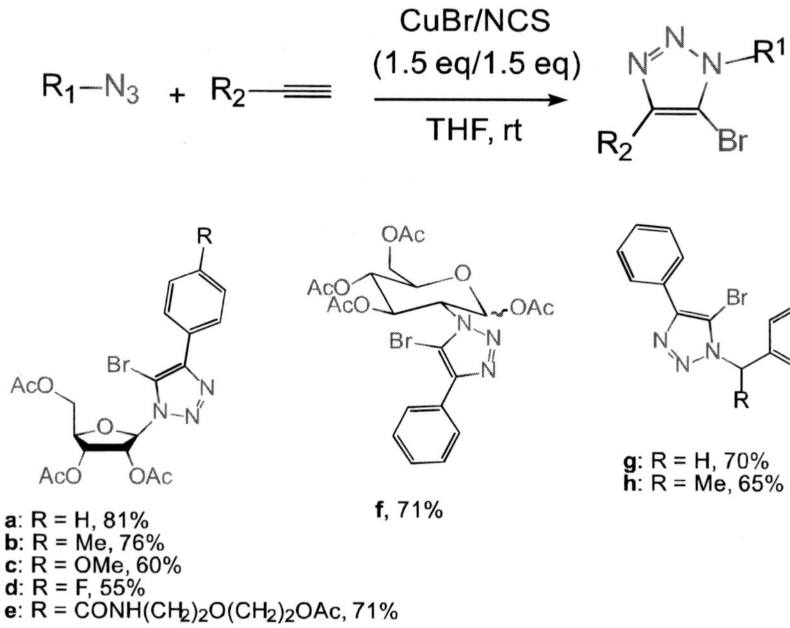

Scheme 20. One-pot procedure for preparation of 5-bromo-1,4-disubstituted-1,2,3-triazoles with CuBr and NCS.

In the presence of CuI, NBS and DIEPA, various organic azides could couple with trimethylsilyl acetylene to produce a number of 4-trimethylsilyl-5-iodo-1,2,3-triazoles in good yields. Catalyzed with CuBr, this iodotriazoles are treated with nucleophiles, such as various phenyloxy, arylthio, and alkynyl groups, affording variant 1,5-disubstitued 1,2,3-triazoles. This procedure provides an attractive alternative for the preparation of 1,5-disubstituted 1,2,3-triazoles (Scheme 22) (Li, Shang et al. 2015).

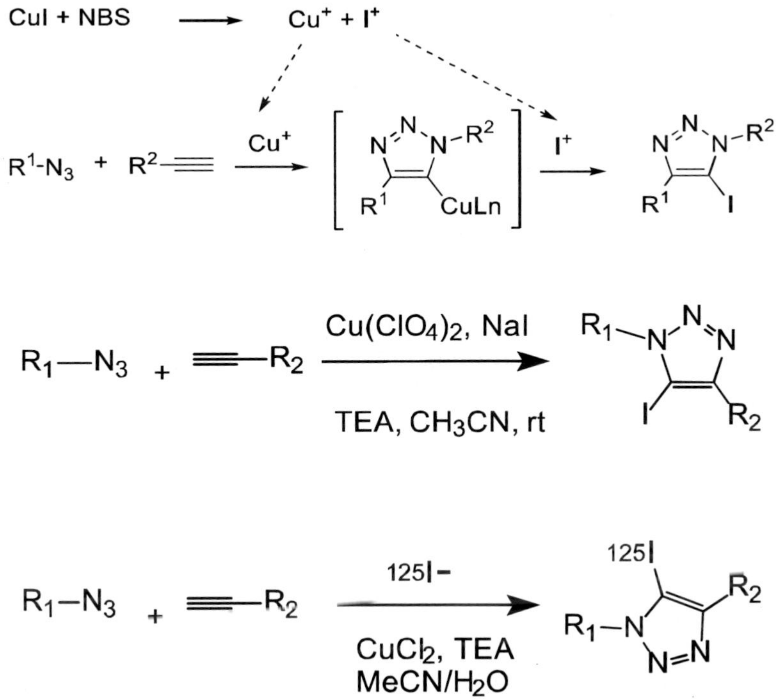

Scheme 21. One-pot procedure for preparation of 5-iodo-1,4-disubstituted-1,2,3-triazoles with $Cu(ClO_4)_2$-NaI or $CuCl_2$-NaI.

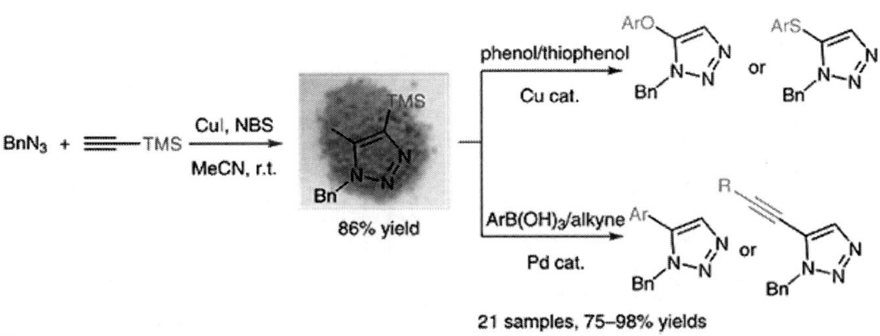

Scheme 22. Copper-mediated cycloaddition reaction between azides and trimethylsilyl acetylene (0.20 mmol alkynes, 0.22 mmol azides and 0.80 mmol $TMSCF_3$ were used as substrates).

tert-Butyldimethylsilyl chloride (TBDMSCl) was found to be able to activate aerobic oxidation of CuX (X = I, Br) to produce X_2 and Cu^+, which could in situ effectively promote one-pot synthesis of 5-halo-1,2,3-triazoles from alkynes, azides, and CuX under O_2 atmosphere at room temperature. The advantages of this method include green O_2 as oxidant, mild conditions, non-oxidative additive, and wide scope of substrates (Scheme 23) (Li, Hao et al. 2013).

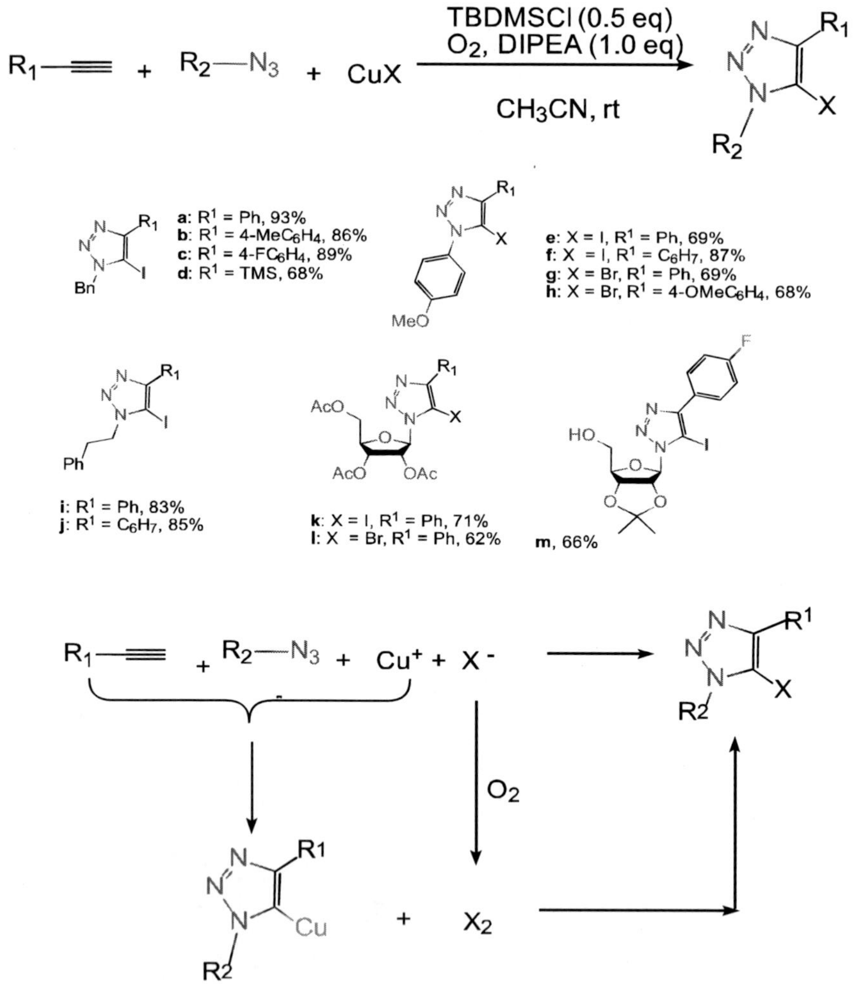

Scheme 23. TBDMSCl-promoted one-pot alkyne-azide-CuX reaction from terminal alkynes, organic azides, and CuI or CuBr in the presence of 0.5 equiv TBDMSCl.

Copper-catalytic aerobic oxidation reaction is a mild and effective reaction condition for the preparation of diverse 5-fluoroalkyl 1,2,3-triazoles (Zhu, Xing et al. 2019). Various 5-fluoroalkyl-1,2,3-triazoles could be synthesized from nucleophilic fluoroalkyl reagents, terminal alkynes and organic azides *via* multi-component one-pot oxidative reactions (Scheme 24). Air was used as the sole oxidant. Alkyl azides or aryl azides and various fluoroalkyl reagents, including trifluoromethyl trimethylsilane (CF_3TMS), pentafluoroethyl trimethylsilane (C_2F_5TMS), and heptafluoropropyl trimethylsilane (C_3F_7TMS), could be used effectively for these reactions. In mechanism, the formed cage-like complexes of glycinamide-type ligands and Cu(I) not only stabilize the triazolyl copper intermediate, but also cause a steric hindrance around the intermediate. The potential nucleophiles, including alkynyl, triazolyl and CF_3^- anions, would react with the copper intermediate via the oxidative coupling reactions. Thus, the cross-oxidation coupling reaction could form the final triazoles product in one-pot.

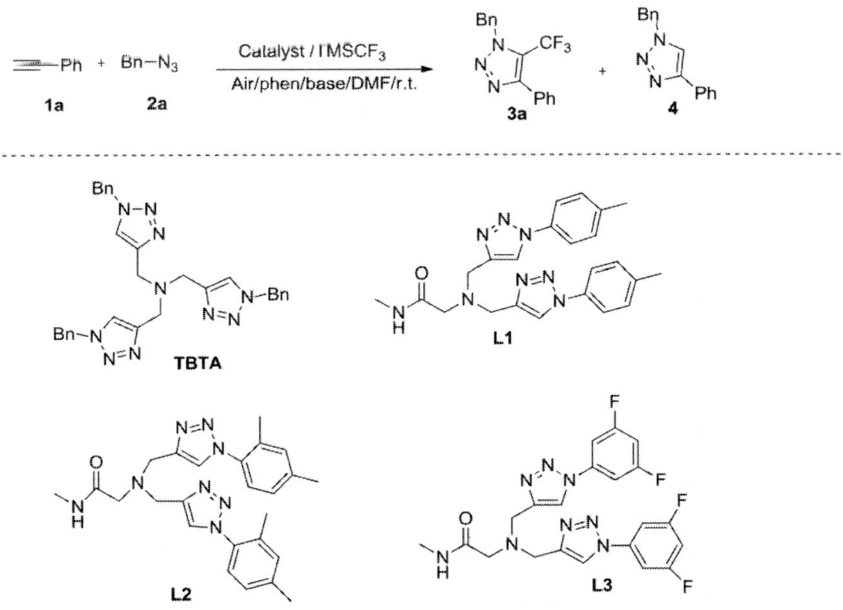

Scheme 24. Copper-catalytic aerobic oxidation reaction of organoazides, alkynes, and fluoroalkyl trimethylsilane (0.20 mmol alkynes, 0.22 mmol azides and 0.80 mmol $TMSCF_3$).

Another aerobic oxidative strategy is the coupling of the highly active anion intermediate, Cu(I)-triazole complex, with nucleophilic *H*-phosphonates, generating a variety of biologically relevant 1,2,3-triazolyl-5-phosphonates (Zhu, Xing et al. 2019). This multicomponent reaction characterizes high regioselectivity and efficiency, mild conditions, and applicability to a broad range of substrates bearing sensitive functional groups (Scheme 25). The mechanism for this three-component domino reaction is proposed. The Cu-catalyzed click-coupling of an azide with an alkyne gives the key intermediate of Cu(I)-triazolide **5'** and followed by aerobic oxidative coupling with *H*-phosphonate (Scheme 26).

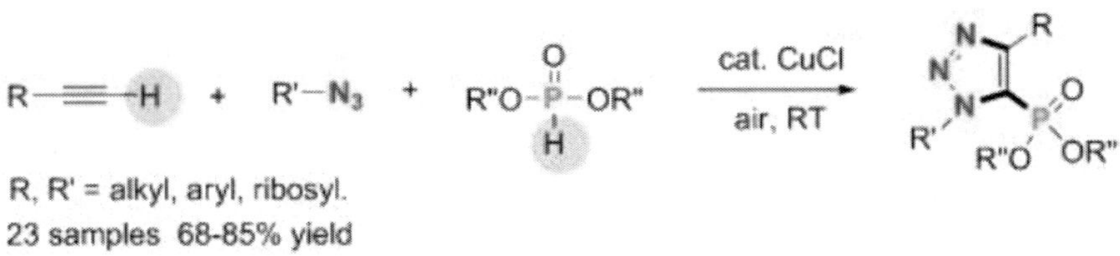

Scheme 25. Copper-catalyzed three-component reaction (CuAA[P]C reaction) of terminal alkynes, azides, and *H*-phosphonates.

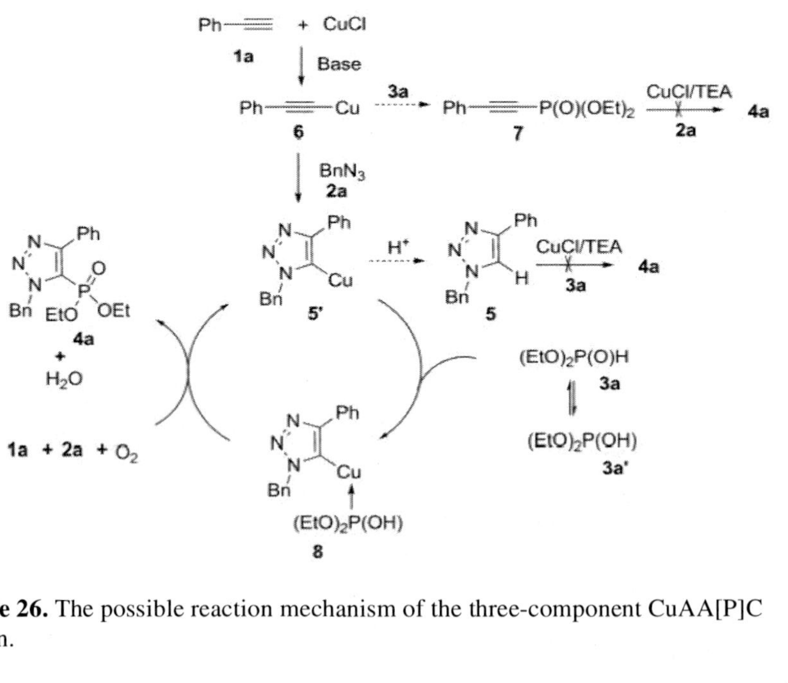

Scheme 26. The possible reaction mechanism of the three-component CuAA[P]C reaction.

A metal-free three-component coupling of α-CF_3 carbonyls, azides and amines for the regioselective synthesis of 1,4,5-trisubstituted 1,2,3-triazoles under air is established (Scheme 27) (Gao, Kuantao et al. 2022). Various substituted 4-amide 1,2,3-triazoles were obtained when the enolizable α-CF_3 ketones were applied in aqueous reaction medium. As described in Scheme 28, the in situ generated β-oxo amide **B** is responsible for the formal oxygen-shift intermediate **D** and the subsequent [3 +2] cycloaddition reaction to form the products 4-amide 1,2,3-triazoles, with amines as both the catalyst and reactant. In the case of non-enolizable α-CF_3 esters, the densely functionalized 5-amino 1,2,3-triazoles **5** were achieved exclusively.

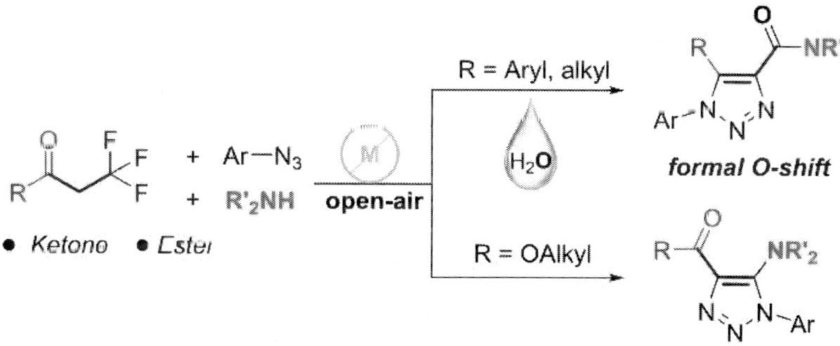

Scheme 27. synthesis of 1,4,5-trisubstituted 1,2,3-triazoles from α-CF_3 ketones, aryl azides and amines.

Terminal alkyne surrogates, propiolic acids, are more convenient to be utilized due to the general solid state. A $CuSO_4{\cdot}H_2O$-catalyzed or Cu_2O-catalyzed decarboxylation of propiolic acids and followed by 1,3-dipolar cycloaddition with sodium azides to form 1,4-disubstituted 1,2,3-triazoles under tandem catalysis conditions have been reported (Scheme 29a and 29b) (Kolarovic, Schnuerch et al. 2011, Shang, Fu et al. 2019). Pan and Xu developed a $Cu(OAc)_2{\cdot}H_2O$-catalyzed decarboxylation/cycloaddition cascade process of propiolic acids, diselenides, and azides to access 5-arylselanyl-1,2,3-triazoles (Scheme 29c) (Cui, Chen et al. 2018). In the presence of MeOLi, O_2, and $Cu(OAc)_2$, decarboxylative cycloaddition of propiolic acids, azides, and arylboronic acids produce a variety of 5-aryl 1,2,3-triazoles (Scheme 29d) (Wang, Xin et al. 2020). A possible mechanism for these transformation is described in Scheme 30. Initially, the decarboxylation of propiolic acid generates the copper(I) acetylide **A** in the presence of copper salt and MeOLi. The [3 + 2] cycloaddition of the copper(I) acetylide with the azide leads to triazolyl-copper intermediate **C**, and followed by a transmetalation with borate complex generated from phenylboronic acid and MeOLi under oxidative conditions to afford the intermediate **D**. Finally, reductive elimination and releasing the copper(I) species of the intermediate **D** provides the triazole product **4a**. O_2 probably acts as an oxidant to produce high-oxidation-state copper intermediates.

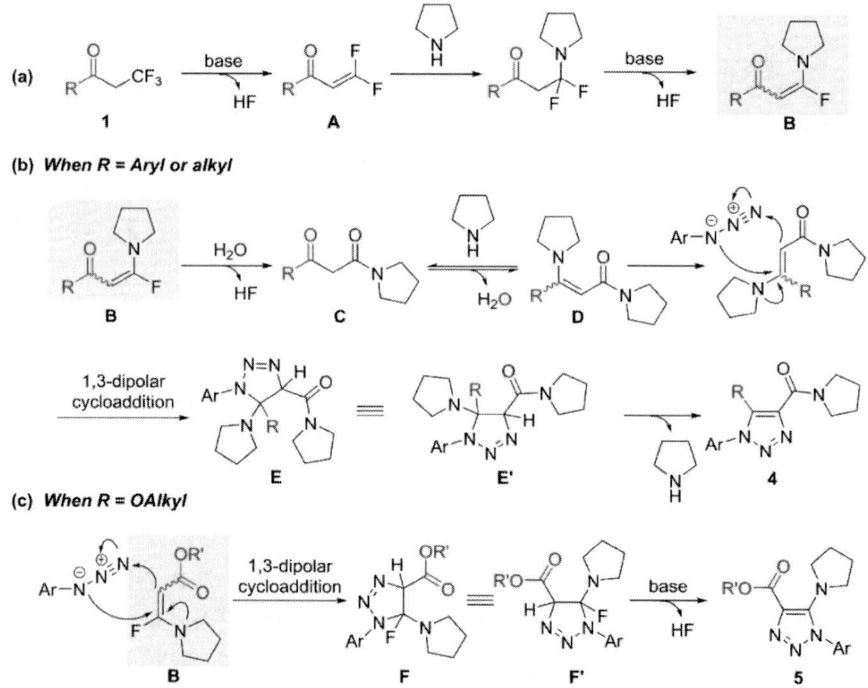

Scheme 28. Proposed mechanism for the synthesis of 1,4,5-trisubstituted 1,2,3-triazoles from α-CF_3 ketones, aryl azides and amines.

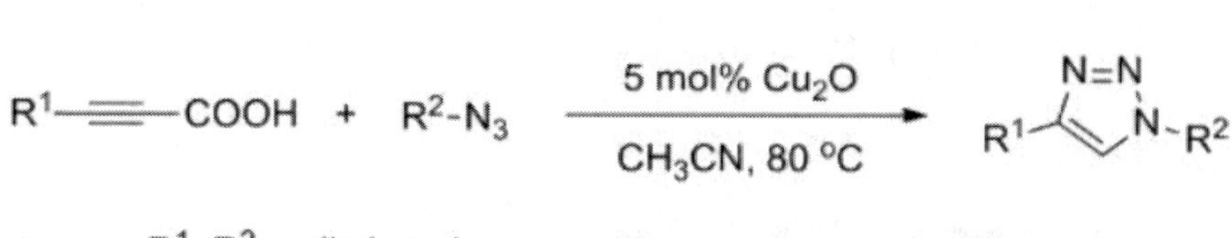

(a)

Conditions: propiolic acids (0.30 mmol), azides (0.30 mmol), and Cu_2O (5 mol%) in MeCN (3 mL) at 80°C for 12 h under an air atmosphere.

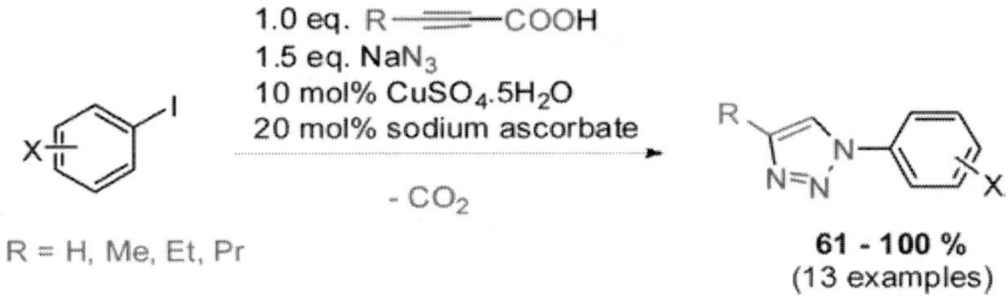

(b)

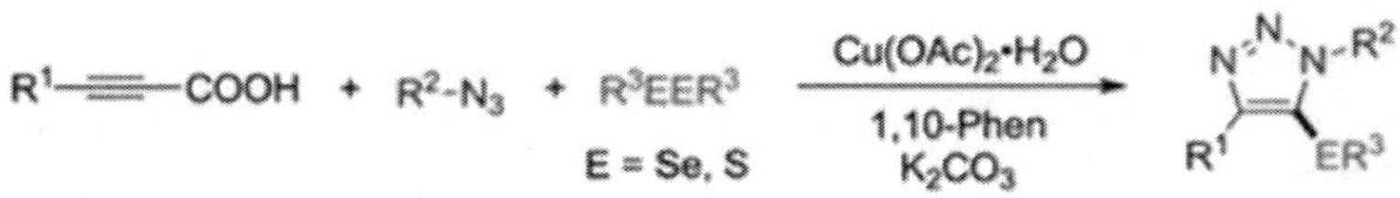

(c)

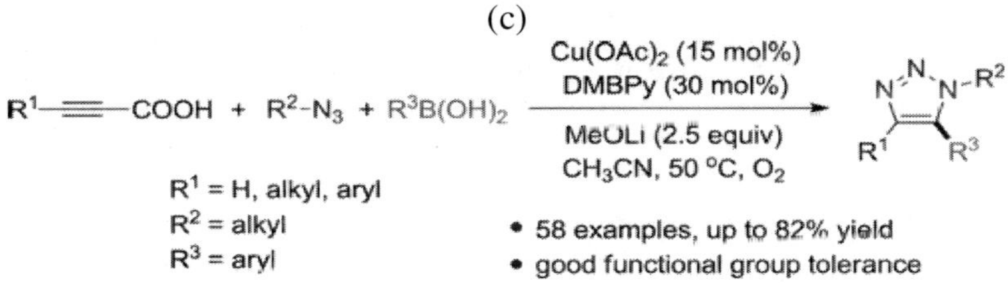

(d)

Scheme 29. Copper-catalyzed decarboxylative 1,3-dipolar cycloaddition.

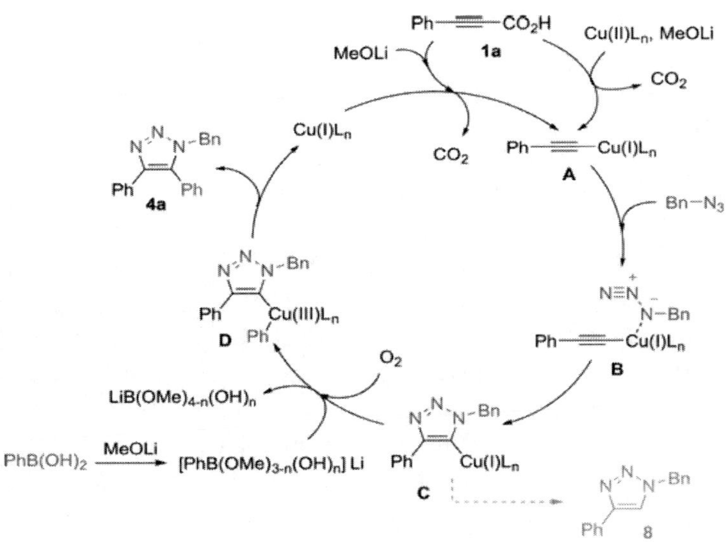

Scheme 30. Proposed mechanism for cascade copper-catalyzed decarboxylative 1,3-dipolar cycloaddition.

Similarly, using CuCl as catalyst, TEA as base, and O_2/chloramine-T as oxidant through one-pot *in situ* oxidative-coupling reaction, a variety of 5-aryl-1,2,3-triazole compounds could be efficiently prepared from the organic azides, terminal alkynes and arylboronic acids under mild conditions (Scheme 31). This tandem reaction proceeds through the Cu-catalyzed click-coupling between an azide and an alkyne and followed by aerobic oxidative coupling with arylboronic acids (Wang, Li et al. 2021).

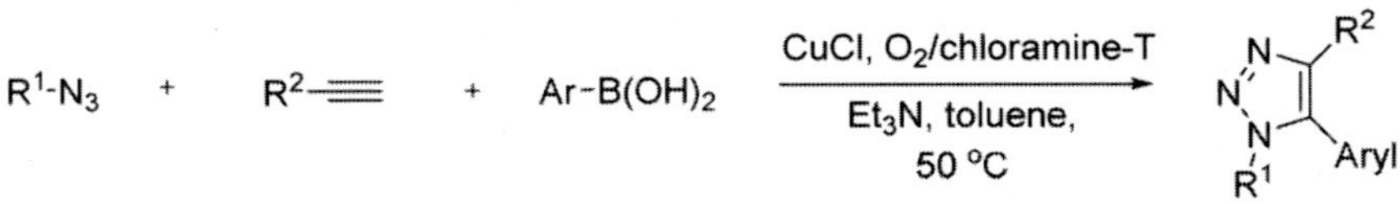

Scheme 31. Multicomponent synthesis of 5-aryl-1,2,3-triazoles from arylboronic acids, organic azides and alkynes (0.15 mmol azides, 0.6 mmol arylboronic acids, 0.17 mmol terminal alkynes, 0.015 mmol CuCl, 0.15 mmol chloramine T and 0.15 mmol triethylamine were used as substrates).

A three-component coupling reaction of alkynes, azides, and diazo compounds produces fully substituted triazoles (Scheme 32). The reactivities of alkyl azides and diazo compounds toward Cu-acetylide were controlled by the introduction of a ligand and the stoichiometry of azides and diazo compounds to suppress the undesired protonation and the alkyne-diazo coupling, maximizing the selectivity for the three-component coupling. Besides, the use of aliphatic alkynes was crucial for achieving high selectivity for the three-component coupling reaction. This method features mild reaction conditions, broad substrate scope, and good functional group tolerance. Alkyl azides are generally more reactive than diazo compounds toward Cu-acetylide formation. The vinyl-Cu species is more reactive than Cu-acetylide toward the diazo compound (Chen, Mamidipalli et al. 2021).

Scheme 33 shows a mechanistic picture of all competing pathways leading to the observed products **1**, **2**, and **4**. The initially formed Cu-acetylide **A** reacts mainly with an azide to form adduct **B**, which undergoes [3 + 2] cycloaddition to generate intermediate **C**. At the same time, Cu-acetylide **A** competitively reacts with a diazo group to generate the minor product **2**. Subsequently, intermediate **C** reacts with ethyl diazolacetate to generate Cu-carbenoid **D**, which transforms to intermediate **E**. Intermediate **C** can also interact with an alkyne followed by proton exchange to generate triazole **1** and Cu-acetylide **A**. Finally, an alkyne coordination with **E** followed by proton exchange would generate **4** and regenerate Cu-acetylide **A**, which enters a new catalytic cycle (Chen, Mamidipalli et al. 2021).

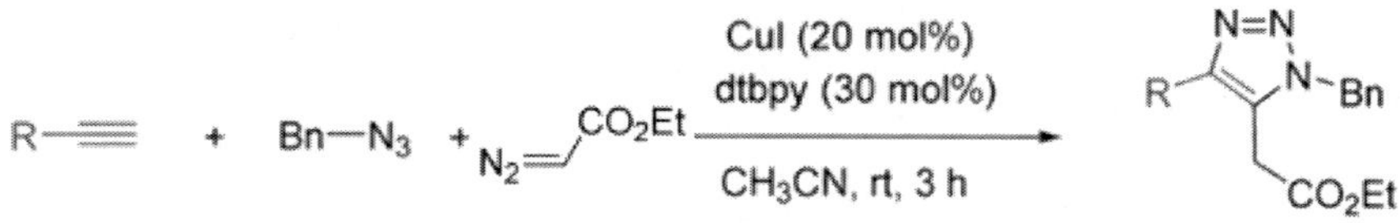

Scheme 32. Tandem three-component coupling reaction of alkynes, azides, and diazo compounds.

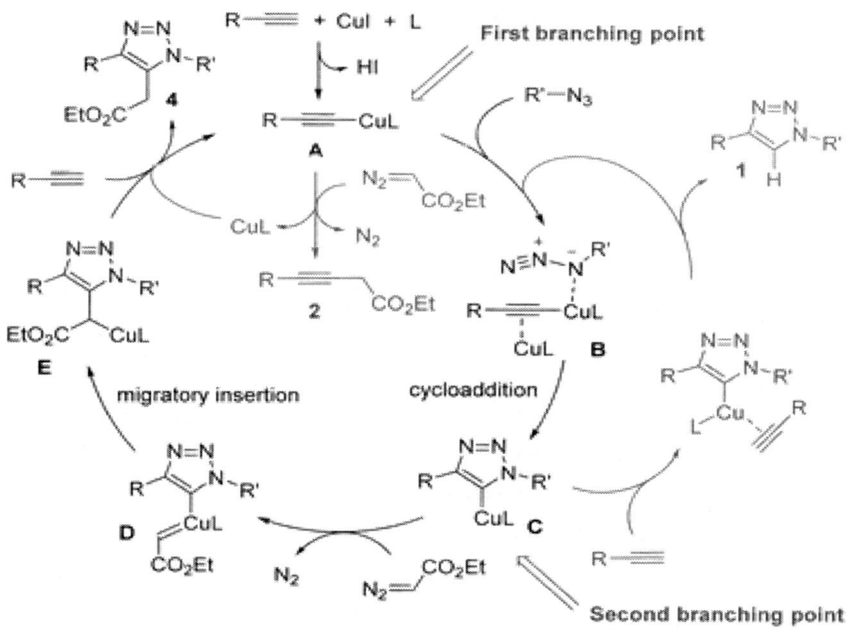

Scheme 33. Proposed mechanism for the three-component coupling reaction of alkynes, azides, and diazo compounds.

Highly efficient microwave-assisted solvent- and catalyst-free multicomponent synthesis of 5-R-imino-[4-trialkylsilyl(germyl)]-1*H*-1,2,3-triazoles from trimethylsilyl azide, primary amines, and Si/Ge-substituted propynals has been developed (Scheme 34). These products are polyfunctional triazoles containing interconnected nucleophilic and electrophilic centers (Medvedeva, Demina et al. 2017). As to this process, the 1,3-dipilar [3 + 2] cycloaddition between the triple bond and the azide first proceeds to form 5-formyl triazoles **A**, which then couples with an amine forming an iminoproduct **B** or intermediate. As a 2-aminoethanethiol substrate is used, an intramolecular cyclization would spontaneously occur to afford a bis-heterocyclic adduct **C** with an 1,3-thiazolan-2-yl group (Scheme 35).

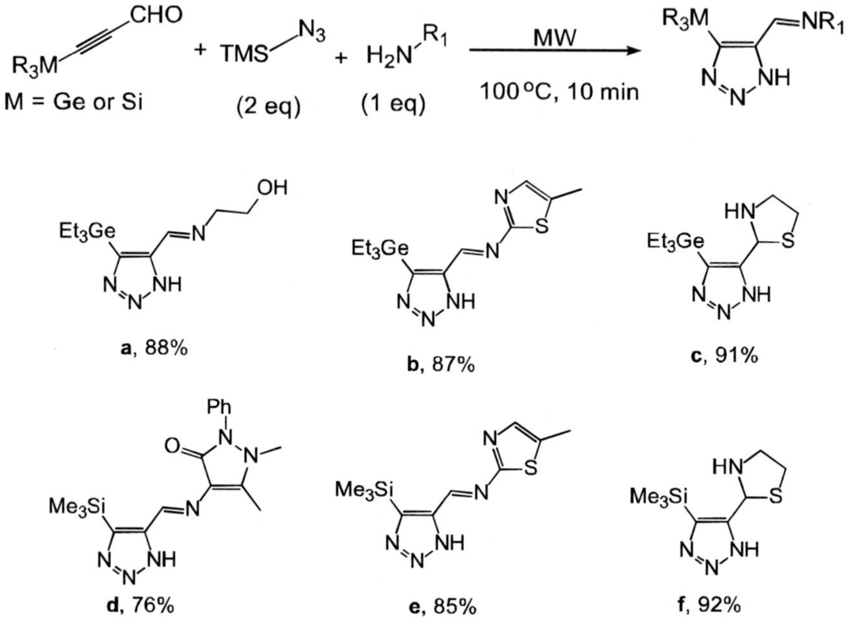

Scheme 34. Microwave assisted three-component reaction to synthesize *NH*-1,2,3-triazolimides without solvents and catalysts.

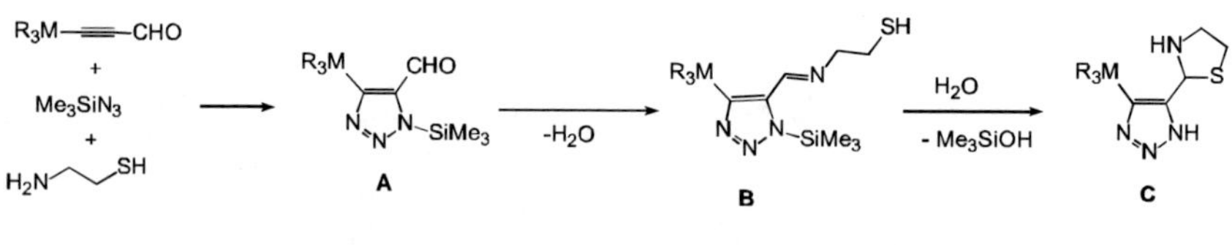

Scheme 35. Process for synthesis of 5-(1,3-thiazolan-2-yl)-1,2,3-triazoles.

Besides 1,2,3-triazoles, organoselenium compounds are also useful in organic synthesis, pharmaceuticals, and biology. In 2009, Benhida et al. reported a three-component reaction of terminal alkynes, ribosyl azides, and phenylselanyl bromide, generating 1,2,3-triazole selenides in the presence of CuI and *N,N*-diisopropylethylamine (Scheme 36). The phenylselanyl bromide acts as electrophilic reagents to trap the 5-copper(I) triazolide intermediate to form the 5-selanyl 1,2,3-triazoles (Malnuit, Duca et al. 2009).

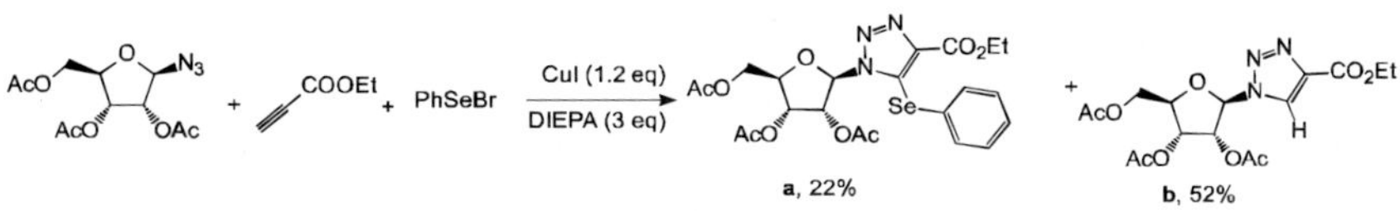

Scheme 36. three-component reaction of terminal alkynes, ribosyl azides, and phenylselanyl bromide.

A copper-catalyzed four-component cascade reaction for the preparation of triazoles bearing *β*-hydroxy chalcogenides from terminal alkynes, azides, epoxides, and Se/K_2S is reported (Scheme 37). This reaction features utilizing sustainable selenium and sulfur reagents, mild reaction conditions and good functional group compatibility (Wang, Sun et al. 2022). Disproportionation reaction of elemental selenium generates Se^{2-} and SeO_3^{2-} species in the presence of KOH (Scheme 38). Meanwhile, the reaction of alkyne **1 a** and copper salt leads to the copper(I) acetylide **A**, followed by a [3+2] cycloaddition with an azide to give regioselective 5-cuprated triazole **C**. This intermediate subsequently undergoes a ligand exchange with Se^{2-}/S^{2-} and oxidation by SeO_3^{2-}/O_2 affording copper (III) species **D**, which takes place a reductive elimination to provide triazolylchalcogenide anion **E** and release the copper(I) catalyst. Finally, nucleophilic attack of triazolylchalcogenide anion to the epoxide forms product **4 a**.

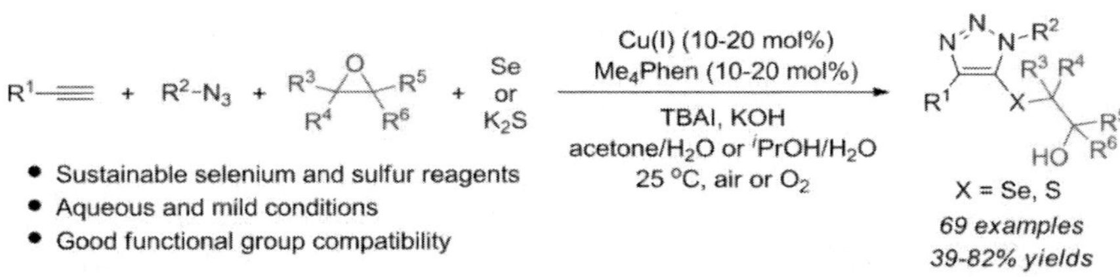

Scheme 37. Reaction conditions: **1** (0.30 mmol), **2** (0.45 mmol), Se (0.75 mmol), **3** (0.9 mmol), CuCl (10 mol%), Me_4Phen (10 mol%), KOH (1.0 equiv.) and TBAI (1.0 equiv.) in 1.0 mL acetone/H_2O [1/1 (v/v)] at 25 °C for 6 h under air.

In contrast to the well-developed Cu-catalyzed AACs of terminal alkynes, the direct reaction of internal alkynes and azides, especially the intermolecular reaction, for the synthesis of fully substituted 1,2,3-triazoles is a great challenge due to the high activation barrier. Most of the intermolecular AACs of internal alkynes require high temperatures and/or heteroatom-functionalized internal alkynes, such as haloalkynes, ynamides, and thioalkynes. As propiolic acids, diselenides and azides were used as reagents, a simple and efficient method for the preparation of various 1,4-disubstituted 5-arylselanyl-1,2,3-triazoles was developed (Scheme 39). The mechanistic study revealed that the alkynyl selenium was formed first via decarboxylative reactions, followed by the intermolecular AAC reaction with azides to afford 5-seleno-triazoles. The resulting multisubstituted 5-seleno-1,2,3-triazoles show potent cancer cell-growth inhibition activities (Cui, Chen et al. 2018) (Scheme 39).

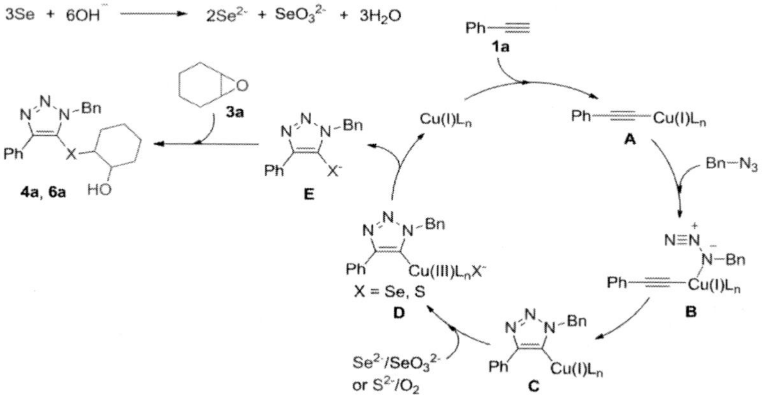

Scheme 38. A plausible mechanism for the copper-catalyzed four-component cascade reaction.

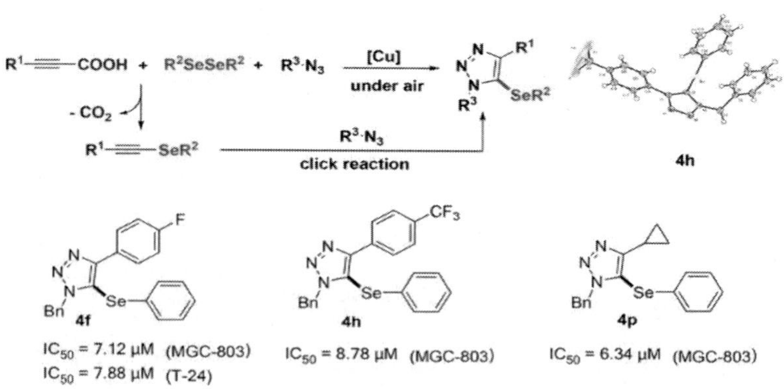

Scheme 39. Synthesis of 5-seleno-1,2,3-triazoles from propiolic acids, diselenides and azides.

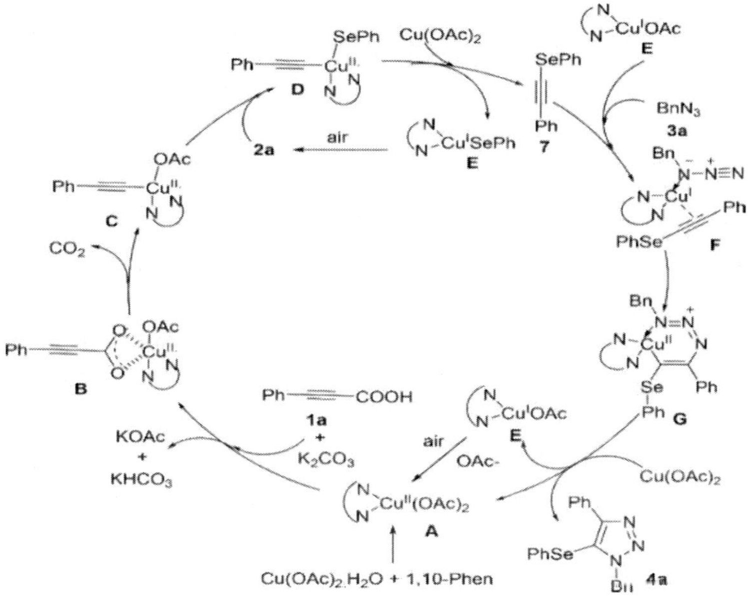

Scheme 40. Plausible mechanism for synthesis of 5-seleno-1,2,3-triazoles from propiolic acids, diselenides and azides.

In details, first, the coordination of 1,10-phenanthroline to $Cu(OAc)_2 \cdot H_2O$ to form the active copper(II) intermediate **A**, which reacts with phenylacetic acid **1a** to afford copper(II) intermediate **B**. The decarboxylation of **B** produces copper (II) intermediate **C** and releases one molecular CO_2 (Scheme 40). The reaction of **C** with diphenyl diselenide **2a** gives intermediate **D**. The reductive elimination of **D** results in phenyl(phenylethynyl)selane **7**, as well as copper(I) species **E**. The complexation of **E** with phenyl(phenylethynyl)selane **7** and azide **3a** generates intermediate **F**, which undergoes oxidative cyclization to form intermediate **G**. The reductive elimination of **G** affords the desired product **4a** and copper(I) species **E**. Finally, **E** is converted into **A** under air atmosphere oxidation (Cui, Chen et al. 2018).

In the presence of CuI (10 mol%) and 1,10-phenanthroline (10 mol%), a three-component reaction of ethynylstibanes, organic azides, and diaryl diselenides forms 5-selanyltriazoles in moderate-to-excellent yields. This reaction proceeds also efficiently under aerobic conditions (Scheme 41) (Yamada, Matsumura et al. 2018). A mechanism is proposed (Scheme 42). Additionally, with selenium powder as a reagent along with ethynylstibanes and organic azides, a cascade reaction catalyzed by CuI occurs to generate either monoselenide 1,2,3-triazoles or diselenide 1,2,3-triazoles via controlling the amount of Se powder under mild reaction conditions (Scheme 43) (Yamada, Matsumura et al. 2019). A proposed mechanism is described in Scheme 44. First, the coupling of ethynylstibane **1** with the azide catalyzed by CuI produces 5-stibanotriazole **A**. Catalytically active complex **B** is generated from CuI and 1,10-phenanthroline. Transmetallation between 5-stibanotriazole **A** and complex **B** forms intermediate **C**, which is attacted by **A** and selenium along with aromatization via elimination of the antimony moiety to form intermediate **E** and Ph_2SbOH. Finally, intermediate **E** undergoes reductive elimination to give the coupling product **3** with releasing Cu(I). Alternatively, the oxidative homo-coupling of intermediate **E** can afford diselenide **4**, which can be transformed to intermediate **F** mediated by bimetallic complex **B**. 5-Stibanotriazole **A** attacks intermediate **F** to form intermediate **H** and selenide anion **G**. The intermediate **H** undergoes reductive elimination and aromatization to give monoselenide **3**, releasing Ph_2SbOH and Cu(I) (Yamada, Matsumura et al. 2019).

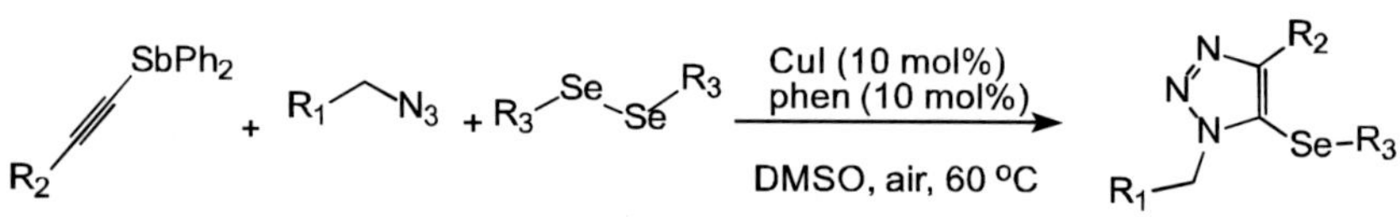

Scheme 41. Three-component reaction of organoantimony, azide, and diaryl diselenides for 5-seleno-1,2,3-triazoles.

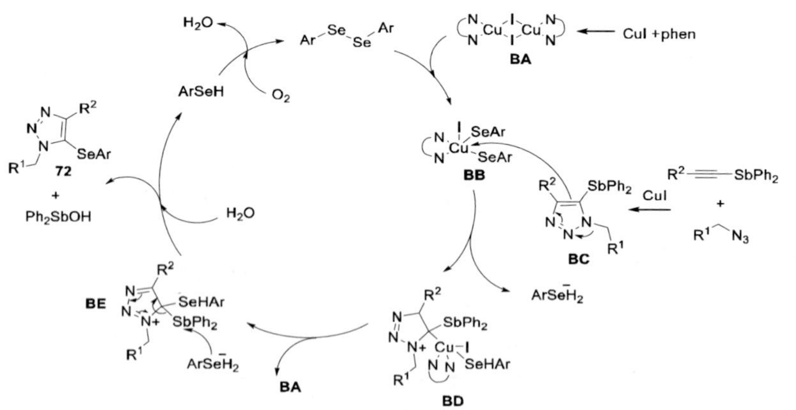

Scheme 42. Proposed mechanism for the three-component reaction of organoantimony, azide, and diaryl diselenides for 5-seleno-1,2,3-triazoles.

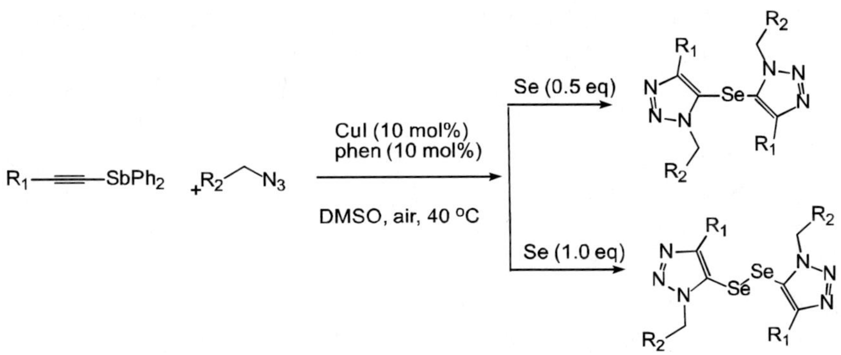

Scheme 43. Three-component reaction of alkynylantimony, azide, and selenium for selenide 1,2,3-triazoles.

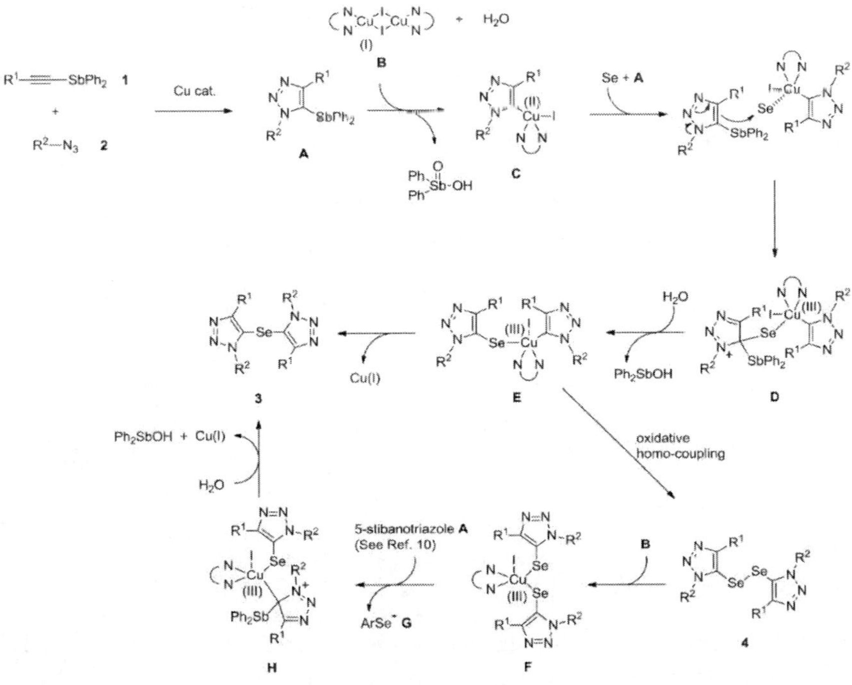

Scheme 44. Proposed mechanism for synthesis of selenide 1,2,3-triazoles from ethynylstibanes organic azides, and selenium.

A microwave-assisted solvent- and catalyst-free multicomponent approach for 5-R-imino-[4-trialkylsilyl(germyl)]-1*H*-1,2,3-triazoles from functionalized primary amines, Si/Ge-substituted propynals, and trimethylsilyl azide has been reported (Scheme 45). A more expedient route to this sequential cascade procedure for bis-heterocyclic adducts starts from the 1,3-dipolar cycloaddition of trimethylsilyl azide to the triple bond of the propynals to afford triazole-5-formaldehyde **Z**, and followed by nucleophilic addition of 2-aminoethanethiol to the aldehyde group, generating the corresponding triazoloimines **AA**. Intramolecular cyclization of intermediates **AA** affords the target (1,3-thiazolan-2-yl)-4-trimethylsilyl(triethylgermyl)-1*H*-1,2,3-triazoles **10a,b** (Scheme 46) (Medvedeva, Demina et al. 2017).

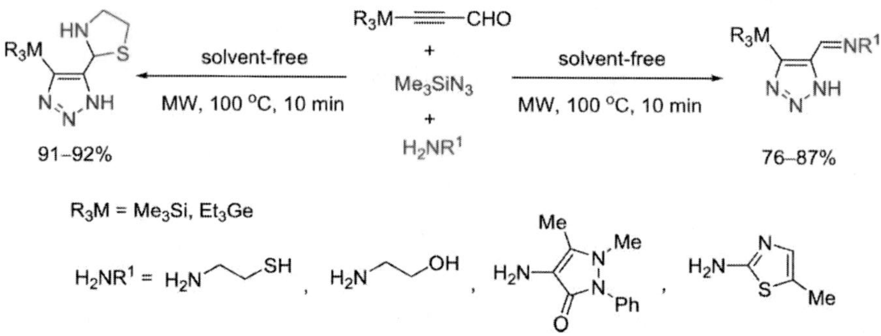

Scheme 45. Three-component synthesis of 1,3-thiazolanyl-1*H*-1,2,3-triazoles and *NH*-1,2,3-triazoloimines.

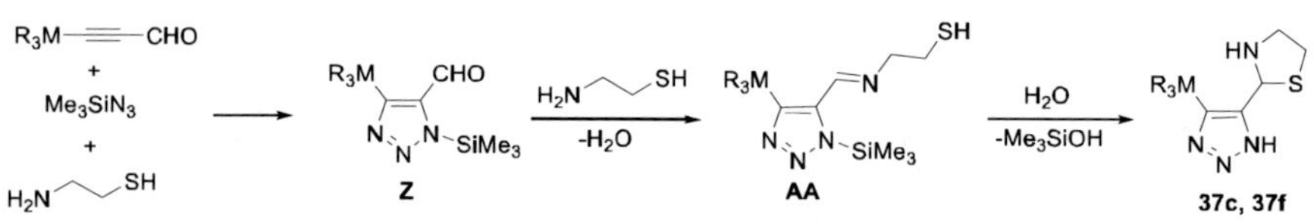

Scheme 46. Proposed mechanism for three-component synthesis of 1,3-thiazolanyl-1*H*-1,2,3-triazoles and *NH*-1,2,3-triazoloimines.

With reagents of internal thioalkynes, organoazides, and *m*-CPBA, regioselective synthesis of both fully substituted 4-sulfonyl-1,2,3-triazoles and fully substituted 5-sulfonyl-1,2,3-triazoles is achieved in good yields (up to 88%) via a tendam process (Scheme 47) (Song, Zheng et al. 2018). One approach, the internal thioalkynes are first oxidized with *m*-CPBA to form internal sulfonyl alkynes, which further cycloadds to azides via the RhAAC process ($[Rh(CO)_2Cl]_2$) to form 5-aryl-4-sulfonyl-1,2,3-triazoles in good yields (80%–88%). Both electron-rich and -deficient aryl azides react smoothly. Alternatively, the RhAAC catalyzing cycloaddition between internal thioalkynes and azides occurrs first to form the 5-sulfur-1,2,3-triazoles and then followed by oxidation *in situ* in a one-pot manner to afford fully substituted 5-sulfonyl-1,2,3-triazoles with excellent yields.

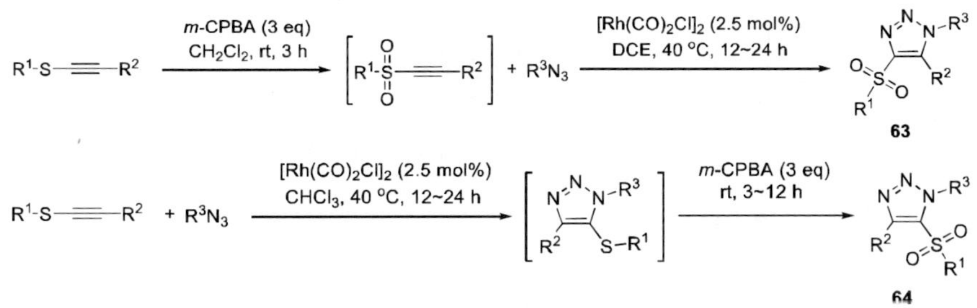

Scheme 47. Tendam process for the fully substituted 5-sulfonyl-1,2,3-triazoles and 4-sulfonyl-1,2,3-triazoles.

The proposed mechanism of the regiodivergent RhAAC reaction is shown in scheme 48 (Song, Zheng et al. 2018). For the sulfone-controlled RhAAC to access 4-sulfonyl-1,2,3-triazoles, the oxidation occurs first followed by the cycloaddition. The high 1,4-regioselectivity is derived from intermediate **A**. The electron-withdrawing sulfonyl group reverses the initial reactivity of the thioalkyne and could not coordinate with Rh(I). Then, Rh-carbene intermediate **B** is formed through the oxidative addition of the β-carbon of the alkyne and azide. The subsequent Rh-assisted isomerization of intermediate **B** to intermediate **C** is followed by the reductive elimination of intermediate **C** to form final 4-sulfonyl-1,2,3-triazole **3**. For the sulfide-controlled RhAAC reaction, the cycloaddition occurs first followed by the in situ oxidation. The high regioselectivity for 5-sulfonyl triazoles **4** is ascribed to the strong coordination of the sulfide of thioalkyne **1** to the rhodium, leading to a key intermediate **B′**.

Wang and co-workers used *O*-benzoyl hydroxylamines as precursors to generate electrophiles to trap Cu(I) triazole intermediates. The synthesis of 1,4,5-trisubstituted 5-dialkylamino-1,2,3-triazoles is performed *via* the tandem reaction of 1-copper(I)-alkynes, azides, and *O*-benzoyl hydroxylamines. The key intermediate of Cu(I) triazoles leads to regioselective 5-dialkylamino-1,2,3-triazoles. Moreover, 1-Cu(I)-alkynes are usually yellow crystals and stable to air and moisture (Scheme 49) (Wang, Liu et al. 2015).

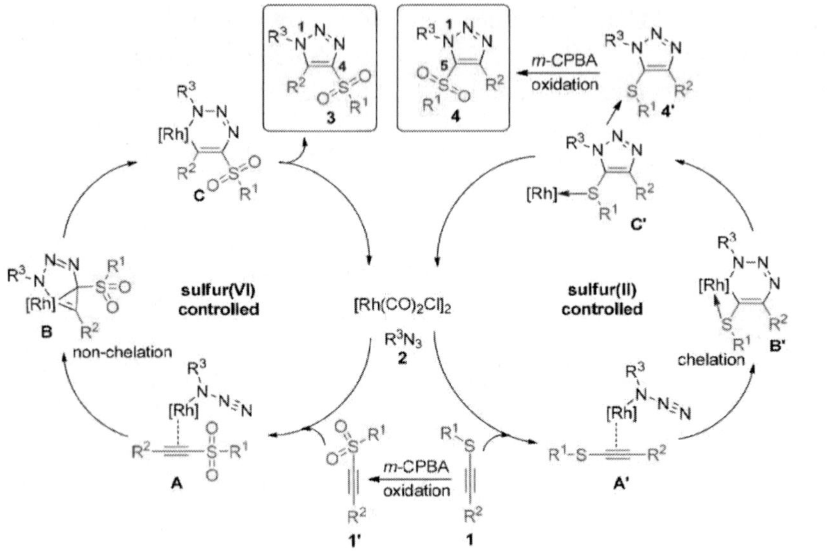

Scheme 48. Proposed mechanism for the regiodivergent RhAAC reaction.

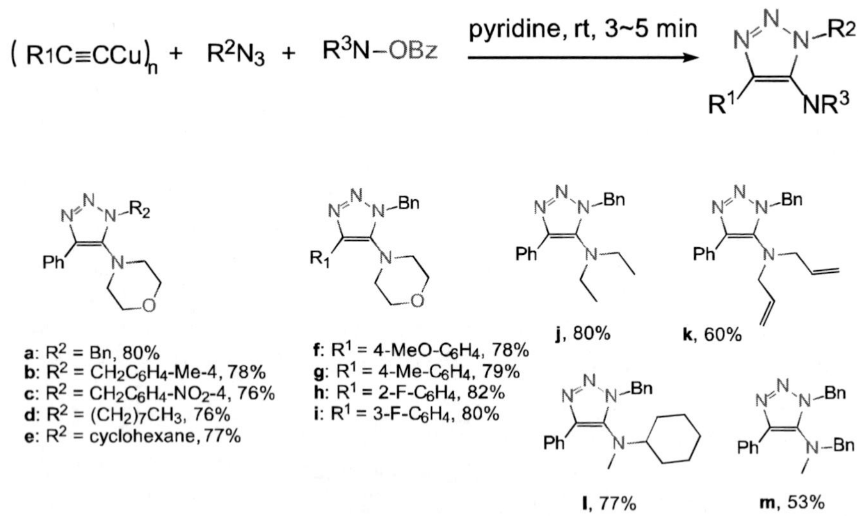

Scheme 49. Regioselective synthesis of 5-dialkylamino-1,2,3-triazoles from 1-copper(I)-alkynes, azides, and *O*-benzoyl hydroxylamines.

In absence of catalyst, the reaction of the bromomagnesium acetylides with azides gives the 4-bromomagnesium-1,5-disubstituted triazoles preferentially, which can be transformed to 1,5-disubstituted triazoles after hydrolysis. Rationally, this magnesium triazoles can also be transformed to trisubstituted triazoles as the magnesium is substituted with electrophiles. The electrophiles, for example, include H_2O, I_2, ClCOOMe, HCOOEt, CO_2, PhCHO, and PhNCO (Scheme 50) (Krasinski, Fokin et al. 2004).

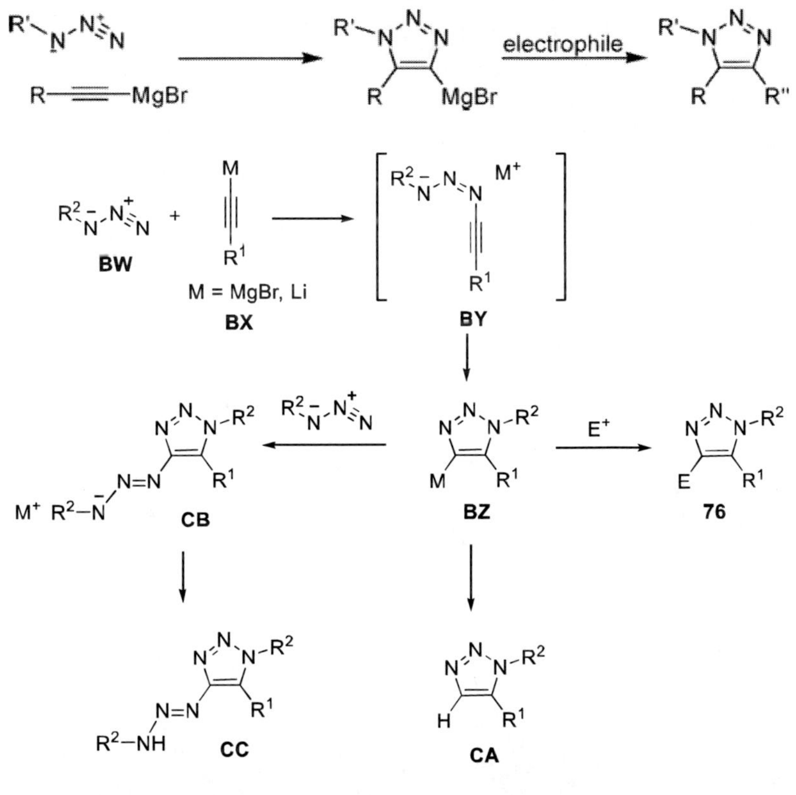

Scheme 50. One-pot synthesis of multisubstituted triazoles from alkynyl Grignard reagents, azides and electrophiles and proposed reaction mechanism.

Synthesis of Functionalized 1,2,3-Triazoles from Non-Alkynes

An efficient three-component domino reaction of aldehydes, amines, and dimethyl α-diazo-β-oxopropylphosphonate (Bestmann-Ohira reagent, BOR) is developed to access full subatituted triazolines and triazoles in methanol solution without any additive. The mechanism of this domino-multicomponent reaction for triazolines and triazoles is proposed as shown in scheme 51. The reaction was initiated by in situ generation of Schiff base **I** from the aldehydes and amines. Next, dimethyl diazomethylphosphonate (DAMP) anion generated in situ from BOR undergoes nucleophilic addition to the intermediate **II**. The **II** would undergo a 5-*endo-dig* cyclization to afford the product **4**. As amine and aldehyde moieties are aromatic, product **4** further undergoes a spontaneous air-assisted oxidative aromatization to provide triazole derivatives **6** (Ahamad, Kant et al. 2016).

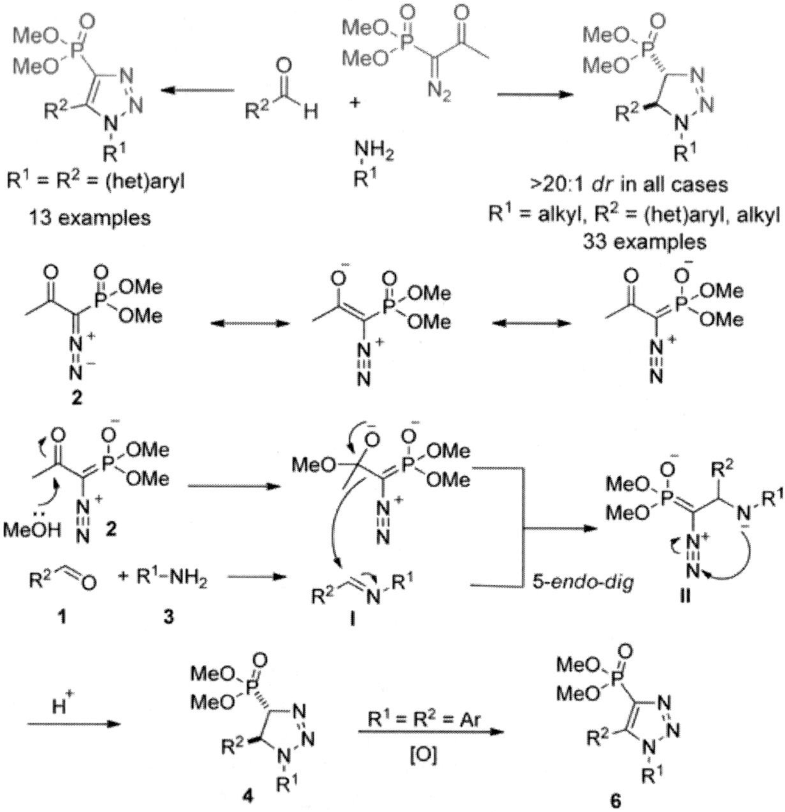

Scheme 51. One-pot synthesis of 4-phosphonate-substituted-1,2,3-triazoles from aromatic aldehydes, aromatic amines (1.5 eq.) and Bestmann-Ohira reagent (1.5 eq.) (K_2CO_3,1.5 eq.; MeOH, rt,12 h) and proposed reaction mechanism.

We developed a metal-free domino reaction for efficient synthesis of 4,5-disubstituted 1,2,3-(*NH*)-triazoles by sequentially coupling sulfur salts with aldehydes and sodium azide in the presence of *L*-proline. The sulfur salts are generated in situ via coupling of dimethylsulfur ether with α-haloacetates, bromomethyl ketone, bromoacetonitrile, or α-bromoacetamide. The coupling of sulfur salts with aldehydes forms in situ the olefinic sulfur salts through proton transfer with *L*-proline. The electron-deficient double bond in olefinic sulfur salts (strong electrophiles like Julia-type reagent, nitroalkene, and acyl acetonitrile) undergoes [2+3]-cycloaddition to azide ions to form a 4,5-disubstituted-1,2,3-(*NH*)-triazole (Scheme 52). This process features mild conditions, high efficiency, commercially available starting materials, and wide substrate scope (Wu and Wu 2018).

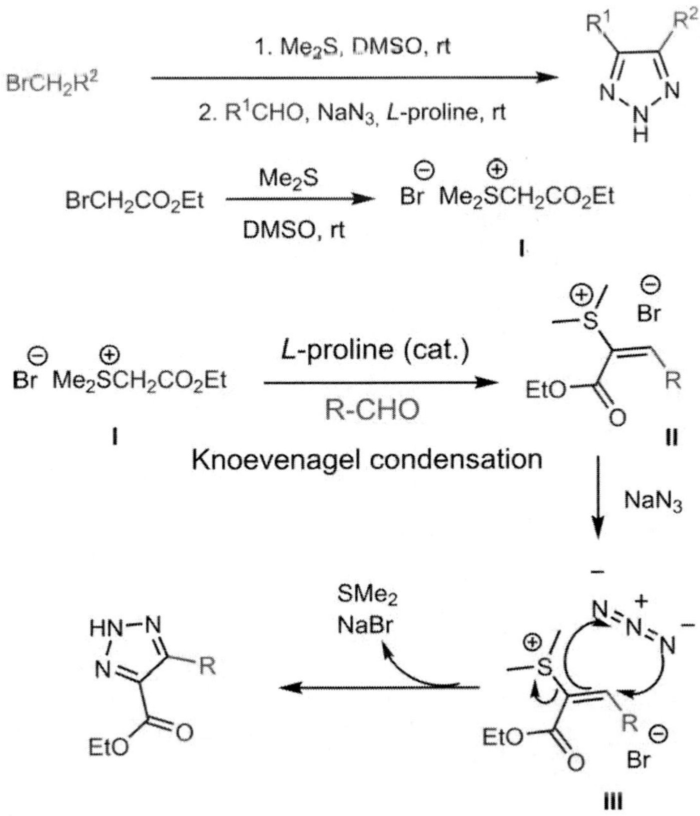

Scheme 52. Domino-multicomponent synthesis of 4,5-disubstituted-1,2,3-(*NH*)-triazoles from organohalides, aldehydes, and sodium azides and proposed reaction mechanism.

Under similar procedure and reaction mechanism, displacement of dimethyl sulfide with pyridine or triphenylphosphine, domino coupling of organohalides, aldehydes, and sodium azide proceeds also efficiently in one-pot to form various 4,5-disubstituted-1,2,3-(*NH*)-triazoles under mild conditions. The olefinic pyridinium salt (Scheme 53a) and the olefinic phosphonium salt (Scheme 53b) are vital intermediates, generated also in situ, respectively (Wu and Wu 2018, Wu and Wu 2018).

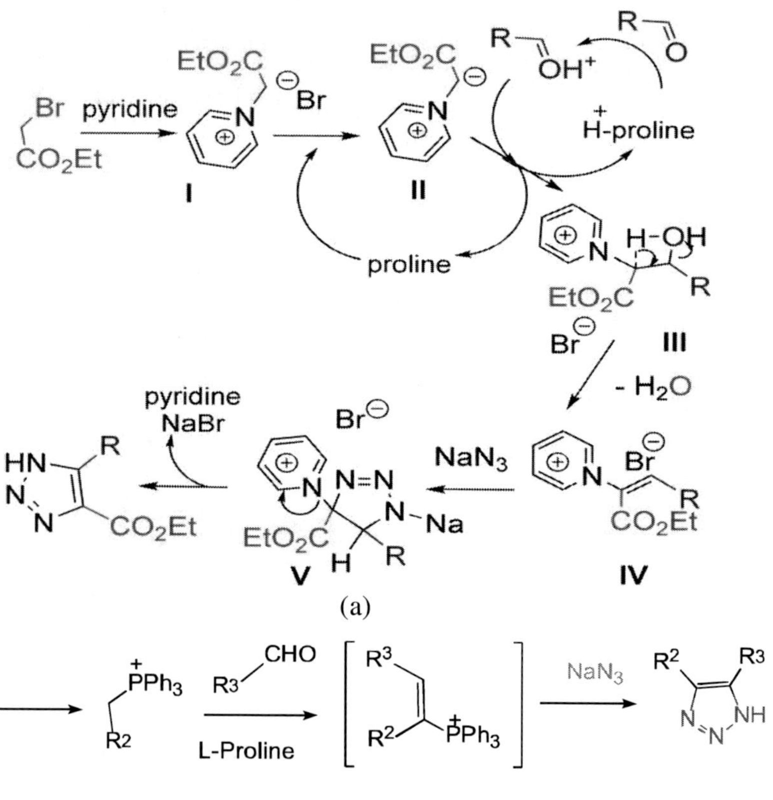

Scheme 53. Proposed mechanism for domino-multicomponent synthesis of 4,5-disubstituted-1,2,3-(*NH*)-triazoles from organohalides, aldehydes, and sodium azide, mediating with pyridine and triphenyl phosphine, respectively.

Furthermore, nitroalkanes can also couple with aldehydes and organic azides to form 1,4,5-trisubstituted 1,2,3-triazoles via a metal-free three-component cascade reaction (Scheme 54). This process starts from a

Knoevenagel condensation of the formyl group with the nitro compound to form nitroalkenes in the presence of amine and acid catalysis. The following 1,3-dipolar cycloaddition of the azide to the activated alkene produces 1,2,3-triazoles with a high yield and regioselectivity.

Notably, the aldehyde is first transformed to an activated iminium intermediate with morpholine salt (Scheme 53). This intermediate is then attacked by a nitroalkane to afford a Knoevenagel adduct, nitroalkenes. The high regioselectivity of this reaction is resulted from the strongly electron-withdrawing nitro group, which lowers the energy of the lowest unoccupied molecular orbital (LUMO) and causes the β carbon atom to be the strongly electrophilic, and further induces 1,3-dipolar cycloaddition reaction to afford triazoles with high regioselectivity (Thomas, John et al. 2014).

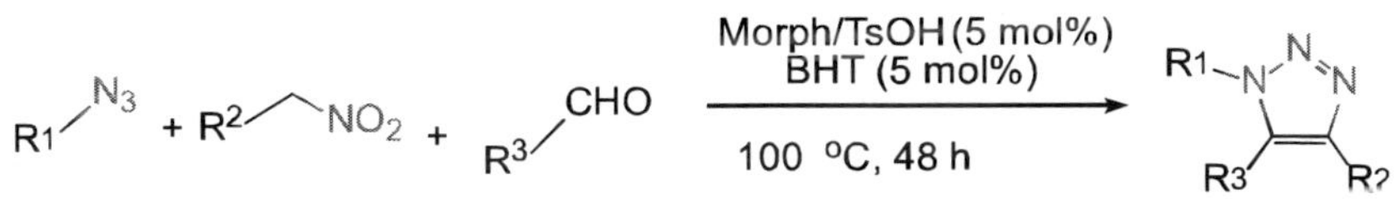

Scheme 54. Synthesis of 4-heteroatom-substituted triazoles from azides, nitro-compounds and aldehydes. (Reaction conditions: aldehydes (1.3 equiv), nitroalkanes (1.0 equiv), organic azides (1.0 equiv), Morph/TsOH (5 mol %), BHT (5 mol %), 4 Å M.S. (50 mg), toluene (0.2 mL), 100°C, 48 h).

Disclaimer

None

References

Ahamad, S., R. Kant and K. Mohanan, Metal-Free Three-Component Domino Approach to Phosphonylated Triazolines and Triazoles. *Organic Letters* (2016)18: 280-283.

Akter, M., K. Rupa and P. Anbarasan, 1,2,3-Triazole and Its Analogues: New Surrogates for Diazo Compounds. *Chemical Reviews* (2022)122: 13108-13205.

Appukkuttan, P., W. Dehaen, V. V. Fokin and E. Van der Eycken, A microwave-assisted click chemistry synthesis of 1,4-disubstituted 1,2,3-triazoles via a copper(I)-catalyzed three-component reaction. *Organic Letters* (2004)6: 4223-4225.

Ashok, D., Reddy, M. R., Thara, G., Dharavath, R., Ramakrishna, K., Nagaraju, N., Gundu, S., & Sarasija, M., A new library of 1,2,3-triazole based benzofuran scaffolds: Synthesis and biological evaluation as potential antimicrobial agents. *Journal of Heterocyclic Chemistry* (2022)59: 1376-1384.

Neha Batra, Vinoth Rajendran, Drishti Agarwal, Ishan Wadi, Prahlad C. Ghosh, Rinkoo D. Gupta, Mahendra Nath. Synthesis and Antimalarial Evaluation of 1,2,3 -Triazole-Tethered Sulfonamide-Berberine Hybrids. *Chemistryselect* (2018)3: 9790-9793.

Beutick, S. E., P. Vermeeren and T. A. Hamlin, The 1,3-Dipolar Cycloaddition: From Conception to Quantum Chemical Design. *Chemistry-an Asian Journal* (2022)17(17): e202200553.

Bock, V. D., D. Speijer, H. Hiemstra and J. H. van Maarseveen, 1,2,3-Triazoles as peptide bond isosteres: synthesis and biological evaluation of cyclotetrapeptide mimics. *Organic & Biomolecular Chemistry* (2007)5: 971-975.

Bonandi, E., Christodoulou, M. S., Fumagalli, G., Perdicchia, D., Rastelli, G., & Passarella, D., The 1,2,3-triazole ring as a bioisostere in medicinal chemistry. *Drug Discovery Today* (2017)22: 1572-1581.

Bozorov, K., J. Zhao and H. A. Aisa, 1,2,3-Triazole-containing hybrids as leads in medicinal chemistry: A recent overview. *Bioorganic & Medicinal Chemistry* (2019)27: 3511-3531.

Brotherton, W. S., R. J. Clark and L. Zhu, Synthesis of 5-Iodo-1,4-disubstituted-1,2,3-triazoles Mediated by in Situ Generated Copper(I) Catalyst and Electrophilic Triiodide Ion. *Journal of Organic Chemistry* (2012)77: 6443-6455.

Chen, F.-J., Mamidipalli, P., Sabbasani, V. R., Liu, H., Xia, Y., & Lee, D., Three-component coupling reaction for the synthesis of fully substituted triazoles: reactivity control of Cu-acetylide toward alkyl azides and diazo compounds. *Organic Chemistry Frontiers* (2021)8: 6095-6107.

Cui, F., Chen, J., Mo, Z., Su, S., Chen, Y., Ma, X., Tang, H., Wang, H., Pan, Y., & Xu, Y. Copper-Catalyzed Decarboxylative/Click Cascade Reaction: Regioselective Assembly of 5-Selenotriazole Anticancer Agents. *Organic Letters* (2018a) 20: 925-929.

Das, U. K., R. K. Jena and M. Bhattacharjee, Synthesis, structure and catalytic properties of Ru(dppp)(2)(CH3CN)Cl BPh4 and isolation of catalytically active Ru(dppp)(2)Cl BPh4: ruthenium catalysed alkyne homocoupling and tandem alkyne-azide cycloaddition. *Rsc Advances* (2014)4: 21964-21970.

Devender, N., Gunjan, S., Chhabra, S., Singh, K., Pasam, V. R., Shukla, S. K., Sharma, A., Jaiswal, S., Singh, S. K., Kumar, Y., Lal, J., Trivedi, A. K., Tripathi, R., & Tripathi, R. P., Identification of beta-Amino alcohol grafted 1,4,5 trisubstituted 1,2,3-triazoles as potent antimalarial agents. European *Journal of Medicinal Chemistry* (2016)109: 187-198.

Feng, L.-S., M.-J. Zheng, F. Zhao and D. Liu, 1,2,3-Triazole hybrids with anti-HIV-1 activity. *Archiv Der Pharmazie* (2021)354(1): e2000163.

Gabba, A., Robakiewicz, S., Taciak, B., Ulewicz, K., Broggini, G., Rastelli, G., Krol, M., Murphy, P. V., & Passarella, D., Synthesis and Biological Evaluation of Migrastatin Macrotriazoles. *European Journal of Organic Chemistry* (2017)2017: 60-69.

Gadzikwa, T., Farha, O. K., Malliakas, C. D., Kanatzidis, M. G., Hupp, J. T., & Nguyen, S. T., Selective Bifunctional Modification of a Non-catenated Metal-Organic Framework Material via "Click" Chemistry. *Journal of the American Chemical Society* (2009)131: 13613.

Gao, G., Kuantao, Mao, L. Lv and Z. Li, Three-Component Coupling of alpha-Trifluoromethyl Carbonyls, Azides and Amines for the Regioselective Synthesis of 1,4,5-Trisubstituted 1,2,3-Triazoles. *Advanced Synthesis & Catalysis* (2022)364: 1402-1408.

Garbarino, S., D. Ravelli, S. Protti and A. Basso, Photoinduced Multicomponent Reactions. *Angewandte Chemie-International Edition* (2016) 55: 15476-15484.

Golas, P. L., and K. Matyjaszewski, Marrying click chemistry with polymerization: expanding the scope of polymeric materials. *Chemical Society Reviews* (2010)39: 1338-1354.

Hakimian, S., A. Cheng-Hakimian, G. D. Anderson and J. W. Miller, Rufinamide: a new anti-epileptic medication. *Expert Opinion on Pharmacotherapy* (2007)8: 1931-1940.

Heravi, M. M., and N. Nazari, Isocyanide-based Cascade Reactions in the Synthesis of Aza-heterocycles: A Marriage of Convenience. *Current Organic Chemistry* (2017)21: 1440-1529.

Jiang, X., Hao, X., Jing, L., Wu, G., Kang, D., Liu, X., & Zhan, P., Recent applications of click chemistry in drug discovery. *Expert Opinion on Drug Discovery* (2019)14: 779-789.

John, S. E., S. Gulati and N. Shankaraiah, Recent advances in multi-component reactions and their mechanistic insights: a triennium review. *Organic Chemistry Frontiers* (2021)8: 4237-4287.

Juricek, M , P. H. J. Kouwer and A. E. Rowan, Triazole: a unique building block for the construction of functional materials. *Chemical Communications* (2011)47: 8740-8749.

Kolarovic, A., M. Schnuerch and M. D. Mihovilovic, Tandem Catalysis: From Alkynoic Acids and Aryl Iodides to 1,2,3-Triazoles in One Pot. *Journal of Organic Chemistry* (2011)76: 2613-2618.

Krasinski, A., V. V. Fokin and K. B. Sharpless, Direct synthesis of 1,5-disubstituted-4-magnesio-1,2,3-triazoles, revisited. *Organic Letters* (2004)6: 1237-1240.

Lauria, A., Delisi, R., Mingoia, F., Terenzi, A., Martorana, A., Barone, G., & Almerico, A. M., 1,2,3-Triazole in Heterocyclic Compounds, Endowed with Biological Activity, through 1,3-Dipolar Cycloadditions. *European Journal of Organic Chemistry* (2014) 2014: 3289-3306.

Li, L., G. Hao, A. Zhu, X. Fan, G. Zhang, Zhang L., A Copper(I)-Catalyzed Three-Component Domino Process: Assembly of Complex 1,2,3-Triazolyl-5-Phosphonates from Azides, Alkynes, and H-Phosphates. *Chemistry-a European Journal* (2013a)19: 14403-14406.

Li, L., G. Hao, A. Zhu, S. Liu and G. Zhang, Three-component assembly of 5-halo-1,2,3-triazoles via aerobic oxidative halogenation. *Tetrahedron Letters* (2013b)54: 6057-6060.

Li, L., R. Li, A. Zhu, G. Zhang and L. Zhang, CuBr-NCS-Mediated Multicomponent Azide-Alkyne Cycloaddition: Mild and Efficient Synthesis of 5-Bromo-1,4-Disubstituted-1,2,3-Triazoles. *Synlett*: (2011)874-878.

Li, L., Shang, T., Ma, X., Guo, H., Zhu, A., & Zhang, G., 4-Trimethylsilyl-5-iodo-1,2,3-triazole: A Key Precursor for the Divergent Syntheses of 1,5-Disubstituted 1,2,3-Triazoles. *Synlett* (2015)26: 695-699.

Li, L., G. Zhang, A. Zhu and L. Zhang, A convenient preparation of 5-iodo-1,4-disubstituted-1,2,3-triazole: Multicomponent one-pot reaction of azide and alkyne mediated by CuI-NBS. *Journal of Organic Chemistry (*2008)73: 3630-3633.

Liebert, T., C. Hansch and T. Heinze, Click chemistry with polysaccharides. *Macromolecular Rapid Communications* (2006)27: 208-213.

Malnuit, V., M. Duca, A. Manout, K. Bougrin and R. Benhida, Tandem Azide-Alkyne 1,3-Dipolar Cycloaddition/Electrophilic Addition: A Concise Three-Component Route to 4,5-Disubstituted Triazolyl-Nucleosides. *Synlett*: (2009)2123-2128.

Medvedeva, A. S., Demina, M. M., Kon'kova, T. V., Nguyen, T. L. H., Afonin, A. V., & Ushakov, I. A., Microwave assisted solvent- and catalyst-free three-component synthesis of NH-1,2,3-triazoloimines. *Tetrahedron* (2017) 73: 3979-3985.

Neu, H. C., and K. P. Fu, Cefatrizine activity compared with that of other cephalosporins. *Antimicrobial Agents and Chemotherapy* (1979) 15: 209-212.

Qin, A., J. W. Y. Lam and B. Z. Tang, Click polymerization. *Chemical Society Reviews* (2010)39: 2522-2544.

Rani, A., Singh, G., Singh, A., Maqbool, U., Kaur, G., & Singh, J., CuAAC-ensembled 1,2,3-triazole-linked isosteres as pharmacophores in drug discovery: review. *Rsc Advances* (2020)10: 5610-5635.

Rostovtsev, V. V., L. G. Green, V. V. Fokin and K. B. Sharpless, A stepwise Huisgen cycloaddition process: Copper(I)-catalyzed regioselective "ligation" of azides and terminal alkynes. *Angewandte Chemie-International Edition* (2002)41: 2596.

Shang, J.-Q., Fu, H., Li, Y., Yang, T., Gao, C., & Li, Y.-M., Copper-catalyzed decarboxylation/cycloaddition cascade of alkynyl carboxylic acids with azide. *Tetrahedron* (2019)75: 253-259.

Smith, C. D., and M. F. Greaney, (2013) Zinc Mediated Azide-Alkyne Ligation to 1,5-and 1,4,5-Substituted 1,2,3-Triazoles. *Organic Letters* 15: 4826-4829.

Song, W., Zheng, N., Li, M., Dong, K., Li, J., Ullah, K., & Zheng, Y., Regiodivergent Rhodium(I)-Catalyzed Azide-Alkyne Cycloaddition (RhAAC) To Access Either Fully Substituted Sulfonyl-1,2,3-triazoles under Mild Conditions. *Organic Letters* (2018)20: 6705-6709.

Thirumurugan, P., D. Matosiuk and K. Jozwiak, Click Chemistry for Drug Development and Diverse Chemical-Biology Applications. *Chemical Reviews* (2013)113: 4905-4979.

Thomas, J., J. John, N. Parekh and W. Dehaen, A Metal-Free Three-Component Reaction for the Regioselective Synthesis of 1,4,5-Trisubstituted 1,2,3-Triazoles. *Angewandte Chemie-International Edition* (2014)53: 10155-10159.

Tornoe, C. W., C. Christensen and M. Meldal, Peptidotriazoles on solid phase: 1,2,3 -triazoles by regiospecific copper(I)-catalyzed 1,3-dipolar cycloadditions of terminal alkynes to azides. *Journal of Organic Chemistry* (2002)67: 3057-3064.

Vala, D. P., R. M. Vala and H. M. Patel, Versatile Synthetic Platform for 1,2,3-Triazole Chemistry. *Acs Omega* (2022)7(42):36945-36987.

Wang, B., Liu, N., Chen, W., Huang, D., Wang, X., & Hu, Y., Minutes Synthesis of 1,4,5-Trisubstituted 5-Dialkylamino-1,2,3-triazoles by 1-Copper(I)-Alkyne Controlled Tandem Process. *Advanced Synthesis & Catalysis* (2015) 357: 401-407.

Wang, C., Li, Q., Wang, S., Zhu, G., Zhu, A., & Li, L., Copper-catalyzed in situ oxidative-coupling for one-pot synthesis of 5-aryl-1,4-disubstituted 1,2,3-triazoles under mild conditions. *Rsc Advances* (2021)11: 38108-38114.

Wang, W., Y. Lin, Y. Ma, C.-H. Tung and Z. Xu, Copper(I)-Catalyzed Three-Component Click/Persulfuration Cascade: Regioselective Synthesis of Triazole Disulfides. *Organic Letters* (2018) 20: 2956-2959.

Wang, W., X. Peng, F. Wei, C.-H. Tung and Z. Xu, Copper(I)-Catalyzed Interrupted Click Reaction: Synthesis of Diverse 5-Hetero-Functionalized Triazoles. *Angewandte Chemie-International Edition* (2016)55: 649-653.

Wang, X., Sun, B., Zhao, Z., Chen, X., Xia, W., Shen, Y., & Li, Y., Copper-Catalyzed Four-Component Cascade Reaction for the Construction of Triazoles Bearing beta-Hydroxy Chalcogenides. *Advanced Synthesis & Catalysis* (2022)364: 165-171.

Wang, X.-X., Xin, Y., Li, Y., Xia, W.-J., Zhou, B., Ye, R.-R., & Li, Y.-M., Copper-Catalyzed Decarboxylative Cycloaddition of Propiolic Acids, Azides, and Arylboronic Acids: Construction of Fully Substituted 1,2,3-Triazoles. *Journal of Organic Chemistry* (2020a)85: 3576-3586.

Wang, Y., Yu, P., Ren, Q., Jia, F., Chen, Y., & Wu, A., Synthesis of 2,1-Benzoisoxazole-Containing 1,2,3-Triazoles through Copper-Catalyzed Three-Component Domino Reactions of o-Bromoacetophenones, Aldehydes, and Sodium Azide. *Journal of Organic Chemistry* (2020b)85: 2688-2696.

Wei, F., T. Zhou, Y. Ma, C.-H. Tung and Z. Xu, Bench-Stable 5-Stannyl Triazoles by a Copper(I)-Catalyzed Interrupted Click Reaction: Bridge to Trifluoromethyltriazoles and Trifluoromethylthiotriazoles. *Organic Letters* (2017)19: 2098-2101.

Worrell, B. T., J. A. Malik and V. V. Fokin, Direct Evidence of a Dinuclear Copper Intermediate in Cu(I)-Catalyzed Azide-Alkyne Cycloadditions. *Science* (2013)340: 457-460.

Wu, G.-L., and Q.-P. Wu, A Domino Reaction of Pyridinium Salts: Efficient Synthesis of 4,5-Disubstituted 1,2,3-(NH)-Triazoles. *Chemistryselect* (2018a)3: 5212-5215.

Wu, G.-L., and Q.-P. Wu, Metal-Free Multicomponent Reaction for Synthesis of 4,5-Disubstituted 1,2,3-(NH)-Triazoles. *Advanced Synthesis & Catalysis* (2018b)360: 1949-1953.

Wu, G.-L., and Q.-P. Wu, A Mild Multi-Component Reaction for the Synthesis of 4,5-Disubstituted 1H-1,2,3-Triazoles from Phosphonium Salts, Aldehydes, and Sodium Azide. *Synthesis-Stuttgart* (2018c)50: 2768-2774.

Yamada, M., Matsumura, M., Sakaki, E., Yen, S., Kawahata, M., Hyodo, T., Yamaguchi, K., Murata, Y., & Yasuike, S., Copper-catalyzed three-component reaction of ethynylstibanes, organic azides, and selenium: A simple and efficient synthesis of novel selenides and diselenides having 1,2,3-triazole rings. *Tetrahedron* (2019)75: 1406-1414.

Yamada, M., Matsumura, M., Takino, F., Murata, Y., Kurata, Y., Kawahata, M., Yamaguchi, K., Kakusawa, N., & Yasuike, S., Synthesis of Fully Functionalized 5-Selanyl-1,2,3-triazoles: Copper-Catalysed Three-Component Reaction of Ethynylstibanes, Organic Azides, and Diaryl Diselenides. *European Journal of Organic Chemistry* (2018)2018: 170-177.

Yan, R., Sander, K., Galante, E., Rajkumar, V., Badar, A., Robson, M., El-Emir, E., Lythgoe, M. F., Pedley, R. B., & Årstad, E., A One-Pot Three-Component Radiochemical Reaction for Rapid Assembly of I-125-Labeled Molecular Probes. *Journal of the American Chemical Society* (2013)135: 703-709.

Zhang, L.-L., M.-T. Li, L.-L. Shen and Q.-P. Wu, Efficient Synthesis of 5-Trifluoromethylthio-1,2,3-Triazoles: One-Pot Multicomponent Reaction from Elemental Sulfur and TMSCF3. *Synthesis-Stuttgart* (2020)52: 304-310.

Zhang, L.-L., Li, Y.-T., Gao, T., Guo, S.-S., Yang, B., Meng, Z.-H., Dai, Q.-P., Xu, Z.-B., & Wu, Q.-P., Efficient Synthesis of Diverse 5-Thio- or 5-Selenotriazoles: One-Pot Multicomponent Reaction from Elemental Sulfur or Selenium. *Synthesis-Stuttgart* (2019) 51: 4170-4182.

Zhang, X., R. P. Hsung and H. Li, A triazole-templated ring-closing metathesis for constructing novel fused and bridged triazoles. *Chemical Communications*: (2007)2420-2422.

Zhang, X., H. Li, L. You, Y. Tang and R. P. Hsung, Copper salt-catalyzed azide- 3+2 cycloadditions of ynamides and bis-ynamides. *Advanced Synthesis & Catalysis* (2006)348: 2437-2442.

Zhu, A., Xing, X., Wang, S., Yuan, D., Zhu, G., Geng, M., Guo, Y., Zhang, G., & Li, L., Multi-component syntheses of diverse 5-fluoroalkyl-1,2,3-triazoles facilitated by air oxidation and copper catalysis. *Green Chemistry* (2019)21: 3407-3412.

Chapter 2

Design and Synthesis of New 1,2,3-Triazole Hybrids Possessing Antitubercular and Antimicrobial Activities

Pramod S. Phatak[1,2]
Adinath D. Badar[1,2]
and Kishan P. Haval[2,*]
[1]Department of Chemistry,
Late Pushpadevi Patil Arts and Science College,
Risod Dist Washim (MS), India
[2]Department of Chemistry,
Dr. Babasaheb Ambedkar Marathwada University,
Sub-campus, Osmanabad (MS), India

Abstract

This chapter deals with synthesis of various triazoles and their uses in medicinal therapeutics. The compounds with triazole moiety are one of the significant heterocycles with broad range of biological activities such as antimicrobial, antitubercular, anticancer, etc. In present chapter, we have described the triazole derivatives synthesized by our research group possessing potential antitubercular and antimicrobial activities. The antitubercular activity is examined against *Mycobacterium tuberculosis* H37Rv (MTB) and most of them have shown better activity with compared to standard drugs. The cytotoxicity of the active moieties were examined against HEK 293 cell and RAW 264.7 cell line by MTT assay and no toxicity was observed even at small concentration. The

* Corresponding Author's Email: havalkp@gmail.com

In: Triazoles and Their Uses
Editor: Karl Roland
ISBN: 979-8-88697-800-1

antimicrobial activities were examined against both antibacterial *(S. aureus, B. cerus, B. subtilis, E. aerogenes, E. coli, S. typhi, P. aeruginosa, S. boydii, S. abony*) and antifungal (*A. niger, C. albicans, S. cerevisiae*) pathogens and have displayed potent activities.

A molecular docking study was performed against *enoyl-ACP-reductase* (InhA) and *DNA gyrase* to gain an insight into the molecular mechanism. The structure-activity relationship (SAR) is also studied. The pharmacokinetic parameters of moieties were studied and displayed acceptable drug-likeness score.

Keywords: 1,2,3-triazoles, click chemistry, antimicrobial activity, antitubercular activity, molecular docking study

Introduction

1,2,3-Triazole scaffold containing three nitrogen atoms and two carbon atoms are one of the significant heterocycles that exhibit broad biological activities. 1,2,3-Triazoles are synthesized by "Click Chemistry" approach. It was firstly reported by Sharpless and research group [1]. Its unique structure facilitates the formation of various non-covalent bonds with enzymes and receptors, inducing various biological activities, such as anticancer [2], antitubercular [3], antibacterial [4], anti-HIV and many more.

In addition, 1,2,3-triazole containing drugs like cefatrizine (**A**) and CAI (**B**) are very successful as anticancer drugs [5].

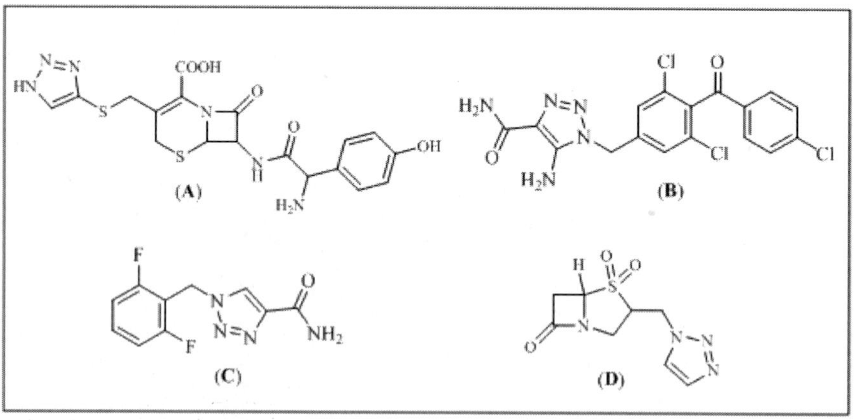

Figure 1. Some 1,2,3-Triazole containing drugs used in clinical trials.

Rufinamide (**C**) was approved by FDA for the treatment of pediatric epilepsy [6, 7]. Tazobactam (**D**) has been successfully used in antibiotic therapy [8, 9] (Figure 1). In this study, we have focused on the various 1,2,3-triazole containing compounds which we have synthesized in our laboratory and discussed their biological activities mainly their antitubercular and antimicrobial activities.

Results and Discussion

Antitubercular Activity

In continuation with our research on synthesis of new antitubercular agents, herein we have synthesized new indanol-1,2,3-triazole derivatives by employing click chemistry approach. The synthesized compounds were screened for their *in vitro* antimycobacterial activity against *Mycobacterium tuberculosis* H37Rv (MTB) strain. These compounds have displayed excellent antitubercular activity compared to first line antitubercular drugs, ciprofloxacin and ethambutol. The compound **1a** has displayed excellent antitubercular activity with MIC value 1.56μM, which is more potent than ethambutol and equivalent to another standard drug, ciprofloxacin. The three compounds **1b**, **1c** and **1d** were shown significant antitubercular activity with MIC value 6.25μM. The compound **1k** has shown good antitubercular activity against *Mycobacterium tuberculosis* H37Rv strain with MIC value 6.25 μM. The four compounds **1e**, **1f**, **1g** and **1h** were displayed noticeable antitubercular activity with MIC value 12.5 μM. The compounds **1i** and **1j** displayed good antitubercular activity with MIC value 25μM [10] (Figure 2).

In another study, we have reported the synthesis of substituted 2-(4-((1,3-dioxoisoindolin-2-yl)methyl)-1*H*-1,2,3-triazol-1-yl)-*N*-phenylacetamides starting from phthalimide *via* click chemistry approach and studied their antitubercular activity. The compounds **2a**, **2b**, **2c**, and **2d** have displayed notable antitubercular activity with MIC values 12.5, 25, 25, and 12.5μg/mL, respectively [11] (Figure 3).

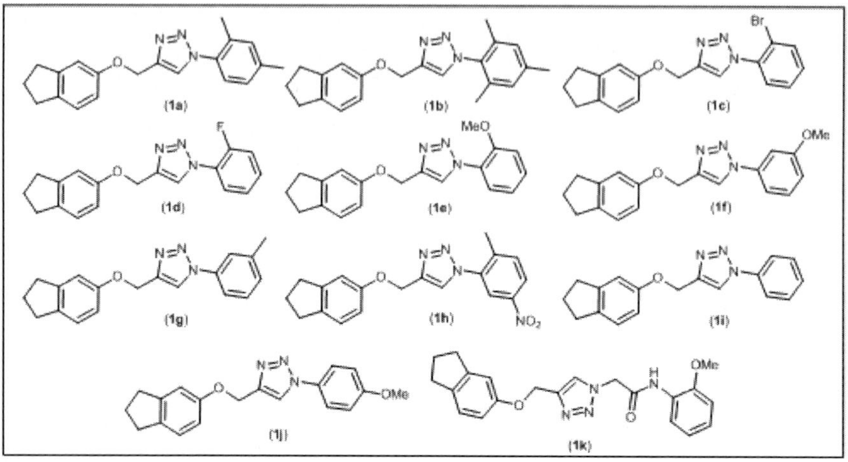

Figure 2. Indanol-1,2,3-triazole derivatives possessing antitubercular activity.

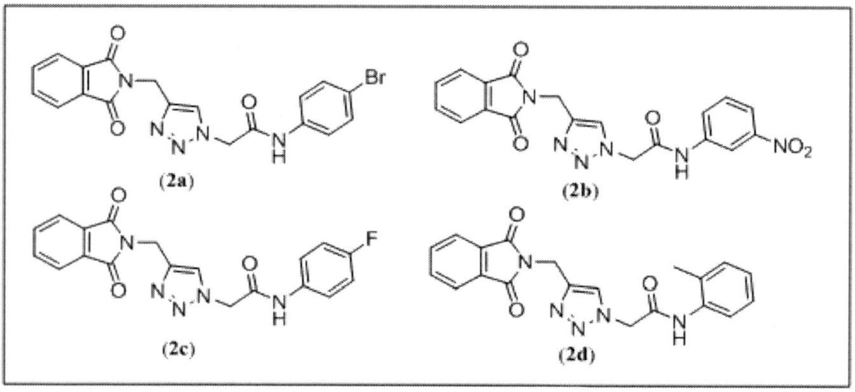

Figure 3. Phthalimide-1,2,3-triazole derivatives possessing antitubercular activity.

A series of new isoniazid-1,2,3-triazole conjugates via click chemistry approach have been reported by our group. The newly synthesized compounds were assessed for their *in vitro* antimycobacterial activities against *M. tuberculosis* H37Rv (Mtb) strain. The four derivatives have exhibited good antitubercular activity. The compound **3a** is equally active to standard antitubercular drug, Ciprofloxacin with MIC value 1.56μg/mL. It has shown better activity than the antitubercular drug, Ethambutol. The compound **3b, 3c** and **3d** have displayed antitubercular activity with MIC values 6.25, 12.5 and 12.5μg/mL, respectively. The *in vitro* cytotoxicity of the most active antitubercular derivatives with lower MIC value were evaluated against growth inhibition of RAW 264.7 cells at 25μg/mL concentration by 3-(4,5-dimethylthiazol-2-yl)-2,5-diphenyltetrazolium bromide (MTT) assay. The results obtained indicate that these compounds are not harmful to the normal cells [12] (Figure 4).

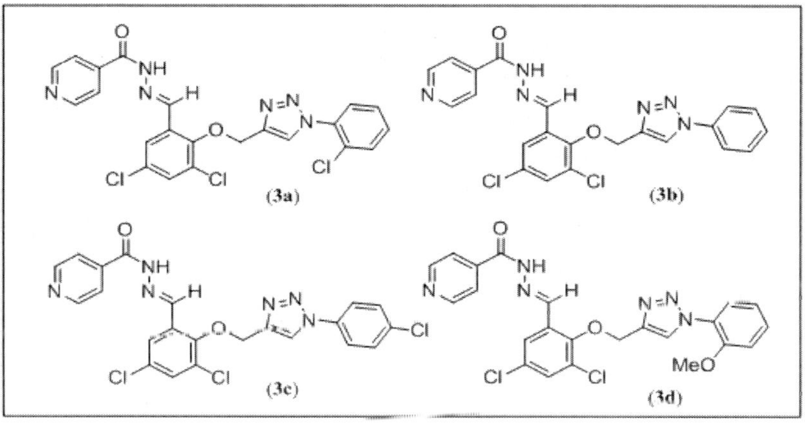

Figure 4. Isoniazid -1,2,3-triazole derivatives possessing antitubercular activity.

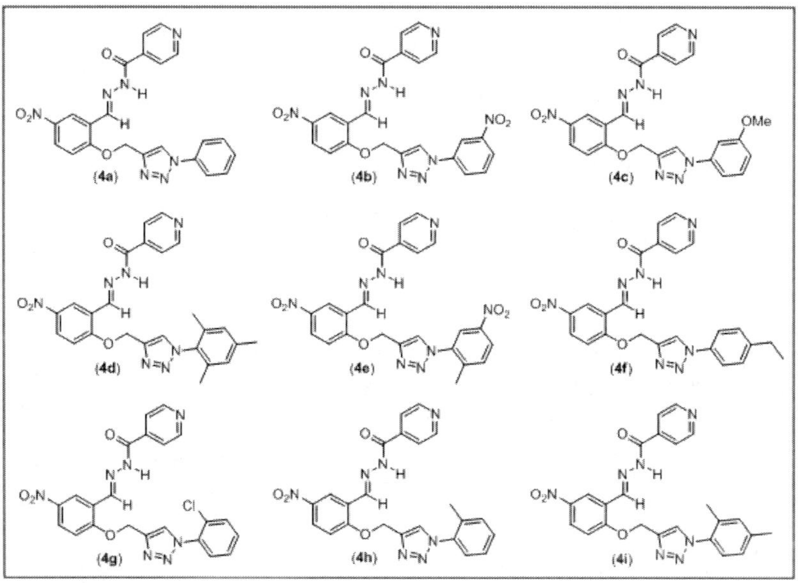

Figure 5. Isoniazid embedded 1,4-disubstituted 1,2,3-triazole derivatives possessing antitubercular activity.

In one of the studies, we have synthesized series of novel isoniazid embedded 1,4-disubstituted 1,2,3-triazole analogues starting from 5-nitrosalicylaldehyde and screened for their antitubercular activity. The compounds **4a**, **4b**, **4c**, **4d**, **4e** and **4f** have shown excellent antitubercular activity with MIC value 0.78μg/mL. While, compounds **4g**, **4h** and **4i** have displayed noticeable antitubercular activity with MIC value 1.56, 1.56 and 3.125μg/mL, respectively [13] (Figure 5).

Antimicrobial Activity

We have synthesized isoniazid embedded triazole derivatives and evaluated for their *in vitro* antimicrobial activities against both Gram-positive and Gram-negative bacteria by agar well diffusion assay. Tetracycline and Fluconazole were used as reference standard for antibacterial and antifungal activities, respectively. The compounds **4g**, **4h**, **4j**, **4k** and **4l** have exhibited good antimicrobial activities against both antibacterial and antifungal pathogens [13] (Figure 6).

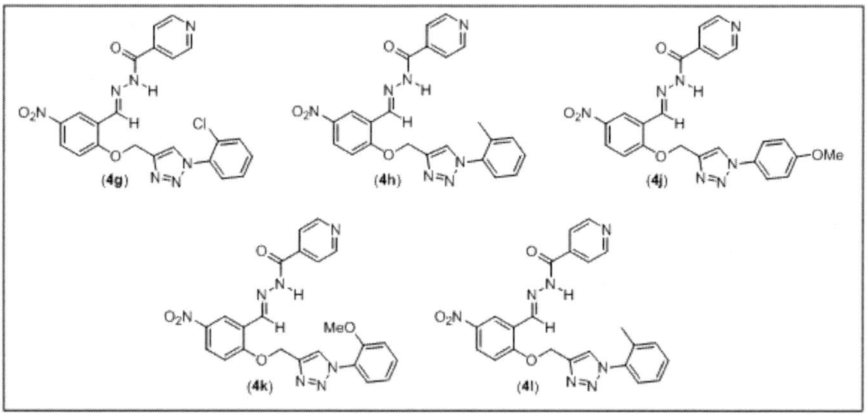

Figure 6. Isoniazid embedded 1,4-disubstituted 1,2,3-triazole derivatives possessing antimicrobial activities.

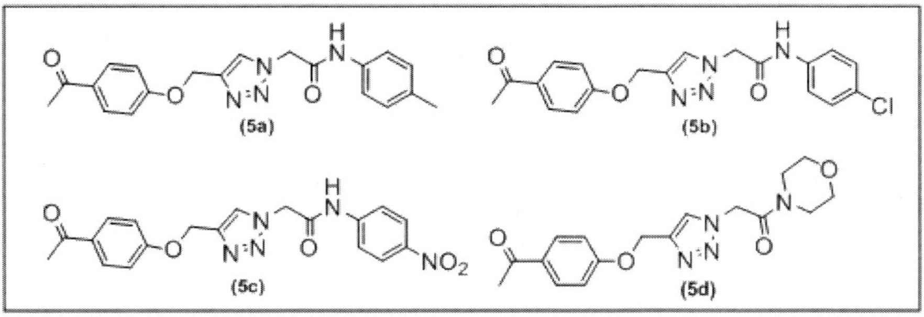

Figure 7. Acetylphenoxymethyl-triazolyl-*N*-phenylacetamide derivatives possessing antimicrobial activities.

A series of substituted acetylphenoxymethyl-triazolyl-*N*-phenylacetamide derivatives were reported. The synthesized compounds were tested for the antibacterial and antifungal activity. Both Gram-positive and Gramnegative bacterial pathogens were used in the antibacterial activity assay. Fluconazole and tetracycline were used as antifungal and antibacterial standard reference compounds, respectively. Among the synthesized compounds, four compounds **5a**, **5b**, **5c**, and **5d** have displayed very good antimicrobial activities. The compound **5b** has displayed a good inhibitory activity against *S. cerevisiae* ATCC 9763 [14] (Figure 7).

In the present study, substituted formylnaphthalenyloxymethyl-triazolyl-*N*-phenylacetamide derivatives have been designed and synthesized employing click chemistry approach and evaluated for their in vitro antifungal and antibacterial activities. Among the screened compounds **6a**, **6b**, **6c**, and **6d** have shown good antifungal and antibacterial activities. Compound **6d** has shown very effective antimicrobial activity [15] (Figure 8).

In another research, we have reported a series of new isoniazid-1,2,3-triazole conjugates. These newly synthesized conjugates were screened for their *in vitro* antimicrobial activities by agar well diffusion assay. Fluconazole and Tetracycline were used as reference standard for antifungal and antibacterial activities, respectively. The compounds **7a**, **7b**, **7c** and **7d** have exhibited good antimicrobial activities against both antibacterial and antifungal pathogens [12] (Figure 9).

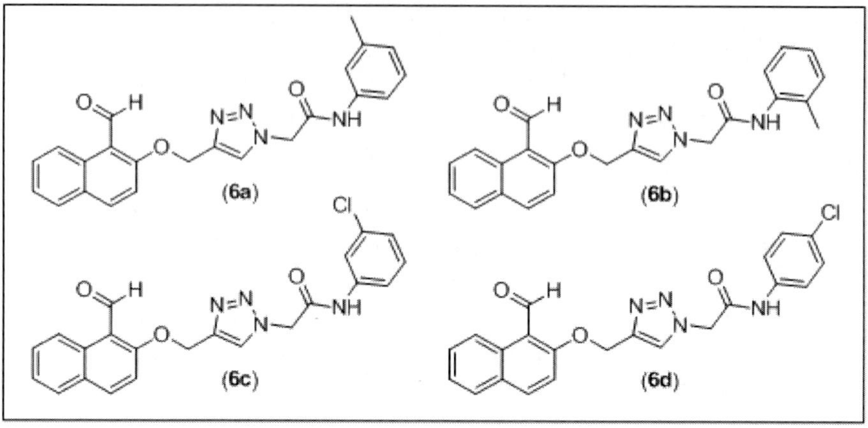

Figure 8. Formylnaphthalenyloxymethyl-triazolyl-*N*-phenylacetamide derivatives possessing antimicrobial activities.

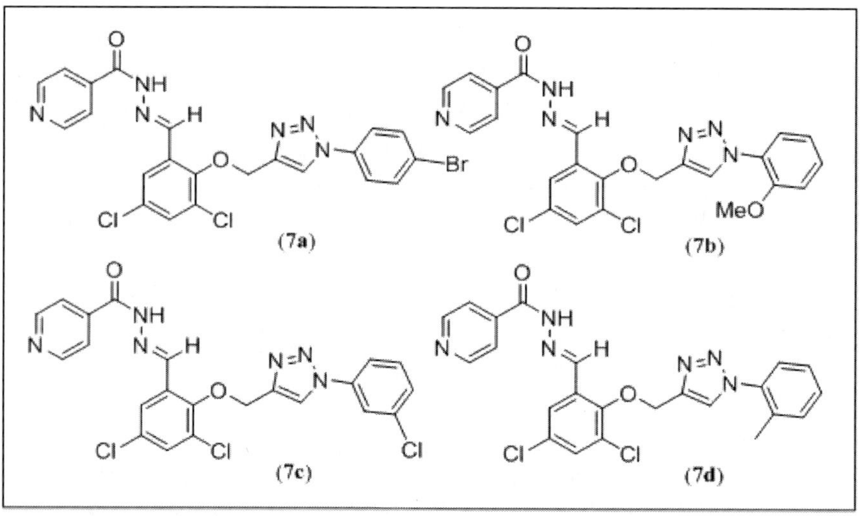

Figure 9. Isoniazid-1,2,3-triazole conjugates possessing antimicrobial activities.

We have synthesized new indanol-1,2,3-triazole derivatives and evaluated for their *in vitro* antimicrobial activities by using agar well diffusion method. The antibacterial screening was executed against Gram-positive *S. aureus*, *B. cereus*, *B. subtilis* and Gram-negative *E. aerogenes*, *E. coli*, *S. typhi*, *P. aeruginosa*, *S. boydii*, *S. abony* pathogens. The antifungal screening was executed against *A. niger*, *C. albicans* and *S. cerevisiae* fungal pathogens.

Among the screened indanol-1,2,3-triazole derivatives, **8a**, **8b**, **8c**, **8d**, **8e**, **8f**, **8g**, **8h**, **8i** and **8j** were shown excellent antimicrobial activities. The minimum inhibitory concentration (MIC) values were determined for the most active indanol-1,2,3-triazole derivatives [10] (Figure 10).

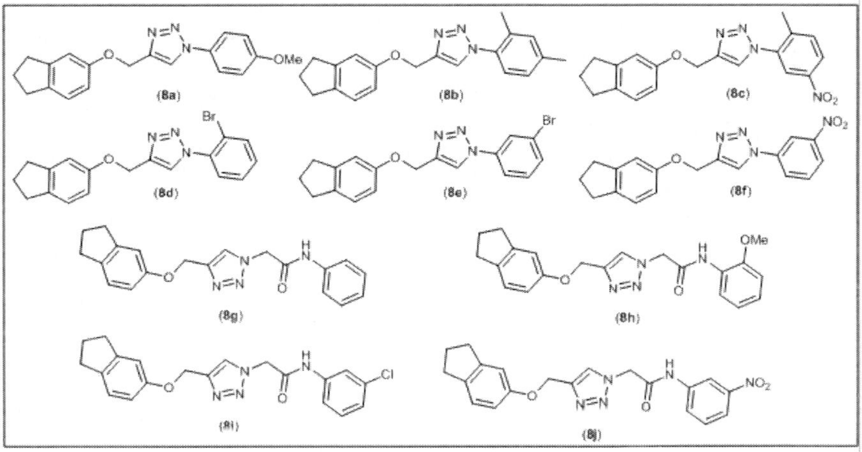

Figure 10. Indanol-1,2,3-triazole conjugates possessing antimicrobial activities.

Molecular Docking Study

The molecular docking analysis was performed to predict possible mode of action of the synthesized triazole derivatives. Crystal structure of the DNA gyrase of crucial antimicrobial target was utilized for docking analysis. The technique involves mainly enzyme 3D structure selection and preparation, identification and assignment of active site (grid generation), ligand 3D structure preparation and optimization, and docking and analysis of docking results (binding affinity). Crystal structure of DNA gyrase (PDB ID: 5l3j) was downloaded from the free protein database www.rcsb.org Crystal structure of the downloaded DNA gyrase was refined prior to the docking analysis via removal of the water and addition of the hydrogen atoms so that native geometry of the protein will be retained. Ligands were prepared in Vlife engine module and optimized via Merck molecular force field. Biopredicta module of the Vlife MDS was utilized for docking analysis.

Conclusion

In conclusion, we have developed a library of novel triazole derivatives *via* click chemistry approach in good to excellent yields. All the newly synthesized compounds were evaluated for their *in vitro* antitubercular and antimicrobial activities. We have identified most of the compounds as potent antitubercular agents (MIC value 0.78 to 12.5μg/mL) against *M. tuberculosis H37Rv* (Mtb). These compounds have displayed low cytotoxicity profile by MTT assay against RAW 264.7 cell line. Among the compounds screened for their *in vitro* antimicrobial activities, most of the compounds have shown good activities against both antibacterial and antifungal pathogens. Furthermore, *in silico* binding affinity data against mycobacterial InhA or DNA gyrase could provide a valuable insight into the plausible mechanism of antitubercular or antimicrobial action for these molecules. The promising results were obtained from this study qualify these new triazole derivatives as a pertinent starting point for structure-based optimization as antitubercular and antimicrobial candidates through design and synthesis approach. The new entities with dual antitubercular and antimicrobial properties are desperately needed. We believe that the analogues reported in this work will be helpful for identification of potential lead molecules for further development.

References

[1] Rostovtsev VV, Green LG, Fokin VV and Sharpless KB. (2002). A Stepwise Huisgen Cycloaddition Process: Copper(I)-Catalyzed Regioselective “Ligation” of Azides and Terminal Alkynes. *Angew Chem*, 114: 2708.

[2] Sharma A, Agrahari AK, Rajkhowa S and Tiwari VK. (2022). Emerging impact of triazoles as anti-tubercular agent. *Eur J Med Chem,* 238: 114454.

[3] Keri RS, Patil SA, Budagumpi S and Nagaraja BM. (2015). Triazole: A promising antitubercular agent. *Chem Biol Drug Des,* 86: 410-423.

[4] Zhang B. (2019). Comprehensive review on the anti-bacterial activity of 1, 2, 3-triazole hybrids. *Eur J Med Chem,* 168: 357-372.

[5] Lucht F, Fauche A and Allaert FA. (1996). Céfatrizine (8 j) versus penicillin V (10 j) dans l’angine à streptocoque *A de l’enfant. Médecine Maladies Infect,* 26:557-561.

[6] Caraballo RH, Pociecha J, Reyes G, Espeche A, Galichio S, Fasulo L, Semprino M. (2020). Rufinamide as add-on therapy in children with epileptic encephalopathies other than lennox–gastaut syndrome: A study of 34 patients. *Epilepsy and Behav,* 108: 107074.

[7] Sharawat IK, Panda PK, Panda P and Dawman L. (2021). Efficacy and safety of rufinamide as adjunctive therapy in patients with lennox gastaut syndrome: A systematic review and meta-analysis. *Seizure,* 91: 296-307.

[8] Pfaller MA, Shortridge D, Harris KA, Garrison MW, Deryke CA, Depestel DD, Moise PA, Sader HS. (2021). Ceftolozane-tazobactam activity against clinical isolates of Pseudomonas aeruginosa from ICU patients with pneumonia: United States, 2015-2018. *Int J Infect Dis,* 112: 321-326.

[9] Ramassamy M, Murris M, Recoche I, Mailhol C, Didier A and Guilleminault L. (2021). Piperacillin-tazobactam-induced fever in chronic respiratory diseases: Safe challenge test under premedication. *Rev Francaise D'allergologie (French Rev of Allergology).*

[10] Phatak PS, Bakale RD, Kulkarni RS, Dhumal ST, Dixit PP, Krishna VS, Sriram D, Khedkar VM and Haval KP. (2020). Design and synthesis of new indanol-1,2,3-triazole derivatives as potent antitubercular and antimicrobial agents, *Bioorg and Med Chem Lett*, 30(22): 127579.

[11] Phatak PS, Bakale RD, Dhumal ST, Dahiwade LK, Choudhari PB, Krishna VS, Sriram D and Haval KP. (2019). Synthesis, antitubercular evaluation and molecular docking studies of phthalimide bearing 1,2,3-triazoles, *Synth Commun*, 49: 2017-2028.

[12] Badar AD, Sulakhe SM, Muluk MB, Rehman NNMA, Dixit PP, Choudhari PB, Rekha EM, Sriram D, Haval KP. (2020). Synthesis of isoniazid-1,2,3-triazole conjugates: Antitubercular, antimicrobial evaluation and molecular docking study. *J Heterocyclic Chem*, 57: 3544-3557.

[13] Patil PS, Kasare SL, Haval NB, Khedkar VM, Dixit PP, Madhu Rekha E, Sriram D and Haval KP. (2020). Novel isoniazid embedded triazole derivatives: Synthesis, antitubercular and antimicrobial activity evaluation, *Bioorg and Med Chem Lett*, 30(19): 127434.

[14] Phatak PS, Sathe BP, Dhumal ST, Rehman NNMA, Dixit PP, Khedkar VM and Haval KP. (2019). Synthesis, Antimicrobial Evaluation, and Docking Studies of Substituted Acetylphenoxymethyl-triazolyl-N-phenylacetamides. *J Heterocycl Chem*, 56: 1928.

[15] Muluk MB, Dhumal ST, Phatak PS, Rehman NNMA, Dixit PP, Choudhari PB, Mane RA and Haval KP. (2019). Synthesis, antimicrobial activity, and molecular docking study of formylnaphthalenyloxymethyl-triazolyl-N-phenylacetamides, *J Heterocycl Chem,* 56: 2411.

Biographical Sketch

Dr. Kishan Prabhu Haval, MSc, PhD, SET/NET/GATE
Department of Chemistry
Dr. Babasaheb Ambedkar Marathwada University, Aurangabad
SubCampus, Osmanabad, Maharashtra, India
Email: havalkp@gmail.com

Education:

2000:	BSc, Phy, Chem, Ind. Chem., S. R. T. M. U. Nanded
2002:	MSc, Organic Chemistry, S. R. T. M. U. Nanded
2003:	SET, Chemical Sciences UGC (Savitribai Phule Pune University, Pune), India
2002, 2003:	NET, Chemical Sciences, UGC-CSIR
2003:	GATE, Chemical Sciences IIT
2009:	PhD, Chemistry, Savitribai Phule Pune University, Pune (NCL,Pune); PhD Thesis: Synthesis of Natural and Unnatural Dialkyl Substituted Maleic Anhydrides and Related Natural products, Dr. N. P. Argade, Sr. Scientist, National Chemical Laboratory, Pune, India

Work Experience:

February 2009- August 2009:	Associate Scientist, Lupin Pharmaceuticals, Pune
04/09/2009- to date:	Assistant Professor, Dr. Babasaheb Ambedkar Marathwada University, Aurangabad

Honors:

	Name of Award	Awarding Agency
2004:	Chancellor's Gold Medal	S. R. T. M. U. Nanded
2004:	Dr. D. S. Deshpande Gold Medal	S. R. T. M. U. Nanded
2002, 2003:	Junior Research Fellowship	CSIR New Delhi
2022:	NSS Best Programme Officer	Dr. Babasaheb Ambedkar Marathwada University, Aurangabad

Grants Received :

One UGC Major Research Project completed with total outlay of Rs. 7,72,800/-

Two Minor Research Projects completed, Dr. Babasaheb Ambedkar Marathwada University, Aurangabad : Rs. 2,50,000/-

PhD students guided : Awarded: 07; Working: 06

Professional Appointments:

- Working as IQAC Coordinator, Dr. B. A. M. U. SubCampus, Osmanabad
- Working as SWAYAM Coordinator Department of Chemistry
- Working as Coordinator for "Green Campus" Dr. B. A. M. U. SubCampus, Osmanabad
- Worked as NSS Osmanabad District Coordinator, Area Coordinator and Program Officer
- Worked as Coordinator Earn & Learn Scheme Worked as Custodian D-CAS Centre
- Worked as Joint Custodian D-CAS Centre
- Chairman and Member of Flying Squad for different university examinations
- Committee Member for various affiliation committees

Publications from the Last 3 Years:

SD Balsure, M Gurav, RH Kadam, KP Haval, RM Tigote, AB Kadam. (2022). X-ray line profile analysis and magnetic and optical properties of FexZn0.95-xCr0.05O nanoparticles fabricated by sol-gel route, *Cerâmica,* 68: 24-32.

SL Kasare, PN Gund, BP Sathe, PS Patil, NNMA Rehman, PP Dixit, PB Choudhari, KP Haval. (2021). Synthesis, antimicrobial screening and docking study of new 2-(2-ethylpyridin-4-yl)-4-methyl-N-phenylthiazole-5-carboxamide derivatives, *J. Chin. Chem. Soc.,* 68: 353-361.

AD Badar, SM Sulakhe, MB Muluk, NNMA Rehman, PP Dixit, PB Choudhari, GM Bondle, KP Haval. (2021). Synthesis, Antimicrobial Evaluation and Docking Study of Novel Thiosemicarbazone clubbed with 1,2,3-Triazoles, *Current Bioactive Compounds,* 17: e010621185856.

PS Phatak, RD Bakale, RS Kulkarni, ST Dhumal, PP Dixit, VS Krishna, D Sriram, VM Khedkar, KP Haval. (2020). Synthesis, biological evaluation and molecular docking study of new indanol-1,2,3-triazole derivatives as potent antitubercular and antimicrobial agents, *Bioorg. Med. Chem. Lett.,* 30: 127579.

PS Patil, SL Kasare, NB Haval, VM Khedkar, PP Dixit, EM Rekha, D Sriram, KP Haval. (2020). Novel isoniazid embedded triazole derivatives: synthesis, antitubercular and antimicrobial activity evaluation, *Bioorg. Med. Chem. Lett.,* 30: 127434.

PS Patil, SL Kasare, AD Badar, RS Kulkarni, PP Dixit, JA Kulkarni, PB Choudhari, KP Haval. (2020). Synthesis, Antimicrobial Evaluation and Molecular Docking Study of New Thiazole-5-Phenylpropenone Derivatives, *Russian J. Gen. Chem.,* 90: 1523-1528.

AD Badar, SM Sulakhe, MB Muluk, NNMA Rehman, PP Dixit, PB Choudhari, EM Rekha, D Sriram, KP Haval. (2020). Synthesis of isoniazid-1, 2, 3-triazole conjugates: Antitubercular, antimicrobial evaluation and molecular docking study, *J. Heterocycl. Chem.,* 57: 3544-3557.

MB Muluk, AS Ubale, ST Dhumal, NNMA Rehman, PP Dixit, KK Kharat, PB Choudhari, KP Haval. (2020). Synthesis of new hydrazones bearing pyridyl and thiazolyl scaffolds and their anticancer and antimicrobial evaluation, *Synth Commun.,* 50: 243-255.

MB Muluk, PS Patil, SL Kasare, RS Kulkarni, NNMA Rehman, PP Dixit, PB Choudhari, KP Haval. (2020). Synthesis and biological evaluation of pyridyl and thiazolyl bearing hydrazones as antimicrobial and antioxidant agents, *Eur. Chem. Bull.,* 9: 184-192.

RS Kulkarni, PN Gund, BP Sathe, PS Phatak, AD Badar, NNMA Rehman, PP Dixit, KP Haval. (2020). Synthesis and antimicrobial evaluation of new tetrazolyl-N-substituted maleimide and phthalimide derivatives, *Der Pharma Chemica,* 12: 41-47.

NV Shinde, AS Dhake, KP Haval, SK Bhosale. (2020). Total Synthesis of Clausenain, a Cyclic Octapeptide and its Analog for Anticancer Activity, *Indian J. Pharm. Educ. Res.,* 54: s330-s336.

MB Muluk, PS Phatak, SB Pawar, ST Dhumal, NNMA Rehman, PP Dixit, PB Choudhari, KP Haval. (2019).

Synthesis, antimicrobial and antioxidant activities of new pyridyl and thiazolyl bearing carbohydrazides, *J. Chin. Chem. Soc.,* 66: 1507-1517.

MB Muluk, ST Dhumal, NNMA Rehman, PP Dixit, KR Kharat, KP Haval. (2019). Synthesis, Anticancer and Antimicrobial Evaluation of New (E)-N′-Benzylidene-2-(2-ethylpyridin-4-yl)-4-methylthiazole-5-carbohydrazides, *Chemistry Select,* 4: 8993-8997.

MB Muluk, ST Dhumal, PS Phatak, NNMA Rehman, PP Dixit, PB Choudhari, RA Mane, KP Haval. (2019). Synthesis, antimicrobial activity, and molecular docking study of formylnaphthalenyloxymethyl-triazolyl-N-phenylacetamides *J. Heterocycl. Chem.,* 56: 2411-2418.

PS Phatak, RD Bakale, ST Dhumal, LK Dahiwade, PB Choudhari, VS Krishna, D Sriram, KP Haval. (2019). Synthesis, antitubercular evaluation and molecular docking studies of phthalimide bearing 1,2,3-triazoles, *Synth. Commun.,* 49: 2017-2028.

PS Phatak, BP Sathe, ST Dhumal, NNMA Rehman, PP Dixit, VM Khedkar, KP Haval. (2019). Synthesis, antimicrobial evaluation and docking studies of substituted acetylphenoxymethyl-triazolyl-N-phenylacetamides, *J. Heterocycl. Chem.,* 56: 1928-1938.

BP Sathe, PS Phatak, NNMA Rehman, PP Dixit, VM Khedkar, SG Vedpathak, KP Haval. (2019). Synthesis, anti-microbial evaluation and molecular docking studies of some novel tetrazole containing azodye derivatives, *Chemistry & Biology Interface,* 9: 96-113.

RS Kulkarni, NB Haval, JA Kulkarni, PP Dixit, KP Haval. (2019). Synthesis, characterization and biological evaluation of substituted 2-phenoxynicotinaldehydes as α-amylase inhibitors. *Eur. Chem. Bull.,* 8: 26-30.

BP Sathe, HS Oman, NNMA Rehman, PP Dixit, AS Kharat, KP Haval. (2019). Synthesis, antimicrobial and β-lactamase enzyme inhibition activity of some new tetrazole containing maleamic and phthaleamic acid derivatives, *Asian J Org & Med Chem,* 4: 166-173.

Books / Reports / Chapters / General Articles etc.

KP Haval, Modern Green Chemistry and Heterocyclic Compounds (Molecular Design, Synthesis and Biological Evaluation). (2020). Apple Academic Press.

KP Haval, The Comparative Study of Biologically Active Agents (Sources and scopes in modern science). (2020). Scholar's Press

Chapter 3

The Distribution of Phytosanitary Products in Sediment Cores from the Bizerte Lagoon (Tunisia)

Mouna Necibi*
and Nadia Mzoughi
High Institute of Environmental Sciences and Technologies of Borj Cedria, Environmental Sciences and Technologies Laboratory, University of Carthage
Hammam Lif, Tunisia

Abstract

The present study investigates the vertical distribution of Triazole pesticides in two sediment cores collected from the Bizerte Lagoon. The sediment core (SC1) is located at the mouth of the channel connecting the lagoon to the Mediterranean Sea and is characterized by the industrial and urban activities of zones A and B. The sediment core (SC2) is situated in zone D, known for its intensive metallurgy and naval construction activities. The Triazole pesticide concentrations ranged from 20.64 ng g^{-1} found at 15 cm depth to 0.07 ng g^{-1} at 6 cm depth in SC1, and varied between 22.41 ng g^{-1} at 9 cm depth and 2.77 ng g^{-1} at 3 cm depth in SC2. Different profiles were observed, making evident that the levels of these pollutants depend strongly on anthropological discharges from industrial, urban, and farming sources. The temporal trends of the studied compounds were influenced by their different historical usages, different properties, or different degradation conditions in the environment.

* Corresponding Author's Email: mouna23necibi@gmail.com.

In: Triazoles and Their Uses
Editor: Karl Roland
ISBN: 979-8-88697-800-1

Keywords: marine environment, triazole pesticides, sediment cores, Bizerte Lagoon

Introduction

Pesticide researchers are students of nature, and every new compound and mechanism turns a page in the ever-expanding encyclopedia of life. Pesticides are instruments for pest management to facilitate food production and improve health, as well as probes to learn about the workings of life.

Pesticides, in contrast to several household and industrial chemicals, are presumed to be harmful to human health and the environment unless proven otherwise (Casida et al. 2017).

Pesticides play an important role in increasing unit outputs because they effectively control crop diseases, insects, and weeds. Triazole pesticides are one of the most important classes of pesticides, which not only have great bacteriostatic action like controlling mildew, rust, sheath blight and other diseases, but also have an effect on plant physiological regulation to achieve insecticidal and herbicidal activities.

People are increasingly concerned about the harmful effects of triazole pesticides on ecosystems and human health, which have been shown to cause tumorigenic and endocrine disrupting effects. Since they have been found in food and water, a rapid, simple and accurate determination method of triazole levels is important for food safety monitoring and regulatory purposes (Xua et al. 2013). Triazole fungicides such as triadimenol, myclobutanile, triazolone, and hexaconazole are systemic pesticides that have been widely applied to avoid fungus diseases in agriculture. However, the triazole pesticides may enter environmental waters such as rivers and lakes or even permeate soil and enter underground waters. They possess high toxicity. At the same time, they are difficult to degrade and belong to persistent organic pollutants. The harm of residual triazole pesticides to aquatic organisms has been demonstrated in previous studies (Lui et al. 2017).

Lagoons are formed and maintained through sediment transport. These ecosystems are highly productive and support a wide range of habitats (Ben Ameur et al. 2013). The Bizerte Lagoon is one of the most economically important areas in Tunisia and is known as a fishing and aquaculture park due to the presence of three mussel sectors. It is subject to intensive industrial activities like oil refineries, metallurgical activities, naval construction, and tire production. According to the literature, diverse types of pollutants have been investigated in surface sediments from the Bizerte Lagoon, especially mercury (Mzoughi et al. 2002), organotin compounds (Mzoughi et al. 2005), polycyclic aromatic hydrocarbons (PAH) (Mzoughi et al. 2002), and synthetic musk compounds (Mzoughi et al. 2002). The OCPs have also been registered in sediment (Necibi et al. 2016), cores (Necibi et al. 2016) and surface water (Necibi et al. 2015).

To the best of our knowledge, there are few studies on the spatial contamination of triazole pesticides in the surface sediment, but no systematic research has been conducted to study the vertical distribution of these pesticides in this area. This paper presents the first study of the vertical distribution of triazole pesticides in two sediment cores taken from the Bizerte Lagoon. The purposes of this study are the identification of the contaminant origin, understanding the sedimentation process, and examination of pollution mechanisms.

Materials and Methods

Study Area

The Bizerte Lagoon is located in the northernmost part of Tunisia, between latitude and longitude; 37°08' and 37°14'N; 9°48' and 9°56'E. The surface area is 128 km^2 and the sea depth is between 3 and 12 m. The lagoon is connected to the Mediterranean Sea and to Lake Ichkeul by straight channels. The exchange of water between the Mediterranean and the lagoon determines the salinity of the lagoon, which varies between 32.5 psu and 38.5 psu. The water temperature is between 10°C during the winter and 29°C during the summer. The sedimentation rates depend on the lithological nature of the deposited sediment and the hydrodynamic parameters within the lagoon. It remains more important in the western and central sectors of the lagoon (2 to 2.5 mm $year^{-1}$) than in the northern and eastern sectors (0 to 0.3 mm $year^{-1}$) (Garali et al. 2013). The selection of the sampling sites was based on different criteria related to the geomorphology of the lagoon, the hydrological regime, and the localization of urban and industrial discharges. SC1 is located at the mouth of the channel connecting the lagoon to the Mediterranean Sea. This channel includes zones A and B, which are highly populated and characterized by industrial activities including cement and metal treatment manufacturing. SC2 is situated in zone D, known for its intensive metallurgy and naval construction activities (Figure 1).

Table 1. Description of sampling sites of sediment cores from the Bizerte Lagoon

	Latitude (N)	Longitude (E)	Depth (m)
Sediment core 1	37°15.11’	09°51.00’	4
Sediment core 2	37°09.06’	09°49.37’	6.5

Sample Collection

The sediment cores (SCs) were collected at a water depth of 4 and 6.5 m by scuba diving during July 2016, using a global positioning system (GPS) (Table 1). The divers used manual acrylic tubes 20 cm in length and 5 cm in diameter (Figure 1). Each SC was sampled three times at each sampling site. The SCs were cut into sub-samples with the thickness of the layers varying between 3 and 5 cm. Sediments from the same layer were homogenized to form a composite sample for each sampling site, then placed in sealed polythene bags and transported to the laboratory. The different sediment layers were freeze-dried, homogenized, and stored in the dark at -20 °C until analysis. Total Organic Carbon (TOC) sediment samples were analyzed with a TOC Analyzer (SHIMADZU H544051).

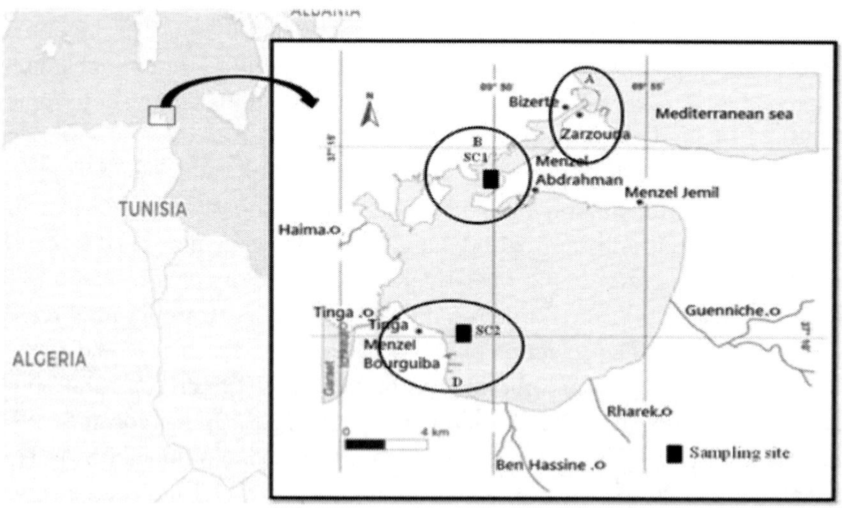

Figure 1. Study areas with sampling locations and different types of industrial areas (A–D) in the Bizerte Lagoon.

Table 2. Analytical performance of the QuECHERS extraction

Compounds	Ions (m/z)	LOD (ng g^{-1})	LOQ (ng g^{-1})	Recovery (%)10 (ng L^{-1})	Recovery (%) 100 (ng L^{-1})	RSD % (n= 5)	RSD% (n= 15)	Correlation coefficient (R^2)
Mesosulfuron-methyl	154.1; 68.0	0.001	0.003	96.19	85	1.07	0.77	0.996
Dimethoate	229;125;87.0	0.008	0.024	85.07	97	1.28	1.34	0.994
Penconazole	248.1; 159.0	0.001	0.003	92	74	1.25	1.09	0.994
Triadimenole	296.1; 168.1; 128.0	0.001	0.003	70.56	79	1.13	1.30	0.997
Hexaconazole	214.0; 83.1	0.008	0.024	76.59	82	1.05	1.63	0.998
Fludioxonile	248.1; 154.1; 127.0	0.007	0.021	76.16	90	1.33	1.13	0.998
Cyproconazole	222.1; 139.0	0.009	0.027	82.61	87	1.52	1.36	0.994
Propiconazole	259.1; 173.0; 69.1	0.004	0.012	75	98	1.85	1.00	0.991
Tebuconazole	250.1; 125.0	0.010	0.030	92	74	1.25	1.91	0.995
Epoxyconazole	206.1; 192.1	0.006	0.018	76.55	79	1.71	1.58	0.994
Fenoxycarbe	301; 186.1; 116.1	0.030	0.090	97.42	86	1.02	1.12	0.997
Bromuconazole	294.9; 214.0; 173.0	0.001	0.003	74.51	99	1.02	1.32	0.995
Triticonazole	299.1; 235.1; 83.0	0.005	0.015	96	98	1.98	1.19	0.993
Myclobutanile	168; 128;112	0.019	0.057	99.38	97	1.05	1.33	0.991
Fluquinconazole	340; 108.0	0.017	0.051	99.55	38	1.20	1.51	0.991

Table 3. Levels of Triazole pesticides (ng g^{-1}), TOC (%) in sediment cores from the Bizerte Lagoon (Mean; expressed in nanograms per gram dry weight ± SD; expressed in %; n = 3)

Depth (cm)	Sediment core 1		Sediment core 2	
	∑triazole	TOC%	∑triazole	TOC%
3	0.50	3.07	2.77	1.39
6	0.07	2.05	7.48	1.20
9	20.01	2.95	22.41	1.10
12	1.04	2.16	12.90	1.13
15	20.64	2.46	17.59	1.70
20	6.53	2.93	5.29	0.05

Analytical Performances

Appropriate blanks were analyzed with a set of analyses; internal standards were analyzed for quality control purposes. Quantification was assured by the calculation of response factors of individual triazole external standards. The reported results were corrected with recoveries of surrogate internal standard fluconazole (Kolberg et al. 2016). The calibration solutions in methanol were prepared daily at five levels: 1; 10; 50; 100; and 200 ng/L. Over the calibration range selected, good linearity was obtained for all the target compounds. The determination coefficients (R^2) were higher than 0.990 for all triazole pesticides. The limits of detection and the limits of quantification have been calculated by multiplying the standard deviations of the calculated amount for each triazole by 3 and 10, respectively. For the validation of the method, the precision and accuracy were determined by analyzing the spiked water samples at two levels of concentrations (10 and 100 ng L^{-1}) on the same day and three different days (Table 3). Relative standard deviations for intraday (n = 5) precisions ranged between 1.02% and 1.98%, and for interday (n = 15) precisions ranged between 0.77% and 1.91%. The extraction recoveries for these selected pesticides were in the range of 70% and 99%, for the two levels of spiked concentrations, respectively (Table 2). These results indicate that this method is satisfactorily accurate, sensitive, and repeatable (Wang et al. 2016; Zhang et al. 2015).

Extraction

The method for extracting triazole pesticides from sediments is based on the method proposed in preliminary studies, with some modifications (Masiá et al. 2013). The QuECHERS method, also previously reported in a variety of formats, was applied to 2 g of lyophilized sediment sample, which was weighed in a 50 mL falcon tube and homogenized with 10 mL of acetonitrile (Anastassiades et al. 2003). Then (1 g of sodium citrate, 0.5 g of sodium hydrogen citrate sesquihydrate, 4 g of magnesium sulfate, and 1 g of sodium chloride) were added for phase-separation adjustment. The mixture was agitated intensively in a vortex for 1 min and centrifuged at 4000 rpm for 5 min. The upper layer (9 mL) was transferred into a new tube which contained the clean up sorbent mixture (400 mg of C18EC, 400 mg of PSA and 1200 mg of magnesium sulfate). After centrifugation (4000 rpm × 5 min), 6 mL of the extract was placed in a 10 mL glass centrifuge tube and evaporated under a gentle stream of azote. Finally, the residues were reconstituted with 50 μL of acetonitrile and the extract was directly transferred into GC vials for the GC–MS analysis.

Gas Chromatography

Analyses of the extracts were performed with an Agilent 7890 A gas chromatography (GC) coupled to a 5975 C inert mass selective detector (Agilent Technologies, USA). The Agilent ChemStation software has been used for instrument control, data acquisition, and data treatment. The temperature program optimization has been realized based on previous work with a few modifications (Kolberg et al. 2016; Xu et al. 2013; Vieira et al. 2017). The chromatographic separation was performed with the HP-5MS capillary column (30 m length x 0.25 mm diameter and 0.25 μm film thickness). The column temperature was held at 50°C for 1 minute, then ramped to 160°C at 7°C min^{-1} for 1 minute, and finally held at 240°C at 10°C min^{-1} for 18 minutes.The target ions have been measured in the selected ion monitoring (SIM) mode. Helium has been selected as the carrier gas with a flow rate of 1 mL min^{-1}. The injector and detector were maintained at 250°C and 300°C respectively. Injection volumes were 1 μL in the splitless mode. The target ions have been measured in the selected ion monitoring (SIM) mode. The mass detector conditions were: transfer line temperature: 250°C; ion source temperature: 180°C; ionization mode-electron impact at 70 eV.

Statistical Analysis

The statistical analysis has been realized by the well-known techniques of principal component analysis (PCA). Correlation analysis has been used to reveal the relationships between triazole pesticides. The difference was statistically significant when $p < 0.05$. For all statistical analyses, the Statistica 6 software package was used.

Results and Discussion

Variations in Triazole Pesticide Concentrations in the Core Profiles

The analysis of the Triazole pesticides in the different sediment layers was carried out in two sediment cores (SCs). The results and average compositions of the studied compounds are presented in Tables 3, 4 and 5. Their Levels for SC1 and SC2 ranged between 0.50 and 2.77 ng g^{-1} at 3cm depth and varied between 6.53 and 5.29 ng g^{-1} at 20 cm depth. The penconazole and myclobutanile were higher than those of other pesticides investigated in this study. The sedimentary profile observation shows high concentrations of the studied pesticides in the deeper sediments, notably in SC2 (Figure 2). The comparison of the contamination in the deep sediments with the superficial sediments in SC1 shows a decrease in the concentrations of the Triazole pesticides down to the bottom of this core. Their concentrations remain relatively high at 9 and 15 cm depth (Table 2). The vertical evolution of the organic matter in sediments shows, in the majority of the SCs, a decrease in the TOC content with depth. Higher TOC contents in SC1 than in SC2 could probably contribute to their accumulation due to their high affinity for organic matter (Table 2).

The increase in the organic matter, content promotes the accumulation of hydrophobic pollutants. Therefore, sediments rich in organic matter have a better ability to retain hydrophilic pollutants. The sediments contained much higher levels of penconazole and myclobutanile than other triazole pesticide isomers. Generally, upper sediment contained much higher average levels of these pesticides than deeper sediment in the two SCs (Bai et al. 2015).

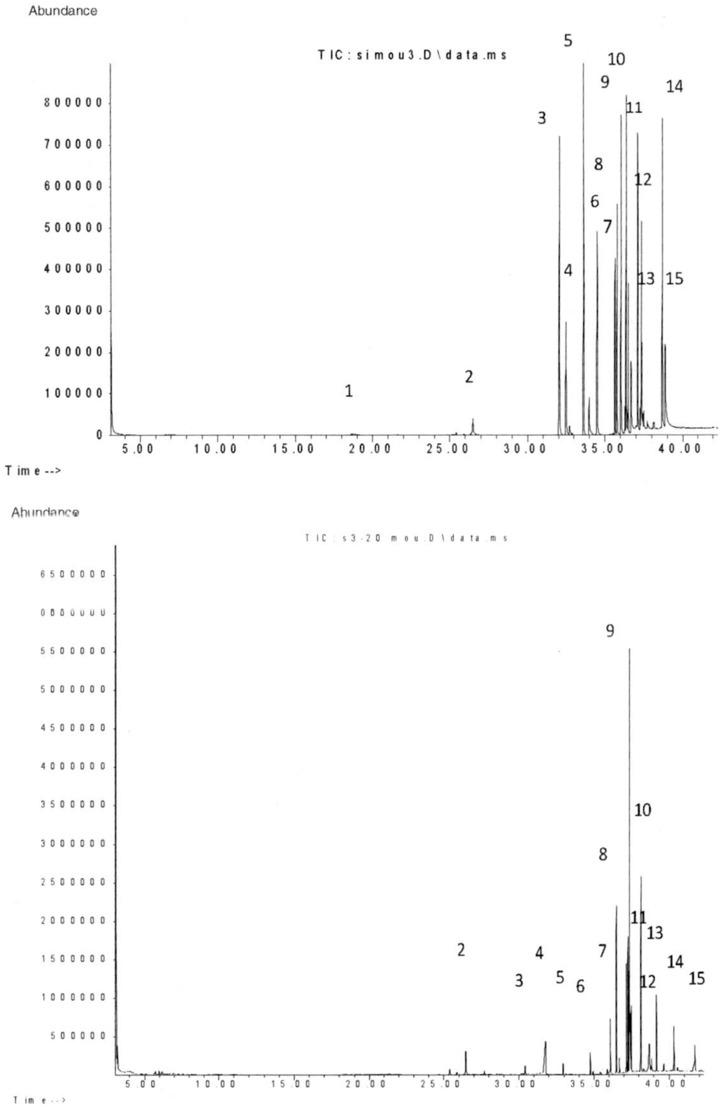

Figure 2. (A) Chromatogram of a standards triazole pesticides mixture at 100 ng g^{-1}(1), (B) chromatogram of the SC2 layer 20 cm: (1) mesosulfuron-methyl, (2)dimethoate,(3) penconazole, (4) tridiminole, (5)fludioxonile, (6) hexaconazole, (7) cyproconazole, (8) propiconazole, (9) tebuconazole, (10) epoxyconazole, (11) fenoxycarbe, (12) bromuconazole, (13) triticonazole, (14) myclobutanile, (15) fluquinconazole.

This might be associated with new sources of Triazole pesticides that were introduced after 1984 or with their continued use for other purposes in addition to their agricultural purposes. The lower detection of some of these pesticides in SCs could be explained by their shorter production histories and lower popularity in usage (Wu et al. 2015). Triazole pesticides could be leached into deeper sediment layers and avoid degradation, which might be associated with their higher concentrations in deeper SCs layers (Wang et al. 2011). Rapid urbanization and industrialization, illegal use, runoff, atmospheric deposition, and wastewater discharges are the main sources of sediment pollution. The southwest and west side's urban areas in the study area experience a very high population concentration and multiple industrial activities. Since the 1950s, around 400 industrial units relating to various fields of activity such as petrochemicals, textiles, iron and steel, and plastic processing have been established on the shores of the Bizerte Lagoon. Consequently, industrial activities, domestic wastewater and open-air municipal landfills scattered around the lagoon can play a major polluting role that could lead to a degradation of the ecological conditions of this body of water.The variation in triazole pesticide concentrations is most likely due to the common effects of tides and water, as well as human activities. Benthic animals have also been suggested to contribute to the identification of older sediments owing to the fact that these animals can accumulate organic compounds in their bodies and these compounds can be transferred to deeper layers when migrating in sediments (Kim et al. 2007). They could even be explained by the variation in the general current, which ends up in great sediment loads, bioturbation, and biogeochemical processes between surface sediment and water.

Vertical Distribution of Triazole Pesticides in the Core Profiles

Triticonazole has only been found in the superficial SC1 and at a depth of 6 cm in the SC1 and the deep layer of the SC2. It presents only 1% and 2% of the total triazole pesticides present in SC1 and SC2, respectively (Miglioranza et al. 1999). Mesosulfuron-methyl and cytoconazole levels were below the limits of detection (LOD) in the SC1 and SC2 sediment layers. As an anthropogenic POP, its use as a fungicide is already banned in most countries due to its high persistence and capacity to disseminate in the environment (Ritteret al. 1995).

Tebuconazole has different profiles in the two SCs; its concentrations were lower than the detection limits in SC1. Levels of Tebuconazole ranged between ND ng g^{-1}obtained at 15 cm depth and 1.49 ng g^{-1}recorded at 9 cm depth in SC2 (Table 4 and 5).

Table 4. Concentrations of Triazole pesticides (ng g^{-1}) in SC1 (Mean; expressed in nanograms per gram dry weight ± SD; expressed in %; n = 3)

Depth (cm)	3	6	9	12	15	20
Mesosulfuron-methyl	ND	ND	ND	ND	ND	ND
Dimitanoate	ND	ND	ND	ND	ND	ND
Penconazole	0.05	ND	1.41	ND	1.43	1.25
Triadimenole	ND	ND	1.15	ND	2.73	ND
Fludioxonile	ND	ND	1.29	ND	ND	ND
Hexaconazole	0.03	ND	ND	ND	2.92	0.21
Cyproconazole	ND	ND	ND	ND	ND	ND
Propiconazole	ND	0.01	2.98	ND	2.22	2.54
Tebuconazole	ND	ND	ND	ND	ND	ND
Epoxyconazole	0.25	ND	ND	ND	0.07	ND
Fenoxycarbe	0.06	ND	2.52	0.18	ND	0.46
Bromuconazole	ND	0.01	ND	ND	2.29	ND
Triticonazole	ND	0.01	ND	ND	ND	ND
Myclobutanile	0.12	0.01	3.70	0.34	4.67	0.75
Fluquinconazole	ND	0.03	6.97	0.52	4.30	1.32

ND *: Not detected

Table 5. Concentrations of Triazole pesticides (ng g^{-1}) in SC2 (Mean; expressed in nanograms per gram dry weight ± SD; expressed in %; n = 3)

Depth (cm)	3	6	9	12	15	20
Mesosulfuron-methyl	ND	ND	ND	ND	ND	ND
Dimitanoate	ND	ND	3.50	ND	ND	1.07
Penconazole	ND	0.68	1.69	0.67	1.74	0.55
Triadimenole	ND	ND	1.18	ND	ND	ND
Fludioxonile	0.03	ND	0.15	ND	ND	ND
Hexaconazole	0.59	0.44	1.07	ND	ND	ND
Cyproconazole	ND	ND	ND	ND	ND	ND
Propiconazole	0.05	1.33	1.19	1.74	ND	0.30
Tebuconazole	0.36	1.03	1.49	0.66	ND	0.76
Epoxyconazole	ND	ND	ND	ND	1.20	ND
Fenoxycarbe	0.06	1.03	3.64	ND	0.06	0.18
Bromuconazole	1.27	ND	2.21	0.90	3.52	0.90
Triticonazole	ND	ND	ND	ND	1.18	0.34
Myclobutanile	0.32	0.15	5.57	0.21	2.63	0.62
Fluquinconazole	0.08	2.82	0.71	8.71	7.27	0.56

ND *: Not detected

Tebuconazole represents 8% of the total triazole pesticides found in SC1 and SC2 collected near the mouth of the channel of Bizerte and near the city of Manzel Bourguiba. Surface sediment contained higher Tebuconazole levels compared with deeper sediments. This result might be associated with the illegal use of Tebuconazole, demonstrating a wider application of Triazole pesticides in this region in the past.

The concentrations of Myclobutanile varied between 0.01 and 4.67 ng g^{-1} for SC1 and between 0.15 and 5.57 for SC2. Epoxyconazole was found only in the subsurface depths of SC1 and in the deep layers of SC2. Myclobutanile presents 20% and 14% of the composition of the total triazole pesticides recorded in SC1 and SC2 respectively. These pesticides have been used in Tunisia for crop protection due to their strong insecticidal properties (Eqani et al. 2012).

The levels of propiconazole ranged between ND and 2.98 ng g^{-1}recorded at 3 cm and 9 cm depth for SC1. However, except for a depth of 15 cm, it has been detected in different layers of SC2 (Tables 4 and 5). Despite this prohibition, propiconazole has been found in the Bizerte Lagoon.This funding could be explained by the historical use of propiconazole and not by current input. The concentrations of fluquinconazole varied between ND and 6.97 ng $g^{-1}$for SC1 and between 0.08 and 8.71 for SC2. Fluquinconazole presents 27% and 29% of the composition of the total triazole pesticides recorded in SC1 and SC2 respectively. Hexaconazole was found only in the subsurface depths of SC2 and in the deep layers of SC1. The highest levels of bromuconazole were 3.52 ng g^{-1}at 15 cm depth in the SC2, and 2.29 ng g^{-1} at the same depth in SC1; this was primarily attributed to industrial and urban activities in zones B and D.

The Fenoxycarbe concentrations in the SC1 varied between ND ng g^{-1} found at 6 cm depth and 2.52 ng g^{-1}obtained at 9 cm depth. The levels of penconazole in the SC2 are higher than those recorded in the SC1. The Triadimenole (1.18 ng g^{-1}) was registered at 9 cm depth in SC2, and at the depths of 9 cm (1.15 ng g^{-1}) and 15 cm (2.73 ng g^{-1}) in the SC1. Dimitanoate was below the detection limits in SC1 and was only detected at 9 and 20 cm in the SC2. Fludioxonile was only detected at a depth of 9cm for the SC1 and at a depth of 9 and 15 cm for the SC2 (Figures 3 and 4).

This finding suggests that triticonazole accumulates in sediment, can leach out of the sediment into the sediment pore water, and then transfers to surface water by processes such as diffusion, which is concentrated within the surface horizon of SCs. This could be due to their high hydrophobicity and lower usage popularity (Jiménez et al. 2008). In fact, wastewater, runoff, and erosion from coastal areas can bring several types of chemical pollutants into marine or lagoon environments, the degree of contamination of which depends on the treatments that have previously been subjected to urban or industrial waste, climate, or an annual period (Bejaoui et al. 2015).

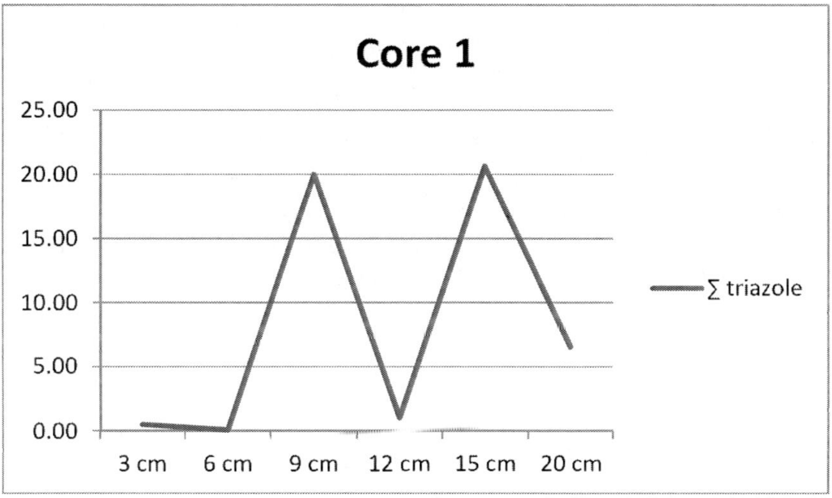

Figure 3. Vertical depth profiles of Triazole pesticides in sediment cores 1.

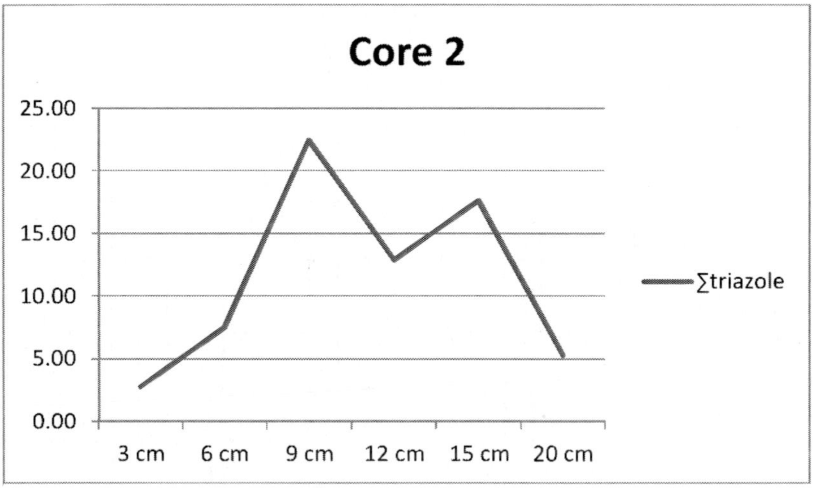

Figure 4. Vertical depth profiles of Triazole pesticides in sediment cores 2.

Table 6. Correlation matrix of Triazole pesticides levels in SC1 of the Bizerte Lagoon

	TOC	Mesosulfuron-methyl	Dimitanoate	Penconazole	Triadimenole	Fludioxonile	Hexaconazole	Cyproconazole	Propiconazole	Tebuconazole	Époxyconazole	Fénoxycarbe	Bromuconazole	Triticonazole	Myclobutanile	Fluquinconazole
TOC	1,00	0.55*	0.52*	0.44*	0.00	0.39	-0.13	0.52*	0.49*	0.52*	0.48*	0.45*	-0.16	-0.62	0.17	0.32
Mesosulfuron-methyl		1.00	1.00	-0.39	-0.26	-0.19	-0.19	1.00	-0.41	1.00	0.96***	-0.22	-0.19	-0.25	-0.33	-0.35
Dimitanoate			1.00	-0.42	-0.28	-0.20	-0.21	1.00	-0.44	1.00	0.96***	-0.24	-0.20	-0.20	-0.35	-0.38
Penconazole				1.00	0.69**	0.48*	0.52*	-0.42	0.98***	-0.42	-0.29	0.52*	0.49*	-0.46	0.82***	0.81***
Triadimenole					1.00	0.22	0.90***	-0.28	0.56*	-0.28	-0.03	0.15	0.91***	-0.28	0.94***	0.71**
Fludioxonile						1.00	-0.22	-0.20	0.58*	-0.20	-0.26	0.98***	-0.20	-0.20	0.50*	0.82***
Hexaconazole							1.00	-0.21	0.35	-0.21	0.07	-0.28	1.00	-0.22	0.73**	0.35
Cyproconazole								1.00	-0.44	1.00	0.96***	-0.24	-0.20	-0.20	-0.35	-0.38
Propiconazole									1.00	-0.44	-0.36	0.64**	0.32	-0.44	0.75**	0.82***
Tébuconazole										1.00	0.96***	-0.24	-0.20	-0.20	-0.35	-0.38
Époxyconazole											1.00	-0.32	0.08	-0.26	-0.15	-0.28
Fénoxycarbe												1.00	-0.27	-0.27	0.46*	0.80***
Bromuconazole													1.00	-0.20	0.74**	0.36
Triticonazole														1.00	-0.38	-0.37
Myclobutanile															1.00	0.90***
Fluquinconazole																1.00

*p<0.05; **p<0.01; ***p<0.001

Table 7. Correlation matrix of Triazole pesticides levels in SC2 of the Bizerte Lagoon

	TOC	**Mesosulfuron-methyl**	**Dimitanoate**	**Penconazole**	**Triadimenole**	**Fludioxonile**	**Hexaconazole**	**Cyproconazole**	**Propiconazole**	**Tébuconazole**	**Époxyconazole**	**Fénoxycarbe**	**Bromuconazole**	**Triticonazole**	**Myclobutanile**	**Fluquinconazole**
TOC	1.00	0.26	-0.28	0.30	0.00	0.06	0.19	0.26	-0.04	-0.38	0.53*	0.00	0.47*	0.27	0.18	0.43*
Mesosulfuron-methyl		1.00	-0.26	-0.63	-0.20	0.00	0.27	1.00	-0.47	-0.34	-0.20	-0.26	-0.08	-0.26	-0.29	-0.43
Dimitanoate			1.00	0.50*	0.95***	0.92***	0.70**	-0.26	0.19	0.75**	-0.26	0.91***	0.23	-0.24	0.85***	-0.47
Penconazole				1.00	0.57*	0.45*	0.16	-0.63	0.06	0.15	0.61**	0.54*	0.74**	0.55*	0.82***	0.31
Triadimenole					1.00	0.98***	0.81***	-0.20	0.28	0.73**	-0.20	0.96***	0.30	-0.26	0.90***	-0.35
Fludioxonile						1.00	0.88***	0.00	0.19	0.68**	-0.24	0.93***	0.29	-0.32	0.86***	-0.44
Hexaconazole							1.00	0.27	0.17	0.67**	-0.39	0.85***	0.01	-0.51	0.61*	-0.61
Cyproconazole								1.00	-0.47	-0.34	-0.20	-0.26	-0.08	-0.26	-0.29	-0.43
Propiconazole									1.00	0.64**	-0.51	0.36	-0.49	-0.61	0.01	0.32
Tébuconazole										1.00	-0.68	0.83***	-0.40	-0.68	0.42	-0.44
Époxyconazole											1.00	-0.26	0.82***	0.96***	0.24	0.51*
Fénoxycarbe												1.00	0.14	-0.33	0.83***	-0.40
Bromuconazole													1.00	0.76**	0.66*	0.26
Triticonazole														1.00	0.18	0.41*
Myclobutanile															1.00	-0.15
Fluquinconazole																1.00

*p<0.05; **p<0.01; ***p<0.001

Correlation and Principal Component Analysis

The distributions of triazole pesticides in SCs were influenced by many factors, such as depth and the TOC, which were observed in this study. Tables 6 and 7 show the correlation matrices of triazole pesticides for SC1 and SC2, respectively. These results showed that there was a significant positive correlation between TOC content and levels of Propiconazole, Tebuconazole, and Cyproconazole ($p < 0.05$) for SC1 and Fluquinconazole for SC2 ($p < 0.05$). This suggests that the distribution of these pesticides may be controlled mainly by the distribution of organic matter in the sediments. Hydrological and sedimentological characteristics of the Bizerte Lagoon, such as water flow, current velocity, and TOC content, can influence the vertical distribution of triazole pesticides in SCs.

The contamination pattern was further studied to estimate the possible sources of triazole pesticides in the catchment. There was a significant positive correlation between Myclobutanile, Penconazole, Triadimenole, Fenoxycarbe, and Fludioxonile ($p < 0.001$), probably reflecting a common origin of these compounds for SC1 (see Table 6). Similarly, the correlation between Fenoxycarbe, Triadimenole, Fludioxonile, Hexaconazole, and Tebuconazole ($p < 0.001$) suggests that these contaminants may have a common source for SC2 (Table 7). In this study, the significant correlation between Tebuconazole, Triadimenole, Fludioxonile Hexaconazole, and Propiconazole ($p < 0.01$), indicated a common source of Triazole pesticides in SC2. A significant correlation between Epoxyconazole, Cyproconazole, and Tebuconazole ($p < 0.001$) was found in SC1, probably reflecting a common origin of these compounds (Table 6).

Health and Environmental Risks of Triazole Pesticides

In general, a large part of the particles suspended in water and that are associated with chemical contaminants end up being deposited on the seabed and are thus integrated into the sediments. These contaminants can accumulat in the tissues of marine populations in direct contact with the sediments, which could pose a major risk to human health by following the trophic chain (Hammami et al. 2005). The triazole pesticides reduced cell viability, disturbed normal cell cycle distribution, and induced cell apoptosis in human trophoblast cells. Most importantly, this work provided evidence that these compounds restrained the invasive and migratory abilities of human trophoblast cells via disrupting protease systems, angiogenic factors, hormones, growth factors, and cytokines. Because trophoblast function is important for successful placentation and fetus development, the inhibitory effects of triazole fungicides on trophoblast invasion and migration may result in a comprehensively adverse influence on human pregnancy (Zhoua et al. 2016) (Durjavaet al. 2013).

Conclusion

Triazole pesticide contamination levels in sediment cores from the Bizerte Lagoon have been studied, and different profiles have been observed. The SCs act as environmental archives that record changes in organic pollutant concentrations. The compositional analysis indicated that myclobutanile and fluquinconazole were the predominant compounds, especially in SC1. Significant positive correlations between TOC contents and levels of triazole pesticides were observed in SC1. Based on the higher percentage of tebuconazole and myclobutanil in SCs samples, there may still be recent illegal use of these compounds. Water flow, current velocity, and TOC content can influence the vertical distribution of triazole pesticides in SCs. Rapid urbanization, industrialization and wastewater discharges are the main sources of organic pollutants in sediments.

Acknowledgments

This work was supported by the Ministry of Higher Education and Scientific Research, Tunisia. The authors have declared no conflict of interest.

References

Anastassiades, M., Lehotay, S. J., Stajnbaher, D., and Schenck, F. J. (2003).Fast and easy multiresidue 623 method employing acetonitrile extraction/partitioning and "dispersive solid-phase extraction" 624 for the determination of pesticide residues in produce, *Journal of AOAC International*, 86, 412-31.

Bai, J., Lu, Q., Zhao, Q., Wang, J., Gao, Z., and Zhang, G. (2015). Organochlorine pesticides (OCPs) in wetland soils under different land uses along a 100-year chronosequence of reclamation in a Chinese estuary *Scientific Reports*, 5, 1–10.

Bejaoui, B., BenCharrada, R., Moussa, M., BenHamadou, R., Harzalla., A., and Chapelle., A. (2005). Caracterisation hivernale de la lagune de Bizerte, *Bulletinde l'Institut National des Sciences et Technologies de la mer*[Winter characterization of the Bizerte lagoon, *Bulletin of the National Institute of Marine Sciences and Technologies*], 32, 79-91.

BenAmeur, W., Trabelsi, S., ElMegdiche, Y., and BenHassine, S. (2013). Concentration of polychlorinatedbiphenyls and organochlorine pesticides in mullet (Mugilcepha et al. lus)and sea bass (Dicentrarchuslabrax) from Bizerte Lagoon (Northern Tunisia), *Chemosphere*, 90, 2372–80.

Castro-Jiménez, J., Deviller, G.,Ghiani, M., Loos, R., Mariani, G.,Skejo, H., and Umlauf, G. (2008).PCDD/F and PCB multi-media ambient concentrations, congener patterns and occurrence in a Mediterranean coastal lagoon (Etang de Thau, France), *Environmental Pollution*, 156, 123-35.

Casida, J. E., and Kathleen, A. (2017). Durkin Pesticide Chemical Research in Toxicology: Lessons from Nature. *Chemical Research in Toxicology*, 17, 30(1), 94-104.

Durjava, M. K., Arnus, B. K. L., Papa, E., Kovarich, S., Sahlin, U., and Peijnenburg, W. (2013). Experimental Assessment of the Environmental Fate and Effects of Triazoles and Benzotriazole *Alternatives to Laboratory Animals*, 41, 65–75.

Eqani, S. A., Malik, R. N., Katsoyiannis, A., Zhang, G., Chakraborty, P., Mohammad, A., and Jones, K. C. (2012). Uptake of organochlorine pesticides (OCPs) and polychlorinated biphenyls (PCBs) by river water fish: the case of River Chenab. *Journal of Environmental Monitoring*, 14, 1645.

Garali, A. B., Ouakad., M., and Gueddari, M. (2013). Episodic Sedimentation of Heavy Metals and Iron in Bizerte Lagoon, Northern Tunisia, *International Journal of Geosciences*, 4, 24.

Hammami, J., Brahim, M., and Gueddari, M. (2005). Essai d'evaluation de la qualite des eaux de ruissellement du bassin versant de la lagune de bizerte,*Bulletin de l'Institut National des Sciences et Technologies de la mer*[Test to assess the quality of runoff water from the watershed of the Bizerte lagoon, *Bulletin of the National Institute of Marine Sciences and Technologies*], 32, 69-77.

Kolberg, D. I. S., Zechmann, S., Wildgrube, C., Sigalov, I., Scherbaum., E., and Anastassiades, M. (2016). Determination of Triazole Derivative Metabolites (TDMs) in Fruit and Vegetables using the QuPPe Method and Differential Mobility Spectrometry (DMS) and Survey of the Residue Situation in Organic and Conventional Produce, *Aspects of food control and animal health*, (2), 2196-3460.

Kim, Y. S., Eun, H., Katase, T., and Fujiwara, H. (2007). Excess copper induced physiological and proteomic changes in germinating rice seeds*Chemosphere*, 67(6), 1182-93.

Liu, C., Liao, Y., and Huang, X. (2017). Extraction of triazole fungicides in environmental waters utilizing poly (ionic liquid)-functionalized magnetic adsorbent *Journal of Chromatography A*, 1524, 13-20.

Martins, M. V. A., Zaaboub, N., Aleya, L., and Frontalini, F. (2015). *Environmental Quality Assessment of Bizerte Lagoon (Tunisia) Using Living Foraminifera Assemblages and a Multiproxy Approach*, 10, 1-24.

Masiá, A., Campo, J., Vázquez-Roig, P., Blasco, C., and Yolanda, P. (2013). Screening of currently used pesticides in water, sediments and biotaof the Guadalquivir River Basin (Spain) *Journal of Hazardous Materials*, 263P, 95-104.

Miglioranza, K. S. B., Moreno, J. E. A. D., Moreno, V. J., Osterrieth,M. L., and Escalante, A. H.(1999). Fate of organochlorine pesticides in soils and terrestrial biota of "Los Padres" pond watershed, Argentina. *Environmental Pollution*, 105, 91-99.

Mzoughi, N., Stoichev, T., Daetchraoui, M., and ElAbed, A. (2002). Inorganic mercury and methylmercury in surface sediments and mussel tissues from a microtidal lagoon (Bizerte, Tunisia), *Journal of Coastal Conservation*, 8, 141-45.

Necibi, M., and Mzoughi, N. (2019a). Distribution of organochlorine pesticides in sediment cores from the bizerte lagoon (Tunisia). *International Journal of Environmental Research*, 99, 1-15.

Necibi, M., Mzoughi, N., Yahia, M.N. D., and Pringault, O. (2015). Distributions of organochlorine pesticides and polychlorinated biphenyl in surface water from Bizerte Lagoon, Tunisia. *Desalination and Water Treatment*, 56 (10), 2663-71.

Ritter, L., Solomon, K. R., and Forget, J. (1995). les polluants organiques persistants:Rapport d'évaluation DDT-aldrine-dieldrıne-endrine-chlordane-Heptachlore -hexachlorobenzène-Mirex-toxaphèneBiphénylespolychlorésDioxines et furanes [*Persistent Organic Pollutants: Assessment Report DDT-aldrin-dieldrin-endrin-chlordane-Heptachlor-hexachloro benzene-Mirex-toxaphene Polychlorinated biphenyls Dioxins and furans*], 7-54.

Wang, H., Yang, X., Hu, L., Gao, H., Lu, R., Zhang, S., and Zhou, W. (2016). Detection of triazole pesticides in environmental water and juice samples using dispersive liquid-liquid microextraction with solidified sedimentary Ionic Liquids. *New Journal of Chemistry*, 40, 4696-04.

Wang, X. Q. (2011). Research on variation of microfibril angle in Phyllostachys pubescens. *Journal of Anhui Agricultural University*, 38, 500-03.

Wu, Q. H., Leung, J. Y. S., Yuan, X., Huang, X. X., Li, H. Y., Huang, Z. Y., and Li, Y. (2015). Biological risk, source and pollution history of organochlorine pesticides (OCPs) in the sediment in Nansha mangrove, South China. *Marine Pollution Bulletin*, 96, 57-64.

Xua, X. Y., Yea, J. Q., Niea, J., Lia, Z. G., and Leeb, M. R. (2013). A new liquid-liquid microextraction method by ultrasound assisted salting out for determination of triazole pesticides in water samples coupled by gas chromatography-mass spectrometry, *Analytical Methods*, 3, 1194-1199.

Zhang, Q. L., Zhou, Y., Yang, X., Hua, H., and Wang, M. (2015). Study on the stereoselective degradation of three triazole fungicides in sediment. *Ecotoxicology and Environmental Safety*, 117, 1–6.

Zhou, J., Zhang, J., Li., F., and Liua, J. (2016). Triazole fungicide tebuconazole disrupts humanplacental trophoblastcell function. *Journal of Hazardous Materials*, 308, 294–302.

Chapter 4

The Distribution of Phytosanitary Products in Sediments from the Bizerte Lagoon, Tunisia

**Mouna Necibi*
and Nadia Mzoughi**
High Institute of Environmental Sciences and Technologies of Borj Cedria,
Environmental Sciences and Technologies Laboratory,
University of Carthage, Hammam Lif, Tunisia

Abstract

The contamination of the marine environment by organic pollutants is of great concern. The Bizerte Lagoon is the second largest lagoon in Tunisia and is known for its industrial and fishing activities. A comprehensive assessment of the spatial distribution and partitioning of fifteen triazole pesticides in surface sediments is essential due to their social and economic impact.

Total organic carbon (TOC) content in sediment ranges from 0.05% found in station S3 to 5.22% found in station S5. The triazole pesticides were investigated in surface sediment from fourteen sampling stations in the Bizerte Lagoon. Levels of these pollutants are in the range of 0.22 and 26.61ng g^{-1}.

The fluquinconazole and hexaconazole are the most dominant compounds in the different studied sites. The distribution of triazoles is different, indicating different sources of contamination. The concentrations of triazole in surface water are generally similar to the background levels from the Mediterranean Sea and pose no threat to human and aquatic living species.

*Corresponding Author's Email:mouna23necibi@gmail.com.

In: Triazoles and Their Uses
Editor: Karl Roland
ISBN: 979-8-88697-800-1

Keywords: triazole pesticides, sediments, distribution, Bizerte Lagoon

Introduction

By design, pesticides are intended to control pests with minimal effects on other environmental organisms by defining the manner and place of their application (Casida et al., 2016). Application of pesticides is a worldwide issue and of most importance for current agricultural productivity standards and the control of vectors and pests of public health relevance. Regarding agrochemicals, pesticides are the group of compounds that cover the largest number of substances, and can be grouped into several classes: insecticides, herbicides, fungicides, rodenticides, molluscicides, insect repellents, nematicides, and plant growth regulators (Maia et al., 2017). Among many classes of fungicides, azole or conazole fungicides represent a large group of substances which are widely used in agriculture for protection of crop plants and in the pharmaceutical industry in the treatment of various fungal diseases. They are synthetic compounds that can be classified into imidazole or triazole, depending on the number of nitrogen atoms in the five-membered ring (Shahinasiet al., 2017).

Triazole fungicides such as triadimenol, myclobutanil, triazolone, and hexaconazole are systemic pesticides that have been widely applied to avoid fungus diseases in agriculture. However, triazole pesticides may enter environmental waters such as rivers and lakes, or even permeate soil and enter underground waters (Lui et al., 2017). They are thoroughly studied and carefully classified as to their potential as carcinogens, mutagens, teratogens, delayed neurotoxicants, and other adverse effects (Casida et al., 2016). The reports on organic pollutants such as SM (Necibi et al. 2016) and OCP occurrence in the marine environment of the Bizerte Lagoon are extensive and regard several matrices, such as cores (Necibi et al. 2019a), suspended particulate matter, sediments (Necibi et al., 2019b), water (Necibi et al., 2015).

The Bizerte Lagoon is the second largest lagoon and one of the most economically important areas in Tunisia. This lagoon is also known as a fishing and aquaculture park due to the presence of three mytiliculture sectors. It is subject to many anthropic pressures, including industrial activities and urbanization (Ameur et al., 2013). The Bizerte Lagoon is delimited by the towns of Bizerte, Zarzouna, Menzel Jemil, and Menzel Abderrahmen on the north side, and Menzel Bourguiba and Tinja on the west and southwest side. Urban areas experience a very high population concentration and multiple industrial activities. Since the 1950s, around 400 industrial units relating to various fields of activity such as petrochemicals, textiles, iron, steel, and plastic processing have been established on the shores of the Bizerte Lagoon. Consequently, industrial activities, domestic wastewater and open-air municipal landfills scattered around the lagoon can play a major polluting role that could lead to a degradation of the ecological conditions of this body of water (Salvator et al., 1900, FAO et al., 2005, Garali et al., 2011, Sahraoui et al., 2013, Bejaoui et al., 2005). In fact, wastewater, runoff, and erosion from coastal areas can bring several types of chemical pollutants into marine or lagoon environments, the degree of which contamination depends on the treatments that have previously been subjected to urban or industrial waste, climate, or annual period. In general, a large part of the particles suspended in water, and which are associated with chemical contaminants end up being deposited on the seabed and are thus integrated into the sediments. These contaminants can be accumulated in the tissues of marine populations in direct contact with the sediments, which could pose a major risk to human health by following the trophic chain (Bejaoui et al., 2010, INS et al., 2008).

The triazole pesticide levels have been widely recorded around the world. However, very little information exists on their environmental levels in surface sediment collected from the Bizerte Lagoon, and there are no previous studies on the distribution of these pollutants in sediment. The goal of the present study was to investigate the partitioning and the spatial distribution of triazole in sediments in the Bizerte Lagoon of Tunisia and to provide useful information for the ecological remediation in this lagoon.

Materials and Methods

Study Area

The Bizerte Lagoon is a Mediterranean lagoon located on the northern coast of Tunisia (latitude: 37°8'- 37°14'N, longitude: 9°46' - 9°56' E). It covers an area of 128km^2 and has an average depth of 7m. It is connected to the Mediterranean Sea by a 6km long canal and to Lake Ichkeul by the Tinja Canal (Figure 1). The water temperature in the lagoon has a seasonal cycle, ranging from 10°C in the winter to 28°C in the summer. The salinity of the water varies from 30 psu in the winter to 38 psu in the summer. The water balance is controlled primarily by the exchange of water due to the tide. The supply from

Lake Ichkeul through the Tinja canal and with a lesser degree of seasonal inflows from the rivers, which mainly cover the southern sector to the east. The general circulation in the Bizerte Lagoon is mainly driven by the West and North-West winds which sweep the surface of the lagoon, accelerating the propagation in the West-East direction along the South and North coast (Bejaoui et al., 2010, INS et al., 2008). There are four main zones of anthropogenic influence (Figure 1):

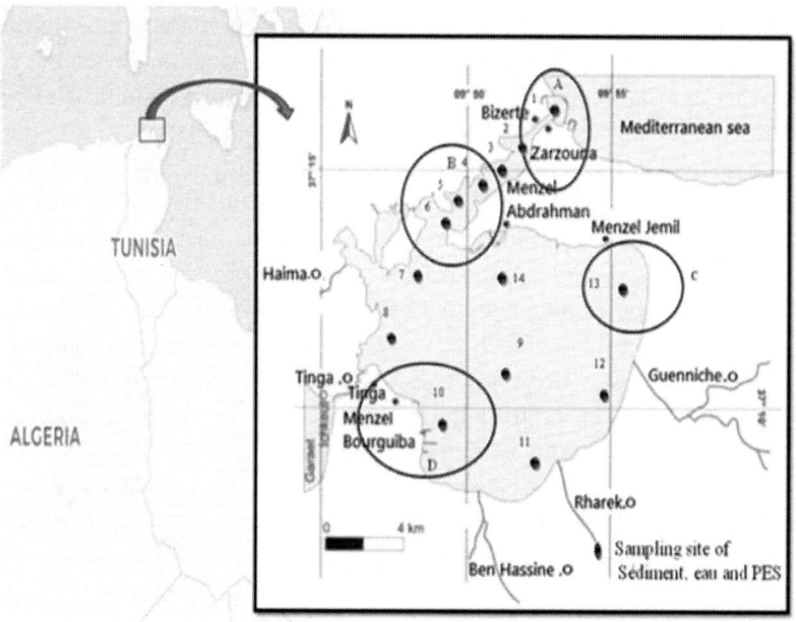

Figure 1. Study areas with sampling stations and different types of industrial areas (A, B, C, and D) in the Bizerte Lagoon.

Zone A: oil refineries, food and ceramic industries.

Zone B: Cement manufacturing, metal treatment (copper craft, asbestos), and beverage sprinkling.

Zone C: ceramics and metallurgy activities.

Zone D: metallurgical activities (Fe, Zn, Cd, Sn, Hg), naval construction and tire production.

Sample Collection

Surface sediment samples were taken from fourteen stations in July 2016 with the aid of a global positioning system (GPS) (Table 1). The same first layer of the surface sediment (1–2cm depth) was collected (about 1–2kg) at all stations using the Van Veen grab. Each sample was homogenized to form a composite sample for each sampling site, and then transported to the laboratory. Sediment samples were freeze-dried and stored at -20°C until analysis. The Total Organic Carbon (TOC) of sediment samples was analyzed with a TOC Analyzer (SHIMADZUH544051).

Table 1. Description of samples collected from the Bizerte Lagoon

Sites	Latitude (N)	Longitude(E)	Water depth (m)
S1	37°17°10	09°54°03	23
S2	37°16°33	09°53°00	9
S3	37°15°52	09°52°05	12
S4	37°15°17	09°51°24	11
S5	37°15°11	09°51°00	4
S6	37°13°37	09°48°56	11
S7	37°11°01	09°48°01	5
S8	37°11°41	09°49°01	7
S9	37°10°52	09°51°00	10
S10	37°09°06	09°49°37	6,5
S11	37°10°10	09°53°05	6
S12	37°11°10	09°54°00	5
S13	37°13°09	09°54°46	5
S14	37°13°04	09°52°00	5

Analytical Performances

Appropriate blanks were analyzed with a set of analyses; internal standards were analyzed for quality control purposes. Quantification was assured by the calculation of response factors of individual triazole external standards. The reported results were corrected with recoveries of surrogate internal standard fluconazole (Kolberg et al., 2016). The calibration solutions in methanol were prepared daily at five levels: 1; 10; 50; 100; and 200ng L^{-1}. Over the calibration range selected, good linearity was obtained for all the target compounds. The determination coefficients (R^2) were higher than 0.990 for all triazole pesticides. The limits of detection and the limits of quantification have been calculated by multiplying the standard deviations of the calculated amount for each triazole by 3 and 10, respectively. For the validation of the method, the precision and accuracy were determined by analyzing the spiked water samples at two levels of concentrations (10 and 100ng L^{-1}) on the same day and three different days (Table 3). Relative standard deviations for intraday (n = 5) precisions ranged between 1.02% and 1.98%, and for interday (n = 15) precisions ranged between 0.77% and 1.91%. The extraction recoveries for these selected pesticides were in the range of 70% and 99%, for the two levels of spiked concentrations, respectively. These results indicate that this method is satisfactorily accurate, sensitive, and repeatable (Wang et al., 2016; Zhang et al., 2015).

Reagents and Standards

High purity reagents of analytical grade are used. Hexane, acetonitrile, and ethyl ether, which have been used for the extraction of triazol pesticides, were obtained from Honeywell Riedel-de Haën (France). Fluka supplied formic acid, which was added to the mobile phase acetonitrile. Riedel-de-Han (Germany) provided purity ranging from 95-100% for the mesosulfuron-methyl, fludioxonil, tebuconazole, cyproconazole, triticonazole, penconazole, bromuconazole, hexaconazole, propiconazole, fluquinconazole, triadimenol, myclobutanil, epoxiconazole, dimethoate, and fenoxycarbe witch were used as an internal standard in this study.

The combined stock solution of the 15 standards was prepared in acetonitrile and diluted with acetonitrile. All solutions are stored in the dark at 20°C. For the linearity and recovery experiments, six different commodities of organic products have been used.

Extraction

The extraction of triazole pesticides from sediments is based on the method proposed in preview works (Masiá et al., 2013).

The QuECHERS method, also previously reported in a variety of formats (Anastassiades et al., 2003), was applied to 2g of lyophilized sediment sample, which was weighed in a 50mL falcon tube, and homogenized with 10mL of acetonitrile.

Then, $MgSO_4$ (6g), NaCl (1.5g), trisodium citrate dehydrate [$Na_3C_6H_5O_7$+ $2H_2O$(1.5g)] and disodium hydrogen citrate sesquihydrate [HOC(COOH) $(CH_2COONa)_2$ $1.5H_2O$ (0.75g)] were added for phase-separation adjustment. The mixture was agitated intensively in a vortex for 1min and centrifuged at 3000rpm for 5min. An aliquot (1mL) of the upper organic phase was cleaned up by dispersive solid phase extraction (d-SPE) using PSA (50mg), $MgSO_4$ (150mg) and C18 (50mg). This mixture was shaken in a vortex for 1min and centrifuged at 3000rpm for 5min. The supernatant was filtered through a 0.45μm PTFE filter and introduced in a vial for GC-MS analysis.

Gas Chromatography

Analyses of the extracts were performed with an Agilent 7890 gas chromatography (GC) coupled to a 5975 C inert mass selective detector (agilent technologies, USA).

The Agilent chemstation software has been used for instrument control, data acquisition, and data treatment. The temperature program optimization has been realized based on previous work with a few modifications (Kolberg et al., 2016; Xu et al., 2013; Vieira et al., 2017). The chromatographic separation was performed with the HP-5MS capillary column (30m length x 0.25mm diameter and 0.25μm film thickness).

The column temperature was initially held at 50°C for 1 minute, then ramped to 160°C at 7°Cmin^{-1}, and finally held at 240°C at 10°C min^{-1} for 18 minutes. The target ions have been measured in the selected ion monitoring (SIM) mode (Table 2). Helium has been selected as the carrier gas with a flow rate of 1mL min^{-1}. The injector and detector were maintained at 250°C and 300°C respectively.

The volumes of injection were 1μL in the splitless mode. The target ions have been measured in the selected ion monitoring (SIM) mode. The mass detector conditions were: transfer line temperature: 250°C; ion source temperature: 180°C; ionization mode-electron impact at 70 eV.

Statistical Analysis

The statistical analyses have been realized by the techniques of principal component analysis. Correlation analyses have been used to reveal the relationships between triazole pesticides. The difference was statistically significant when $p < 0.05$. All statistical analyses were performed with the Statistica 6 software package.

Results and Discussion

Distribution of Triazole Pesticides

Fourteen sampling sites in the Bizerte Lagoon revealed a total of 15 pesticides with the aid of a global positioning system (GPS) (Table 1). Levels of triazole compounds in surface sediment samples are represented in Table 3. The concentration of total triazole pesticides ranged from 0.22 to 26.61ng g^{-1}. The less contaminated sampling sites in the lagoon were S3 and S7 with total triazole pesticide concentrations of 0.94 and 0.22ng g^{-1}, respectively. The most polluted stations were S4, S9, and S12, with total triazole pesticide concentrations of 23.56, 12.30, and 26.61ng g^{-1}, respectively.

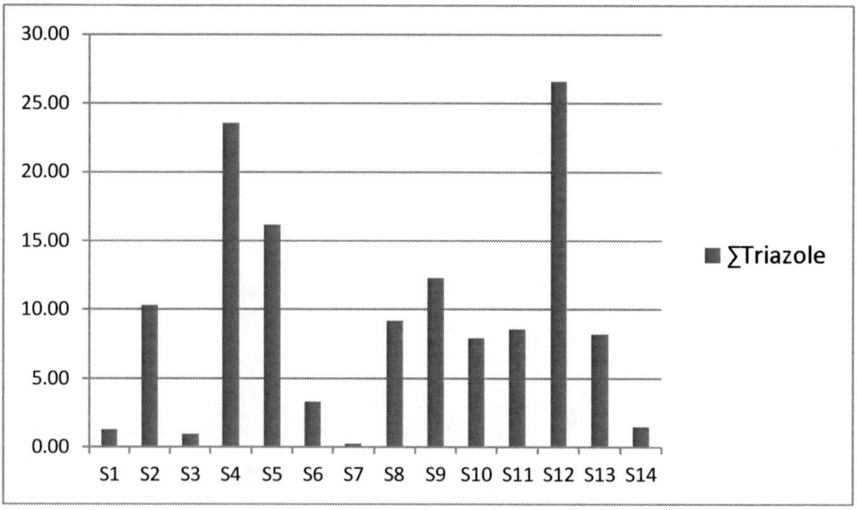

Figure 2. Distribution of triazole in surface water samples collected from Bizerte Lagoon.

Table 2. Analytical performance of the QuECHERS extraction

Compounds	**Ions (m/z)**	**LOD (ng g^{-1})**	**LOQ (ng g^{-1})**	**Recovery (%) 10 (ng L^{-1})**	**Recovery (%) 100 (ng L^{-1})**	**RSD % (n= 5)**	**RSD% (n= 15)**	**Correlation coefficient (R^2)**
Mesosulfuron-methyl	154.1; 68.0	0.001	0.003	96.19	85	1.07	0.77	0.996
Dimethoate	229;125;87.0	0.008	0.024	85.07	97	1.28	1.34	0.994
Penconazole	248.1; 159.0	0.001	0.003	92	74	1.25	1.09	0.994
Triadimenole	296.1; 168.1; 128.0	0.001	0.003	70.56	79	1.13	1.30	0.997
Hexaconazole	214.0; 83.1	0.008	0.024	76.59	82	1.05	1.63	0.998
Fludioxonile	248.1; 154.1; 127.0	0.007	0.021	76.16	90	1.33	1.13	0.998
Cyproconazole	222.1; 139.0	0.009	0.027	82.61	87	1.52	1.36	0.994
Propiconazole	259.1; 173.0; 69.1	0.004	0.012	75	98	1.85	1.00	0.991
Tebuconazole	250.1; 125.0	0.010	0.030	92	74	1.25	1.91	0.995
Epoxyconazole	206.1; 192.1	0.006	0.018	76.55	79	1.71	1.58	0.994
Fenoxycarbe	301; 186.1; 116.1	0.030	0.090	97.42	86	1.02	1.12	0.997
Bromuconazole	294.9; 214.0; 173.0	0.001	0.003	74.51	99	1.02	1.32	0.995
Triticonazole	299.1; 235.1; 83.0	0.005	0.015	96	98	1.98	1.19	0.993
Myclobutanile	168; 128;112	0.019	0.057	99.38	97	1.05	1.33	0.991
Fluquinconazole	340; 108.0	0.017	0.051	99.55	88	1.20	1.51	0.991

Table 3. Concentrations of triazole pesticides in surface water (ng g^{-1}) collected from the Bizerte Lagoon

Compounds	S1	S2	S3	S4	S5	S6	S7	S8	S9	S10	S11	S12	S13	S14
Mesosulfuron-methyl	0.00	0.00	0.00	0.00	1.37	0.00	0.00	0.00	0.00	0.00	1.09	0.00	0.00	0.00
Dimethoate	0.00	0.00	0.00	0.00	0.00	0.00	0.00	0.04	0.00	2.02	0.00	0.00	0.00	0.00
Penconazole	0.00	0.18	0.94	0.00	0.00	0.70	0.00	0.00	1.63	0.63	1.08	3.13	0.46	0.26
Triadimenole	0.00	1.11	0.00	0.00	0.00	0.00	0.00	0.01	0.00	0.00	0.74	0.00	0.00	0.00
Hexaconazole	0.00	0.09	0.00	2.08	5.04	0.00	0.00	0.01	0.07	0.33	0.66	0.00	0.00	0.00
Fludioxonile	0.00	1.24	0.00	15.20	2.92	0.26	0.00	0.01	0.00	0.00	0.99	9.58	0.43	0.80
Cyproconazole	0.00	0.17	0.00	0.00	0.26	0.00	0.00	0.06	0.20	0.10	0.47	0.00	0.00	0.00
Propiconazole	0.00	1.53	0.00	0.00	0.00	0.74	0.03	0.03	1.79	0.02	1.11	2.33	1.32	0.08
Tebuconazole	0.00	0.48	0.00	0.00	0.00	0.00	0.02	0.06	0.19	0.00	0.07	0.11	0.00	0.00
Epoxyconazole	0.00	0.09	0.00	0.00	0.00	0.00	0.00	0.10	0.00	0.00	0.30	0.12	0.00	0.00
Fenoxycarbe	0.00	0.01	0.00	0.00	0.24	0.62	0.00	3.50	0.11	0.00	0.28	4.26	1.02	0.06
Bromuconazole	0.00	0.15	0.00	0.00	0.00	0.00	0.01	0.51	0.09	4.05	0.64	0.00	0.00	0.00
Triticonazole	0.00	0.09	0.00	0.00	0.00	0.00	0.01	0.16	0.23	0.32	0.23	0.00	0.00	0.00
Myclobutanile	0.09	0.27	0.00	4.27	4.30	0.99	0.01	0.19	0.12	0.06	0.21	6.46	1.81	0.11
Fluquinconazole	1.17	4.88	0.00	2.01	2.03	0.00	0.14	4.51	7.86	0.37	0.72	0.61	3.18	0.19
Total triazole	1.25	10.30	0.94	23.56	16.16	3.30	0.22	9.19	12.30	7.91	8.59	26.61	8.21	1.50

The pollution for the other sampling sites was medium. Quite high contamination, with 16.16 and 9.19ng g^{-1}, was found to be in the north of the lagoon (S5 and S8). Concerning the water flow, this contamination can be explained by the drainage waters coming from agriculture activities through the rivers. The highest levels of contamination were observed in sites close to or far from urban zones, agricultural and industrial activities (Figure 2).

Concentrations of Different Triazole Pesticides

Tebuconazole was found in a few samples. It ranged from ND to 0.48ng g^{-1}.

The highest levels were found in S2 (0.48ng g^{-1}) located near the mouth of the Canal of Bizerte.Myclobutanile was detected in most of the sampling sites. The concentrations ranged from ND to 6.64ng g^{-1}. The highest levels were found in S12 (6.46ng g^{-1}) located near the mouth of the Ghennich River. Except for stations S8 and S10, dimethoate was below the detection limits in all sampling sites. Results show that levels of triadimenole varied between ND and 0.32ng g^{-1}. The mesosulfuron-methyl was detected only in S5 (1.37ng g^{-1}) and S11 (1.09ng g^{-1}) and was below the detection limits in all the other sampling sites. The highest levels of Fludioxonile were detected in S4 and S5, with concentrations ranging between ND-5.04ng g^{-1}. The highest concentrations of bromuconazole were observed for S10 (4.05ng g^{-1}). Pesticides may move from their initial planned area of application. Investigators have shown that measurable concentrations of pesticides have been registered in air, water, plants, and animals up to several hundred meters from the application sites (Gatto et al. 2009). Except for S3 and S6, fluquinconazole was detected at high concentrations in all of the samples analyzed, with concentrations ranging between ND and 7.86ng g^{-1}. Levels of cyproconazole varied between ND and 0.47ng g^{-1} (Table 3). The highest hexaconazole levels were found in S4 at 15.20ng g^{-1}, which could be explained by the use of astrol and conazol pesticides on vines, melons, and citrus fruits. Levels of epoxyconazole were observed only for four stations: S2 (0.09ng g^{-1}), S8 (0.10ng g^{-1}), S11 (0.30ng g^{-1}), S12 (0.12ng g^{-1}). Penconazole was detected at high concentrations in S9, S11, and S12, with levels of 1.63ng g^{-1}, 1.08ng g^{-1}, and 3.13ng g^{-1}respectively. The highest levels of propiconazole were observed in stations S2 (1.53ng g^{-1}), S9 (1.79ng g^{-1}),S11 (1.11ng g^{-1}),S12 (2.33ng g^{-1}), S13 (1.32ng g^{-1}). Tridimenole was only detected in S2 (1.11ng g^{-1}) and S11 (0.74ng g^{-1}).The penconazole varied between 0.18ng g^{-1} in S2 and 3.13ng g^{-1} in S12.

Residual Levels of Triazole Pesticides

The percentage of fluquinconazole varied between 2.3 and 93.1%. It accounts for 93.1%, 64.0%, 63.9%, and 47.4% of total triazole pesticides obtained at sampling sites S1, S7, S9, and S2 (Figure 3). The percentage of hexaconazole varied from 0.1 to 64.5% of total triazole pesticides in S8 and S4 respectively. Surface water quality is threatened by point source pollution, including municipal sewage discharges, industrial wastewater loads, and nonpoint source pollution from agriculture. The percentage of penconazole varied from 1.7 and 100% of total triazole pesticides in S2 and S3 respectively. Fenoxycarbe accounts for 18.8%, 38.1%, and 16.0% of total triazole pesticides found at sampling sites S6, S8, and S12, respectively (Figure 3).

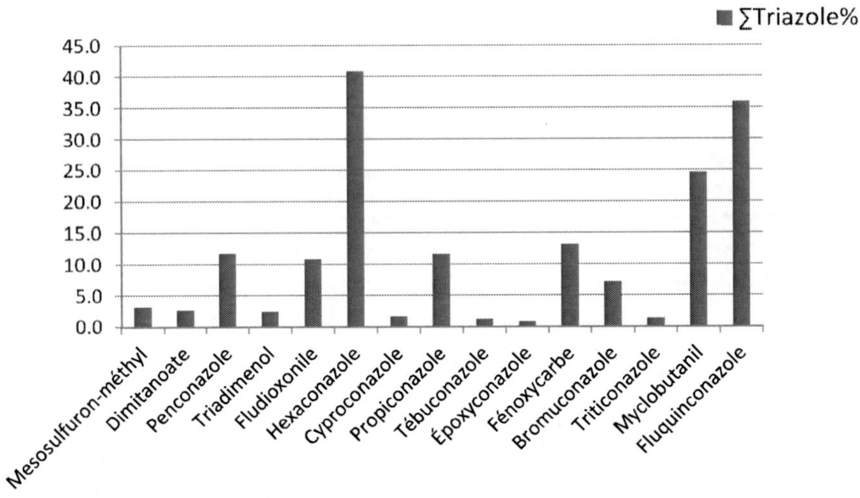

Figure 3. Percentage of individual triazole pesticides in surface water samples collected from Bizerte Lagoon.

Hexaconazole and fluquinonazole were the most abundant triazole pesticides detected in this survey of the Bizerte Lagoon surface sediment, accounting for 24.17% and 21.27% of total triazole pesticides, respectively.

The other triazole pesticides accounted for less than 40% of the total triazole pesticides. Dimethonate, mesosulfuron-methyl (0.94%), tebuconazole, cyproconazole, and triticonazole had respective concentrations of 1.59%, 0.73%, 0.98%, and 0.80%.

Correlation and Principal Component Analysis

The distribution of pesticides in surface water samples was influenced by many factors. The correlation matrix of triazole pesticides in surface sediment collected from the Bizerte Lagoon is listed in Table 4. These results showed that there was a significant positive correlation between Mesosulfuron-methyl and Fludioxonile contents and total oragnic carbon of surface sediment samples ($p < 0.05$) could indicate that TOC % influences the spatial distribution of these compounds.

There was a significant positive correlation between Mesosulfuron-methyl and Cyproconazole ($p < 0.01$) due mainly to the same source of contamination. Similarly, the correlation between dimitanoate and bromuconazole ($p < 0.001$) suggests that these contaminants may have a common source of pollution (Table 4). Hydrological and sedimentological characteristics of the Bizerte Lagoon such as water flow, current velocity, and total organic carbon content, can influence the distribution of triazole pesticides in the Bizerte Lagoon. There was a significant positive correlation ($p <0.05$) between Cyproconazole, Triadimenol, Mesosulfuron-methyl, and Epoxyconazole ($p < 0.05$) indicating the same pollution source.

The principal compound analyses were carried out to further identify the possible sources of triazole pesticides in surface sediment from the Bizerte Lagoon (Figure 4). The triazole pesticides were divided into three groups. The first group, G1, included Bromuconazole and Dimitanoate. The second group G2 included triadimenole, fluquinconazole, tebuconazole, penconazole, fenoxycarbe, and the third group G3, hexaconazole, myclobutanil, and propiconazole (Figure 4). The contamination by triazole pesticides in all groups can be attributed mainly to agricultural activities and atmospheric deposition. The contamination of G1 could be explained by the industrial and agricultural activities of zones A and B while common sources for G2 can be produced by urban and industrial activities. The contamination by group G3, which can be attributed to areas C and D.

Table 4. Correlation matrix of triazole pesticides levels in sediments of the Bizerte Lagoon

	TOC%	**Mesosulfuron-methyl**	**Dimitanoate**	**Penconazole**	**Triadimenol**	**Fludioxonile**	**Hexaconazole**	**Cyproconazole**	**Propiconazole**	**Tebuconazole**	**Epoxyconazole**	**Fenoxycarbe**	**Bromuconazole**	**Triticonazole**	**Myclobutanil**	**Fluquinconazole**
TOC%	1.00	0.50*	0.06	-0.35	0.03	0.59*	-0.21	0.26	-0.22	-0.01	-0.18	-0.22	0.04	-0.05	-0.03	-0.04
Mesosulfuron-methyl		1.00	-0.11	-0.07	0.23	0.75**	-0.01	0.78**	-0.07	-0.11	0.43*	-0.14	-0.04	0.10	-0.14	-0.09
Dimitanoate			1.00	-0.01	-0.11	-0.05	-0.14	0.02	-0.22	-0.14	-0.14	-0.14	0.98***	0.63*	-0.18	-0.20
Penconazole				1.00	-0.05	-0.27	0.23	0.07	0.76**	0.15	0.33	0.52	-0.08	0.13	0.52*	0.02
Triadimenol					1.00	-0.08	-0.10	0.58*	0.36	0.78**	0.63*	-0.18	-0.01	0.26	-0.22	0.22
Fludioxonile						1.00	0.35	0.37	-0.29	-0.19	-0.10	-0.18	-0.07	-0.16	0.08	-0.01
Hexaconazole							1.00	-0.17	0.12	-0.05	0.04	0.25	-0.18	-0.30	0.73**	-0.08
Cyproconazole								1.00	0.22	0.30	0.68	-0.19	0.15	0.57*	-0.33	0.25
Propiconazole									1.00	0.58*	0.40	0.40	-0.20	0.12	0.42	0.41
Tébuconazole										1.00	0.29	0.02	-0.11	0.23	-0.10	0.59*
Époxyconazole											1.00	0.33	0.02	0.38	0.02	-0.01
Fénoxycarbe												1.00	-0.10	-0.05	0.51*	0.08
Bromuconazole													1.00	0.73**	-0.23	-0.15
Triticonazole														1.00	-0.41	0.35
Myclobutanil															1.00	-0.21
Fluquinconazole																1.00

*p < 0.05; **p < 0.01; ***p < 0.001.

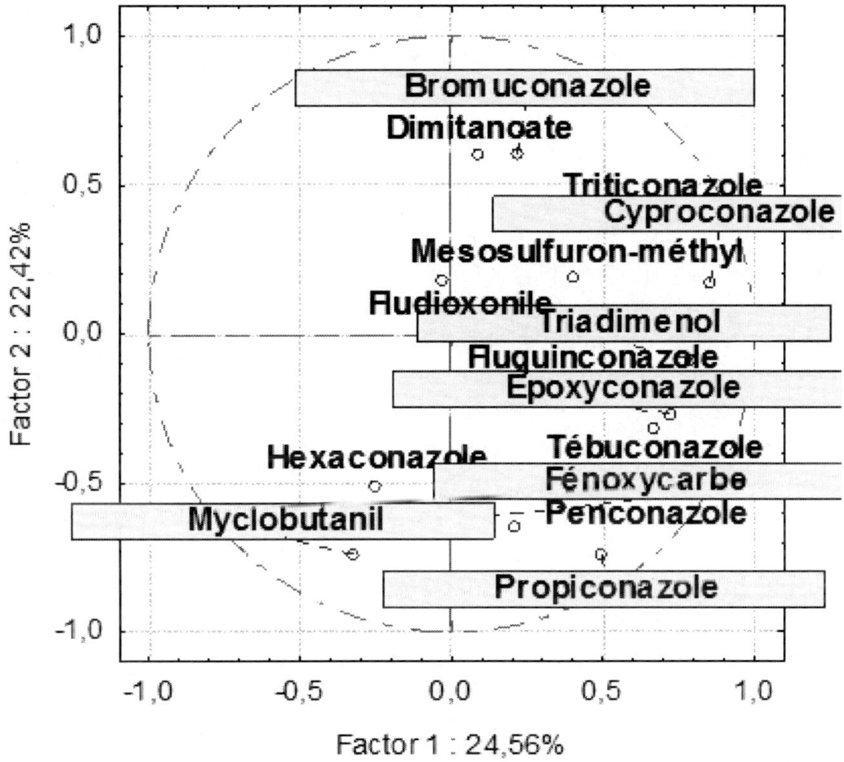

Figure 4. Principal component analysis for individual triazol in sediment collected from the Bizerte Lagoon.

Health Risk of Triazole Pesticides

The intensive or extensive use of these compounds can generate a lot of residues that may potentially lead to substantial plant and environmental contamination. Thus, triazole residues or triazole metabolites may occur in plants or the environment and should be considered as a risk for human health due to their possible transport into the food chain. Triazole pesticides are considered potential water pollutants. It presents a threat to terrestrial and aquatic animals due to its potential for dispersion in the environment. Unfortunately, in mammals, they also impair steroidogenesis, therefore affecting reproduction and development. There is substantial literature on the impacts of triazole pesticide exposure on the endocrine system of marine mammals and fish. In humans, the TOXNET database contains reports of accidental overexposures to triazole pesticides in which exposed people reported, among many other symptoms, hyperactivity (Masiá et al., 2013).

Conclusion

In this study, levels of triazole pesticides have been assessed in the Bizerte Lagoon, which is the most important lagoon in Tunisia. Even though many countries have imposed strict regulations on the concentrations of pesticide residues, those compounds are still detected in the water. The average concentrations of triazole were 0.22 and 26.61ng g^{-1}. The predominant triazole pesticides in the surface sediment were hexaconazole, tebuconazole, and fluquinconazole. The highest levels of contamination were observed in sites close to or far from urban zones, as well as agricultural and industrial activities. Quite high contamination with triazole pesticides was found to be in the south and north of the lagoon. An effective control must take into account both urban and industrial discharges. Results show that levels of triazole pesticides obtained in this study pose no threat to human and aquatic living species.

Acknowledgment

This work was supported by the Ministry of Higher Education and Scientific Research, Tunisia. The authors have declared no conflict of interest.

Disclaimer

None

References

Anastassiades M, Lehotay SJ, Stajnbaher D, Schenck FJ et al. Fast and easy multiresidue 623 method employing acetonitrile extraction/partitioning and "dispersive solid-phase extraction" 624 for the determination of pesticide residues in produce, *Journal of AOAC International,* (2003) 86: 412-431.

Bejaoui B,Ferjani D,Zaaboub N,Chapelle A,Moussa M. Caractérisation hydrobiologique saisonnière de la lagune de bizerte (Tunisie)(Seasonal hydrobiological characterization of the Bizerte Lagoon (Tunisia)), *Journal of Water Science,* (2010) 23:215-232.https://www.erudit.org/en/journals/rseau/2010-v23-n3-rseau3946/ 044686ar/abstract/

Bejaoui B,Ben Charrada R, Moussa M, Ben Hamadou R,Harzalla A, Chapelle A. Caracterisation hivernale de la lagune de Bizerte, *Bulletin de l'Institut National des Sciences et Technologies de la mer* (2005) 32: 79-91. https:// docplayer. fr/215359969-Caracterisation-hivernale-de-la-lagune-de-bizerte.html

Ben Ameur W,Trabelsi S,El Megdiche Y,Ben Hassine S,Barhoumi B, Hammami B, Eljarrat E, Barceló D, Driss MR. Concentration of polychlorinated biphenyls and organochlorine pesticides in mullet (Mugil cephalus)and sea bass (Dicentrarchus labrax) from Bizerte Lagoon (Northern Tunisia), *Chemosphere,* (2013) 90: 2372-2380.https://pubmed.ncbi.nlm.nih.gov/23149188/

Casida JE,Durkin KA. Pesticide Chemical Research in Toxicology: Lessons from Nature. *Chemical Research in Toxicology,* (2016)30(1): 94-104.https:// pubmed.ncbi. nlm nih.gov/27715053/

INS, Rapport annuel. Institut National des Statistique, (2008).http://www.ins.tn/ publication

FAO, Organisation des Nations Unies pour l'alimentation et l'agriculture (2005) 1-11.https://www.fao.org/home/fr

Gatto NM, Cockburn M, Bronstein J, Manthripragada AD and Ritz B. Well-water consumption and parkinson's disease in rural California. *Environmental Health Perspectives,* (2009) 117 (12): 1912-1918. https://www.ncbi.nlm.nih.gov/ pmc/articles/PMC2799466/

Garali AB, Ouakad M, Gueddari M. The Bizerte Lagoon is classified in the category of medium salinity environments Geochemistry and ionic interaction in the Bizerte Lagoon waters (Northern Tunisia),*Journal of Oceanography and Marine Science,* (2011)2: 1–9. https://academicjournals.org/journal/JOMS/article-stat/751F92C4449

Kolberg DIS, Zechmann S., Wildgrube C, Sigalov I, Scherbaum E, Anastassiades M. 2016. Determination of Triazole Derivative Metabolites (TDMs) in Fruit and Vegetables using the QuPPe Method and Differential Mobility Spectrometry (DMS) and Survey of the Residue Situation in Organic and Conventional Produce. Aspects of food control and animal health *Free Journal,* (2016) (2): 2196-3460.https://www.eurl-pesticides.eu/userfiles/file/EurlSRM/EurlSrm_meth_ Triazole Derivative Metabolites.pdf

Liu C, Liao Y,Huang X. Extraction of triazole fungicides in environmental waters utilizing poly (ionic liquid)-functionalized magnetic adsorbent.*Journalof Chromatography A,* (2017)1524: 13-20.https://pubmed.ncbi.nlm.nih.gov/29033150/

Mateo-Sagasta J, Zadeh SM, Turral H. Water pollution from agriculture: A global review, (2017) 1-35. https://www.fao.org/3/i7754e/i7754e.pdf

Maia AS, Ribeiro AR, Castro PML, Tiritan ME. Chiral Analysis of Pesticides and Drugs of Environmental Concern: Biodegradation and Enantiomeric Fraction, *Symmetry,* (2017) 9(9): 196-225. https://www.mdpi.com/2073-8994/9/9/196

Masiá A, Campo J, Vázquez-Roig P,Blasco C, Pico Y.Food Screening of currently used pesticides in water, sediments and biotaof the Guadalquivir River Basin (Spain).*Journal of Hazardous Materials,* 263P (2013) 95-104.https://pubmed.ncbi.nlm.nih.gov/24140087/

Necibi M, Lanceleur L, Mzoughi N, Monperrus. M. Determination of synthetic musks in surface sediment from the bizerte lagoon by QuEChERS extraction followed by GCMS. *Bulletin of Environmental Contamination and Toxicology,* (2016) 97 (5): 659-69. https://pubmed.ncbi.nlm.nih.gov/27743038/

Necibi M, Mzoughi N. Distribution of organochlorine pesticides in sediment cores from the bizerte lagoon (Tunisia).*International Journal of Environmental Analytical Chemistry,* (2019a) 99:1-15.https://www.researchgate.net/publication/ 335124786_Distribution_of_organochlorine_pesticides_in_sediment_cores_from_the_Bizerte_Lagoon_Tunisia

Necibi M, Mzoughi N. Distribution of organochlorine pesticides in suspended particulate matter and sediment from the Bizerte Lagoon. *International Journal of Environmental Analytical Chemistry,* (2019b) 12 (11): 1-27. http://www.iosrjournals.org/iosr-jac/papers/vol12-issue11/Series-1/C1211011827.pdf

Necibi M, Mzoughi N.Yahia MND, Pringault O.Distributions of organochlorine pesticides and polychlorinated biphenyl in surface water from Bizerte Lagoon, Tunisia, *Desalination and Water Treatment,* (2015) 56 (10): 2663-71. https://hal.science/hal-01920850/

Sahraoui I, Bouchouicha D, Mabrouk HH, Hlaili AS. Driving factors of the potentially toxic and harmful species of Prorocentrum Ehrenberg in a semi-enclosed Mediterranean lagoon (Tunisia, SW Mediterranean), *Mediterranean Marine Science,* (2013) 14: 353-362.https://ejournals.epublishing.ekt.gr/index.php/hcmr-med-mar-sc/article/view/12310/12319

Shahinasi E, brahushi F, devolli A, kodra M.The ecotoxicology of pesticides group of triazole and their use to control apple scab (venturia inaequalis), *Journal of Hygienic Engineering and Design,*(2017) 634:11-248https://keypublishing.org/jhed/wp-content/uploads/2020/07/04.-Full-paper-Edlira-Shahinasi.pdf

Salvator L. Bizerte, son passé, son présent et son avenir. Edt. J Rotschild. Paris. (1900) 79.https://www.worldcat.org/title/bizerte-en-son-passe-son-present-et-son-avenir/oclc/23412798

Vieira, A, Santos, M. Figueiredo E. Solid-phase extraction of triazole fungicides from water samples using disks impregnated with carbon nanotubes followed by GC-MS analysis. *International Journal of Environmental Analytical Chemistry,* (2017) 97(1): 29-41.https://www.tandfonline.com/doi/abs/10.1080/03067319.2016.1272679

Wang H, Yang X, Hu L, Gao H,Lu R,Zhang S,Zhou W. Detection of triazole pesticides in environmental water and juice samples using dispersive liquid-liquid microextraction with solidified sedimentary Ionic Liquids. *New Journal of Chemistry*, (2016)40: 4696-4704.https://pubs.rsc.org /en/content/articlelanding /2016/nj/ c5nj03376d

Xu XY, Ye JQ, Nie J, Li ZG and Lee MR. A new liquid-liquid microextraction method by ultrasound assisted salting out for determination of triazole pesticides in water samples coupled by gas chromatography-mass spectrometry. *Analytical Methods,* (2015)7: 1194-1199.https://pubs .rsc.org/en/content/ articlelanding/2015/ ay/c4ay 02448f

Zhang QL, Zhou Y, Yang X, Hua H,Wang M, Shi H. Study on the stereoselective degradation of three triazole fungicides in sediment. *Ecotoxicology and Environmental Safety,* (2015)117: 1-6.https://pubmed.ncbi.nlm.nih.gov/25814463/

Chapter 5

Tautomerism in 1,2,3-Triazoles

Fernanda Petzold Pauli[1]
Adriane Antônia Pereira Amaral[2]
Acácio Silva de Souza[1]
Ruan Carlos Busquet Ribeiro[1]
Luana da Silva Magalhães Forezi[2]
Fernando de Carvalho da Silva[2]
and Vitor Francisco Ferreira[1,*]
[1]Universidade Federal Fluminense, Faculdade de Farmácia, Departamento de Tecnologia Farmacêutica, Niterói, RJ, Brasil
[2]Universidade Federal Fluminense, Instituto de Química, Niterói, RJ, Brasil

Abstract

Triazoles are heterocycles with a privileged structure with important chemical and biological properties. They behave as non-hydrolyzable bioisoters of amides and due to these properties are widely used by Medicinal Chemistry in the search for new substances with relevant therapeutic applications. A large number of drugs are already in the pharmaceutical marketed and among these are 1,2,3- and 1,2,4-triazoles. Additionally, these heterocyclyls are present in a wide range of polymers used as in several types of materials. This chapter will focus on the tautomerism of 1,2,3-triazoles, covering properties such as prototropic tautomerism and structural rearrangements, which are fundamental to understanding the chemistry involved in the main reactions in which these heterocycles are involved.

---

* Corresponding Author's Email: vitorferreira@id.uff.br

In: Triazoles and Their Uses
Editor: Karl Roland
ISBN: 979-8-88697-800-1

Keywords: 1,2,3-triazol, synthesis, tautomerism, heterocycles

1. Introduction

The compounds containing five-membered unsaturated aromatic rings and containing three nitrogen atoms belong to the class of heterocyclics called triazoles or pyrrodiazole. They are composed by three nitrogen atoms and two carbon atoms forming a pair of isomeric compounds [1, 2, 3, 4, 5, 6]. If the nitrogen atoms are arranged sequentially, the heterocycles are known as 1,2,3-triazoles. The 1,2,3-triazoles are important class of compounds among the heterocycles, as they exhibit notable medicinal activities due their great structural diversity and biological activities, transiting as drugs and bioactive compounds between many diseases [7, 8]. Triazoles without group bonded at the one nitrogen atom may present the tautomerism phenomenon between the structures that are in equilibrium in solution (Figure 1). If the triazole has an alkyl or aryl substituent on nitrogen it is possible to have two isomeric series depending on the position of the other substituent on the carbons: 1,4- and 1,5-disubstituted.

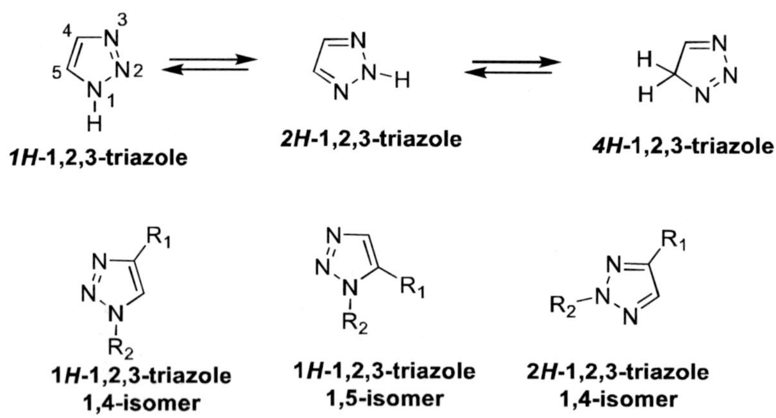

Figure 1. General structures of 1,2,3- and 1,2,4-triazoles.

The 1,2,3-triazoles series are widely studied today for various applications, in particular looking for new biological activities [9]. Its dissemination in the literature is due to its stability, pharmacophore property in medicinal chemistry, relatively simple preparation methods for the 1*H*- or

2*H*-, 1,4- or 1,5-disubstituted isomers [10], mimic a non-isomerizable double bond, ligands in organometallic chemistry, dipolar similarity with amide[11, 12, 13], metabolic degradation, similarity with a peptide bond (can accept hydrogen bonds), amide bioisoster [14] solubility in the polar biological environment, resistance to hydrolysis by acidic and basic conditions, resistance to and reduction and oxidation and the possibility of many structural variations around the heterocycle rings [15]. Indeed, in several biologically active compounds the amide linker has been replaced by 1,2,3-triazoles [16]. There are also many other reports in the literature on the use of triazoles in other fields of study of chemistry, for example, organocatalysis [17] and ionic liquids [18].

It is important to highlight that the preparation of 1,4-disubstituted 1*H*-1,2,3-triazoles are the ones that have the most robust synthetic methods available in the literature and, therefore, the 1,2,3-triazoles are the substances that have been most evaluated in several biological targets [19].

There are several drugs on the pharmaceutical market that contain the 1,2,3-triazole heterocycles. Figure 2 shows those drugs that arc released for the treatment of various diseases as β-lactam-1,2,3-triazole hybrids, tazobactam and cefatrizine, which are used as antibiotics. The first drug came into clinical use in 1992 in the form of sodium salt as an inhibitor of bacterial action acting on β-lactamases enzymes and is currently used as an adjunct to other antibiotics to prevent bacteria resistance [20]. Its association with the antibiotic piperacillin is used to treat infections by *Pseudomonas aeruginosa* or in combination with cephalosporin for the treatment of urinary and intra-abdominal infections [21]. Cefatrizine is also an antibiotic with a broad spectrum of action against many Gram positive and negative microorganisms [22]. Rufinamide [23] is a drug used to treat various seizure disorders by modulating sodium channels that are used as a target for the development of antiepileptic drugs [24].

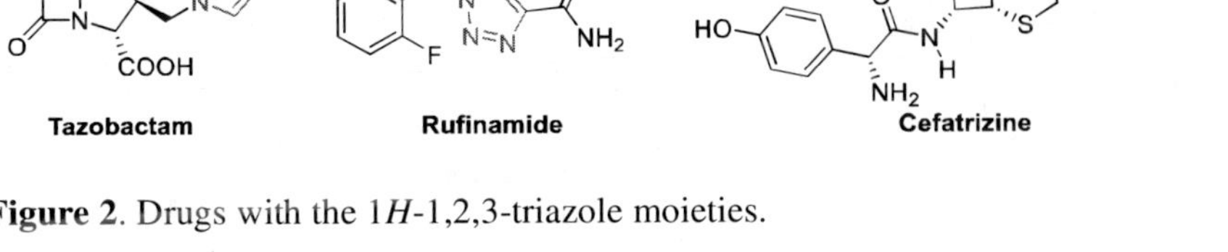

Figure 2. Drugs with the 1*H*-1,2,3-triazole moieties.

2. Prototropic Tautomerism of Triazoles

Prototropic tautomerism is characteristic of isomers that easily interconvert through the reallocation of a hydrogen atom [25]. Concerning the triazole nucleus, the prototropic tautomerism is classified as an annular tautomerism, where the proton can occupy different positions of the heterocycle. Some general aspects of the structural diversity of substituted and unsubstituted 1,2,3-triazoles (Figure 3). The substituted compounds with R_1, R_2 and R_3 can be prepared as 1*H*- or 2*H*-triazole as isomers 1,4, or 1,5. When R_3 = H, the 1,2,3-triazoles can be in tautomeric equilibrium via prototropic tautomeris and them can be subdivided into three groups depending on the positions of R_1 and R_2: 1*H*, 2*H* and e 3(1)*H*-1,2,3-triazole. The latter isomer was only observed in rare cases. Indeed, they exist as an equilibrium mixture of 1*H*- and 2*H*-tautomers of not different stability.

Quantum-chemical investigations and experimental techniques, mainly nuclear magnetic resonance, were used to understanding the factors associated with structural aspects and relative stability of triazole tautomers. In this context, interesting investigations have been dedicated to the study of prototropic tautomerism in triazole systems [26, 27, 28, 29].

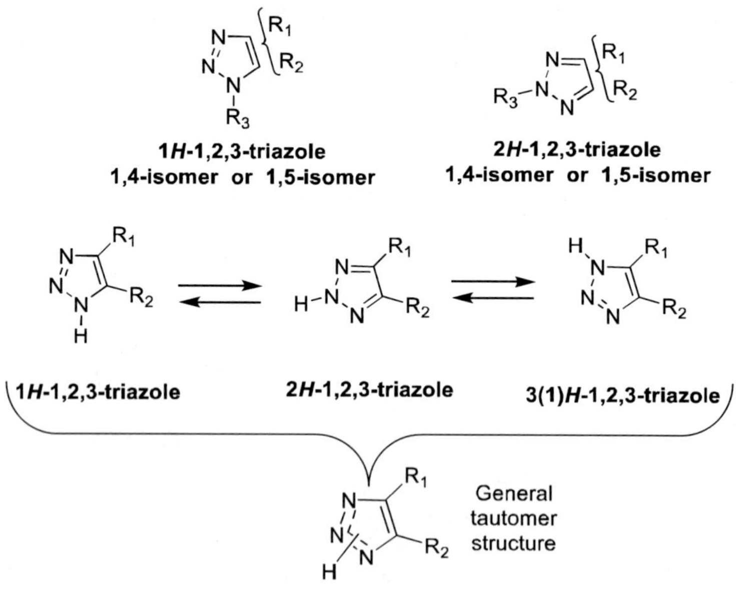

Figure 3. General aspect of triazole diversity and prototropic tautomerism.

Elguero and Soresen studied the prototropic tautomerism of triazole in gas phase. Soresen and coworkers published experimental investigation using microwave and photoelectron spectroscopy, which revealed that in the gas phase, the 2*H*-tautomer dominates strongly over the 1*H*-tautomer in a 1:1000 ratio [30]. Moreover, Elgueiro's group evaluated the tautomerism by theoretical calculations and has shown that in the gas phase the 2*H*-tautomer is more stable than the 1*H*-tautomer by 4.5 kcal.mol^{-1} [31, 32].

Prototropic tautomerism of triazole nucleus were also investigated in solvent medium, Lunazzi and co-workers reported that the relative proportions of each tautomer of 1,2,3-triazole using NMR analysis. According to the authors, in deutered dichloromethane and toluene solutions, at room temperature, the 2*H*-tautomer was assigned as the more stable specie. Moreover, the proportions of each tautomer are dramatically dependent on temperature, concentration, and solvent [33].

Taylor and Albert, for instance, studied the tautomerism of 1,2,3-triazole in aqueous solution using kinetics parameters. The authors identified the 2*H*-tautomer as a favored specie in aqueous solution, which could be explained by the repulsion of free electron pairs [34].

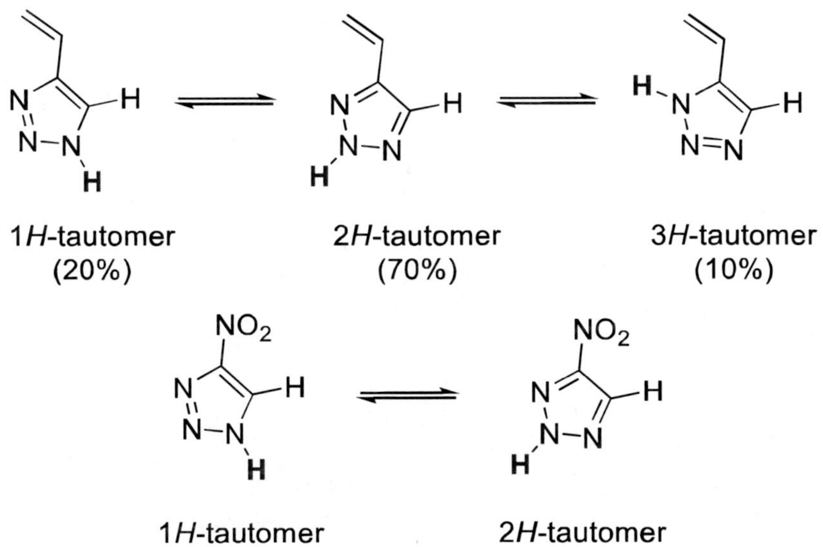

Figure 4. Prototropic tautomerism of 4(5)-vinyl-1,2,3- triazole and nitrotriazoles.

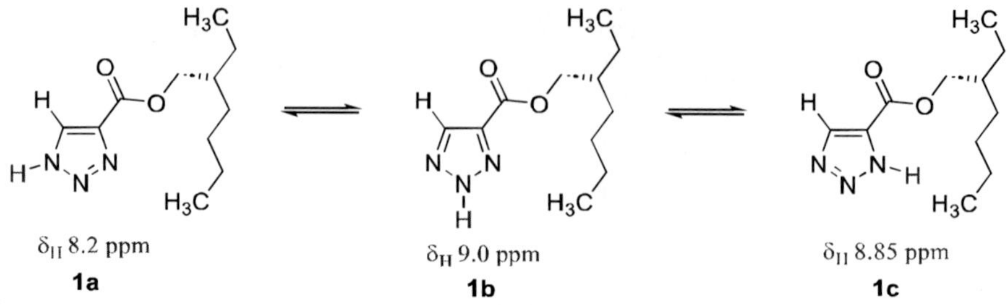

Figure 5. Tautomerism of triazolylcarboxylate by Nguyen and co-workers.

The prototropic tautomerism of C-substituted triazoles were also investigated by Toppet's group, which described the NMR studies of 4(5)-vinyl-1,2,3-triazole in dimethylformamide as solvent. At 40ºC, the ^{13}C spectrum showed an average tautomer, whereas at -55°C allowed the direct observation of the three separate tautomers in calculated proportion of 2:7:1 for 1*H*:2*H*:3*H* tautomer, respectively [35]. Other C-substituted triazoles were investigated and depending of the nature of the substituent and experimental conditions, variations on the preferred tautomer in solution media could be observed. Indeed, some studies in solution assume that the preferred tautomer is 1*H*-1,2,3-triazole. An example was reported by Licht and co-workers which combined NMR data and calculated chemical shifts to demonstrate the predominance of 1*H*-tautomer of 4-NO_2-1,2,3-triazole in solution (Figure 4) [36].

A mixture of the three tautomers of 1,2,3-triazole was reported by Nguyen and co-workers (Figure 5). The triazole **1**, which was obtained by a $CHCl_3$-soluble extract of the stems of *Paramignya trimera*, showed its NH protons in the range of 8,20-9,25 ppm at a ^1H NMR spectra [37].

A similar case was observed by Semenov and co-workers, which published an NMR study to identify the tautomerism of silicon-containing 1,2,3-triazole **2** (Figure 6). With the nitrogen-15 NMR spectra HMBC (^1H-^{15}N), they could unambiguously assign the atoms by comparison of calculated (CFOUR package of CCSD) and experimental NMR chemical shifts. Moreover, the ratio of the prototropic tautomers were determined by comparison between calculated and experimental chemical shifts in ^1H, ^{13}C and ^{15}N of each tautomer, and the results showed that the tautomer **2c** was predominant in solution [38].

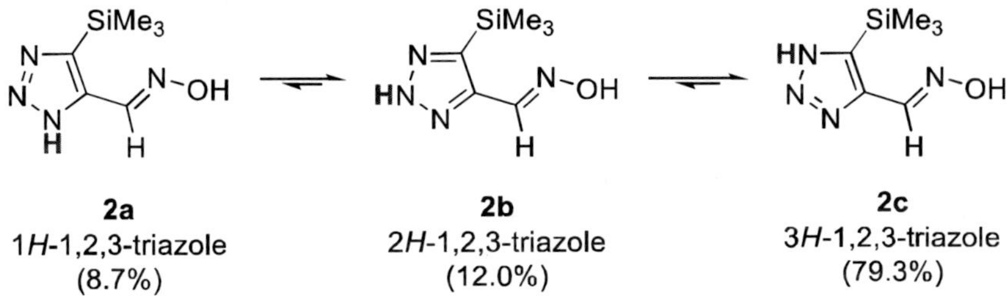

Figure 6. Tautomerism of silicon-containing 1,2,3-triazole **2**.

Considering to prototropic tautomerism of 1,2,3-triazole in solution, although sometimes there is one favored tautomer over the other, depending on the experimental conditions, in the literature there seems to be a consensus regarding the presence of a mixture of tautomers in solution [39]. Another important consideration on this field is about the analysis of the 1,2,3-triazole tautomers in the solid state, which is not very trivial, since the interactions in the crystalline crystal structure depend on many factors. One of the most important interactions is the N-H bond which can be a hydrogen bond donor and also the sp^2 nitrogen which can be hydrogen bond acceptors. In this context, Goddard and co-workers determined the crystal structure of unsubstituted 1,2,3-triazole and found that the crystal is a 1:1 mixture of 1*H*- and 2*H*-tautomers [40]. However, when substituted with the phenyl and nitro group, the triazoles crystallize in the 1*H*-tautormer and 2*H*-tautomer, respectively [38].

The porphyrin compounds complexed with Zn(II) and having a 1,2,3-triazolic ring, crystallized in a dimeric arrangement with control of the 2*H*-tautomer (Figure 7) [41]. The **7** and **8** triazoles were obtained through a Dimroth rearrangement and both crystallized in the form of the 2*H*-1,2,3-triazole conformer tautomer, being confirmed by XRD analysis [42]. Some studies show that it is possible to co-crystallize the **3a** and **3b** triazoles with bulky compounds [43]. Muthuraja and co-workers using the slow evaporation technique prepared a co-crystal of 1*H*-1,2,3-triazole with 3,5-dinitrobenzoic acid in a 1:1 complex [44].

In this context, the investigation studies of prototropic tautomerism have been shown that 1,2,3-triazoles in the solutions, solid state and gas phase can exist as 1*H*-,2*H* and 3*H*-tautomers. The next step is to evaluate the behavior of 1,2,3-triazole tautomers and their reactions in N1, N2 and N3. It is important to emphasize that the reactivity of these nitrogens can give some information about the reactivity of the nitrogens, but other stereoelectronic factors can interfere in the distribution of the products.

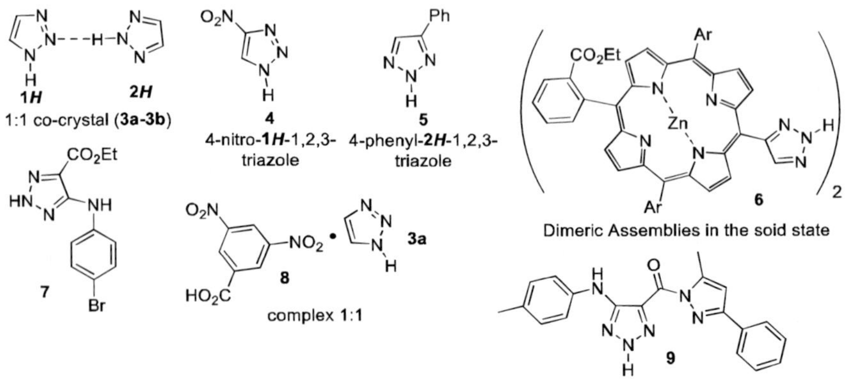

Figure 7. Tautomerism of 1,2,3-triazoles in solid state.

3. Types of Reaction with 1,2,3-Triazoles Affected by Prototropic Tautomerization

The effect of prototropic tautomerization in 1,2,3-triazoles can be clearly observed in some types of reaction that involves the nitrogen atoms of the triazole nucleus. Indeed, the NH-triazoles are considering good nucleophiles, due to their high electron density, which can react with many electrophiles under proper conditions.

Specifically, the *N*-functionalization of 1,2,3-triazoles can lead to mixtures of *N*-1, *N*-2, and *N*-3-regioisomers. It is known that the N1-atom has higher electron density when compared to the N2-atom (Figure 8). Thus, the functionalization of 1,2,3-triazoles is expected to occur in the N1-atom. Despite that, good N-2 regioselectivity for the functionalization of 1,2,3-triazoles has been observed.

Several regioselective methods have been developed based on steric and inductive effects. In addition, different catalytic systems have also reported to promote these reactions. Regarding this topic, the alkylation, arylation, amidation, acetylation, vinylation, silylation and sulfonylation reactions of 1,2,3-triazoles, which are influenced by prototropic tautomerism, as will be demonstrated in this chapter.

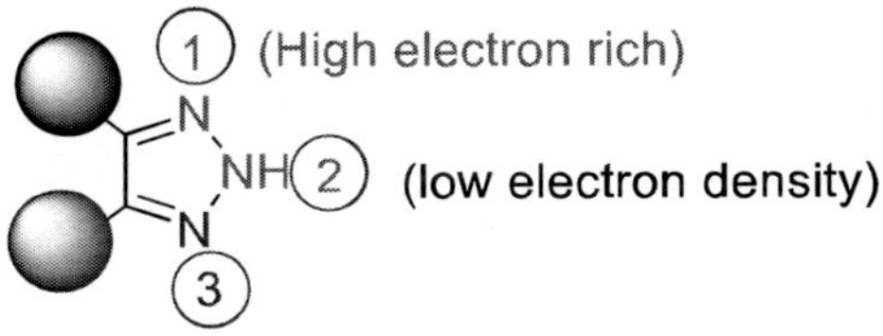

Figure 8. 1,2,3-triazoles structure and electronic description of the N1 an N2 atoms.

3.1. Alkylation Reactions

N-alkylation reactions with different triazole compounds have been extensively studied over the years. The main challenge of these reactions is the control of regioselectivity. In this regard, several regioselective and some non-selective approaches exploring different experimental techniques were used, as well as the use of theoretical tools to explain regioselectivity in the results.

3.1.1. Regioselective N1-Alkylation of 1,2,3-Triazoles Under Metal Catalysis

The lack of regioselectivity in N1, N2 and N3 causes serious problems in the purification of the products in addition to providing products in low yields. In order to provide a methodology that addresses these problems, many metals have been investigated as catalysts for the selectivity of these reactions and preferably for easier recycling and more sustainable solvents.

*N*1-Alkylation of triazoles often result in a mixture of *N*1-, *N*2- and *N*3-products, result of an equilibrium between the 1*H*, 2*H* and 3*H* tautomers. However, the use of catalytic systems including the use of nanoparticles, organocatalysis and photocatalysis has promoted good N1-regioselectivity. Furthermore, it is very common to obtain N1-products when benzotriazoles are used. Conversely, the degradation of N1-alkyl-1,2,3-benzotriazoles due to the equilibrium in different extents of *N*1 and *N*2-isomers in organic solvents such as DMSO, chloroform and others was also pointed out.

The use of ZnO nanoparticles (ZONPs) as catalyst enabled efficient regioselective *N*1-alkylation (Scheme 1). For example, the reaction between 1,2,3-triazole-4-carboxylate and alkyl halides using cesium carbonate as base and NMP solvent resulted in a mixture of isomers **11**, **12** and **13** (Scheme 1A). However, when the reaction was catalyzed by ZnO nanoparticles in the presence of potassium hydroxide as base and DMSO as solvent only *N*1-

isomers were obtained. It is important to mention that alkyl iodide, bromide, and chloride were evaluated, however, only alkyl iodide was active [45]. According to the mechanistic proposal, regioselective alkylation occurs at N1 because the highly resonant structure of substance **10** converts to **10a** by tautomerism (Scheme 1B). The use of ZOPNs increases regioselective alkylation because the coordination between ocurrs of II with ZONPS, followed by regioselective alkylation by the alkyl halide, in the presence of KOH as base.

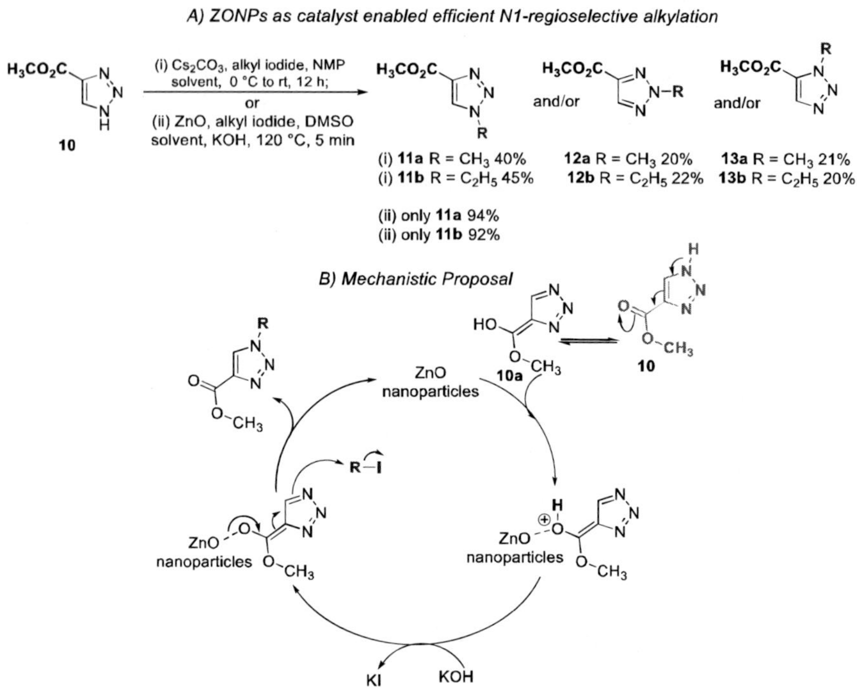

Scheme 1. ZnO nanoparticles as a selective catalyst.

Another example using nanoparticles was reported by Jain's group. The Cu_2S nanocatalyst was successfully used in the regioselective *N*1- and *N*2-alkylation reactions of 1,2,3-triazoles and benzotriazoles with *N*,*N*-dimethylbenzylamine **14** in presence of TBHP (Scheme 2). According to the authors, the reaction is initiated by copper-catalyzed decomposition of TBHP to generate a *tert*-butoxyl radical, and a single electron transfer (SET) from amine **14** generates a radical cation. Interestingly, only *N*1-products were obtained from benzotriazoles. Although only the *N*1 alkylated products were obtained from benzotriazoles, the authors alert to the possible degradation of the compounds when dissolved in organic solvents such as DMSO, chloroform and others. This degradation comes from the equilibrium in different extents of *N*1 and *N*2 benzotriazoles in these solvents. In this approach, compound **17**, an important intermediate in the synthesis of kinase inhibitors, was achieved. According to mechanistic studies, the reaction involves the formation of an iminium ion intermediate [46].

The regioselective *N*-alkylation of benzotriazoles **18** with diazo compounds **19** was accessed through a visible light-induced method catalyzed by *p*-benzoquinone (PBQ) and catalytic amount of *tert*-butyl nitrite (TBN), which proceeds by a radical pathway (Scheme 3). Various triazoles and diazo compounds were employed and *N*1-alkylation products were obtained in moderate to excellent yields with excellent *N*1-selectivities. In the optimization studies carried out, it was observed that PBQ and TBN are responsible for the high selectivity since in the absence of one or both PBQ and TBN, the *N*1 selectivity decreases considerably [47].

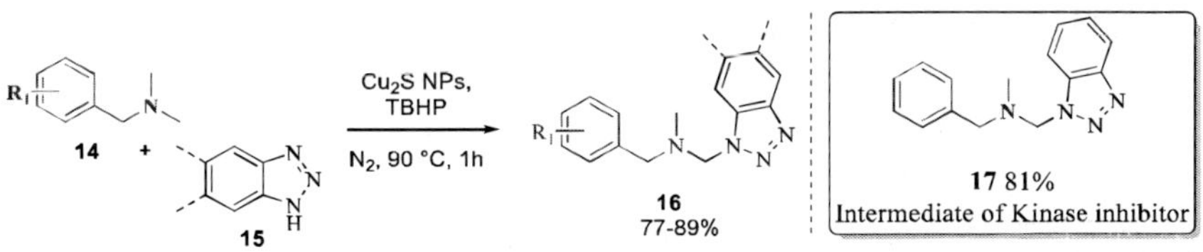

Scheme 2. N1-regioselective alkylation under Cu_2S nanocatalyst as catalyst.

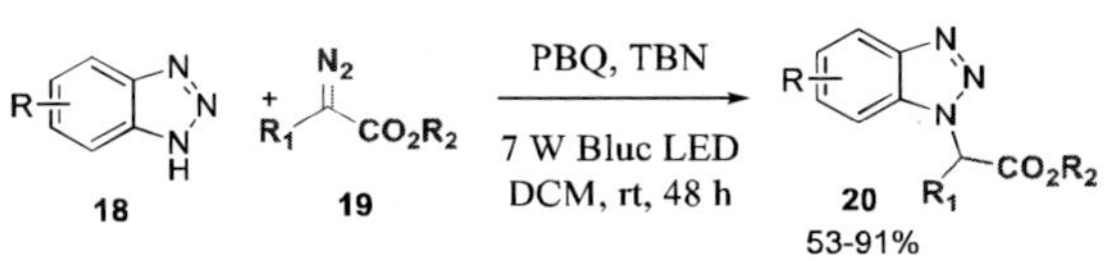

Catalyst	N1/N2
1. DDQ and TBN	93/4
2. DDQ instead of PBQ	71/22
4. without PBQ and TBN	56/30
5. without PBQ	63/29
6. without TBN	71/21

Scheme 3. Regioselective *N*-alkylation of benzotriazoles with diazo compounds.

Alkylation can also be achieved via Michael addition to α,β-unsaturated olefins. An elegant Michael addition method involving the use of benzotriazoles and nitroolefins was also developed. In this transformation, an organocatalyst derived from chinchona was successfully employed (Scheme 4). Michael's adducts were obtained in moderate to high enantioselectivities. According to the authors, only N1 addition products were formed. Even 1*H*-1,2,3-triazole gave good yield (65%), high ee (84% ee) and excellent *N*1-regioselectivity [48].

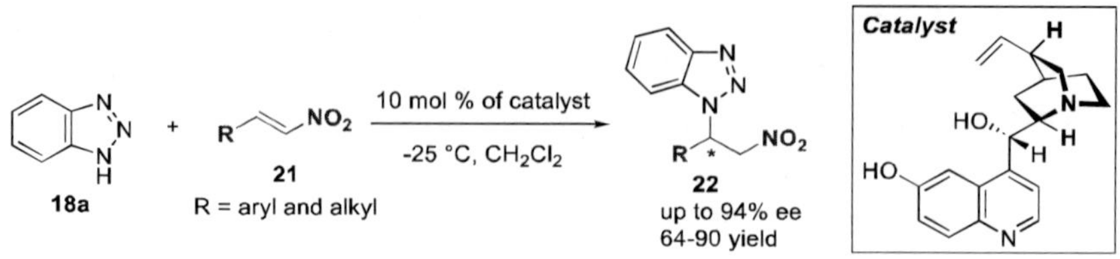

Scheme 4. Michael addition catalyzed by a chinchona-derived catalyst.

Unlike the one-pot approaches presented so far, an interesting systematic study published by Begtrup and Larsen showed the *N*-alkylation of triazoles in two steps. The fact that *N*-alkylation requires a reaction medium that dissolves the base, the nucleophile and the electrophile, without decreasing the efficiency of each substrate, makes the choice of the solvent very difficult. Thus, the authors suggest the use of a solvent for the deprotonation step and another solvent for the alkylation step. For example, the first step of *N*-benzylation of unsubstituted 1,2,3-triazole **3** could be conveniently performed using sodium hydride in acetonitrile. After stirring for 24 hours, the acetonitrile was removed and the azole sodium salt **23** was submitted to the second reaction step of *N*-benzylation, which was carried out with benzyl chloride in methanol at 20°C for 72 h, giving the isomers **24** and **25** in 78% and 22% yield, respectively (Scheme 5). Furthermore, the authors also suggest different methodologies for the deprotonation and *N*-alkylation steps [49].

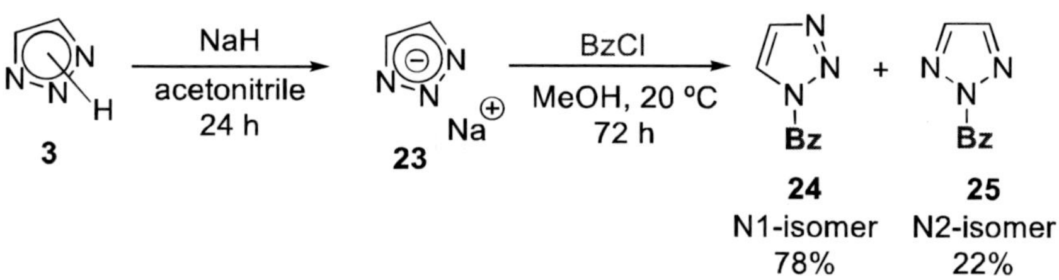

Scheme 5. *N*-alkylation of 1,2,3-triazole in two septs approach by Begtrup and Larsen..

3.1.2. Regioselective N2-alkylation of 1,2,3-triazoles

As mentioned previously, good *N*2 regioselectivity for the functionalization of 1,2,3-triazoles has been observed, usually for triazoles with sterically bulky substituents. In fact, the N1- and N3-atoms would be less available to react than the N2-atoms. Thus, the researchers evaluated the influence of certain linked groups in the C4 and C5 positions aiming to modulate the selectivity of the alkylation reactions at the N2-position.

However, the use of bulky groups in the structure of triazoles is not the only determining factor in the regioselective *N*2-alkylation of 1,2,3-triazoles. Electronegative substituents act by decreasing the reactivity of the neighboring N-atom, favoring the formation of 2-alkyl-1,2,3-triazoles. It is important to emphasize that some groups, such as bromine, can act both by decreasing the reactivity by electronic effect and by promoting steric hindrance. In this case, the group is an electronegative atom that has a large radius. Regioselectivity can also be achieved by varying the structure of the alkylating agent or even by choosing an appropriate solvent. Furthermore, it was observed that regioselectivity can be affected by the base. Typically, weak bases favor *N*2-isomer while bases like sodium hydride favor *N*1-isomer formation. More details will be discussed in the examples below.

A pioneering work in this field was published by Tanaka e Miller, which the regioselectivity study concern on the *N*-alkylation reactions using different 1,2,3-triazoles scaffolds and organic halides. Here, the highest regioselectivity i.e., 84:16 of *N*2:*N*1, was achieve for *N*-alkylation of 4-phenyl-1,2,3-triazole with ethyl chloroacetate in the solvent-base pair of DMF-triethylamine. According to the authors, steric effects of the 4-substituient can be observed, which some "large" substituent (aryl and triphenylphosphonium) directs the attack to the N2. Whereas "smaller" substituents (bromine, hydrogen) all three nitrogens could be attacked [50].

Subsequently, Bulychev and co-workers reported the *N*-alkylation of 4-chloro-5-cyano-1,2,3-triazole **26** with methyl orthoformate (**27**) and ethyl orthoformate (**31**) (Scheme 6). *N*-methylation reaction resulted in three *N*-isomers, which *N*-2-methyl triazole **28** was isolated in 55% yield, and the *N*1-isomer **29** and *N*3-isomer **30** were obtained as a mixture in 29% overall yield. In this case, the unseparated mixture showed a 3:2 ratio of *N*1:*N*3 isomers. Interestingly, the *N*-ethylation reaction gave only the *N*2-isomer **32** in 68% yield [51].

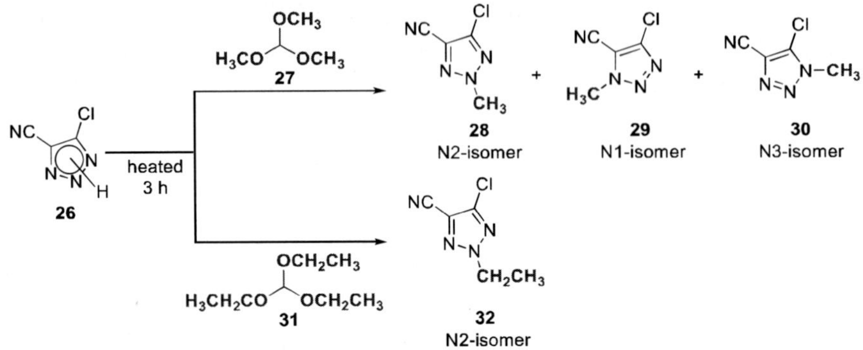

Scheme 6. N-alkylation of 4-chloro-5-cyano-l,2,3-triazole with methyl and ethyl orthoformate.

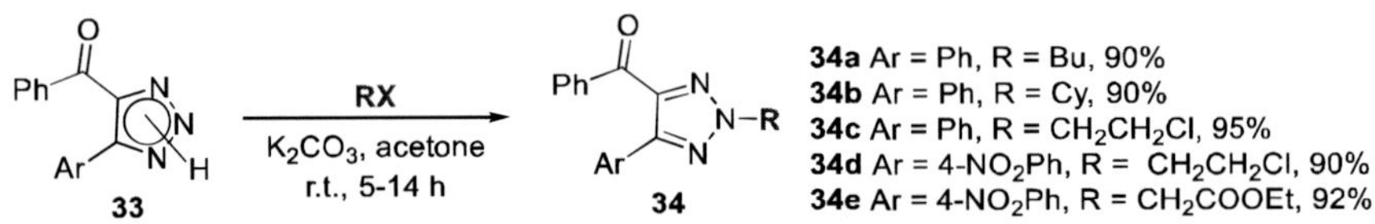

Scheme 7. Selective *N*2-alkylation of 1,2,3-triazoles by Shi's group.

Another approach for the selective *N*2-alkylation of 1,2,3-triazoles was described by Shi's group through investigation of the optimal reaction conditions (Scheme 7). The authors applied different C-4 and C-5 groups at the triazole nucleus and conclude that the regioselectivity increased with the size of this groups, due to sterically hindrance. Indeed, the highest regioselectivity was achieved with a phenyl group at C-4 and a ketone at C-5. Besides that, a screening of the reaction condition demonstrated that the solvent had no significant influence on the regioselectivity. Regarding the base, the application of a strong base, such as sodium hydride, favoring the *N*1-isomer formation. In addition, the electrophile seems to had more influence on the regioselectivity, which hindered electrophiles resulted in results in high yields of *N*2-isomer. Therefore, the authors applied different alkyl and aryl halides in the *N*-alkylation reaction of triazoles **33** with an aryl group in C4 and a ketone in C5 of the triazole nucleus, and an excellent regioselectivity was obtained for the *N*2-isomers **34** (90-92%) (Figure 1) [52].

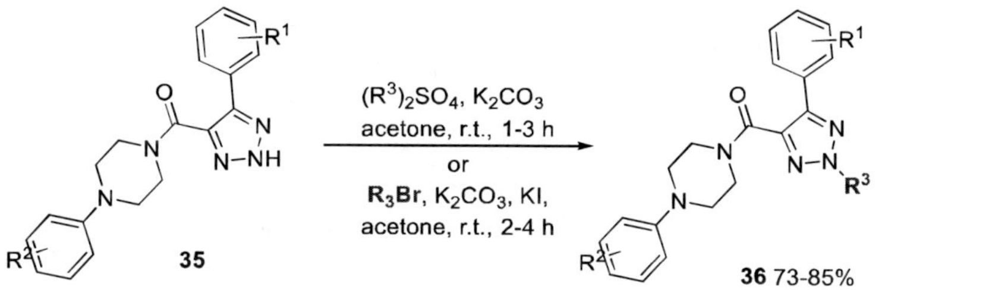

R^1 = H, 2-fluoro, 4-fluoro or 3-chloro
R^2 = H, 3-chloro, 3,5-dimethoxy or 3,5-dimethyl
R^3 = methyl, ethyl, allyl, isopropyl or butyl

Scheme 8. Regioselective *N*2-alkylation of 5-aryl-4-(4-arylpiperazine-1-carbonyl)-2*H*-1,2,3-triazole by Wu and co-workers.

Wu and co-workers reported the regioselective *N*2-alkylation of 5-aryl-4-(4-arylpiperazine-1-carbonyl)-2*H*-1,2,3-triazole **35** (Scheme 8). The authors predicted the alkylation by a density functional theory (DFT) study of the three tautomers, where the energy of 2*H*-tautomer was lower than 1*H*- and 3*H*-tautomers. Moreover, an unambiguous synthetic route could confirm the *N*2-alkylation, forming **36** [53].

In 2009, an interesting regioselective approach to directed de *N*-alkylation to the *N*2-isomer formation (Scheme 9). First, the authors evaluated the reaction conditions, when observed that the solvent has significant impact on regioselectivity. The highest regioselectivity was achieved in DMF as a solvent in -10°C for 6 hours. With the best conditions in hand, the scope of alkylation reaction was tested with five 4-bromo-*N*H-1,2,3-triazole **37** with alkyl bromides and in all cases the *N*2-substituted products 38 could be isolated in good to excellent yields. The results showed that for 4-bromo-5-phenyl 1,2,3-triazole **37** generally gave 10:1 mixture of *N*2:(*N*1-*N*3) isomers. Meanwhile, for the triazoles **37d** or **37e**, which the triazole rings were more electron deficient than the analogue 1, the *N*2 regioselectivity was improved to 13:1 of *N*2 in relation to the mixture of *N*1 and *N*3 [54]. Subsequently, the same group also published this reaction using 4,5-dibromo-1,2,3-triazoles, 4-bromo-5-trimethylsilyl-1,2,3-triazoles, and 4,5-bis(trimethylsilyl)-1,2,3-triazoles, where the respective *N*-alkylated triazoles were obtained in good to excellent regioselectivities and satisfactory yields [55].

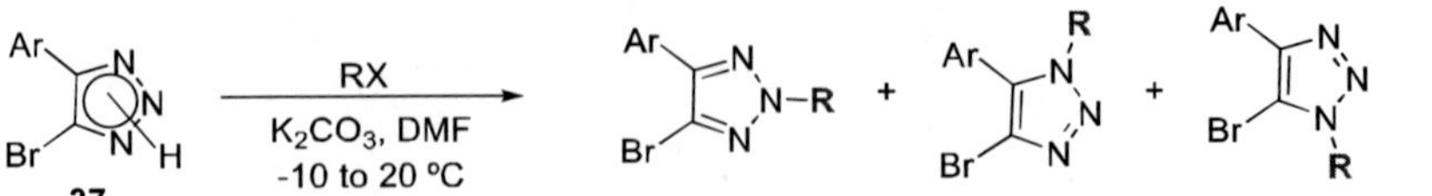

37

	38 *N*2-isomer	39 *N*1-isomer	40 *N*3-isomer
a Ar = Ph, R = Et	88	8	4
b Ar = Ph, R' = phenetyl	91	6	3
c Ar = Ph, R' = 3,5-FPh	92	6	2
d Ar = 3-FPh, R = Et	93	5	2
e Ar = 3-FPh, R' = phenetyl	94	6	<1

Scheme 9. Bromo-Directed *N*2-Alkylation of N*H*-1,2,3-Triazoles by Wang's group.

Jagerovic and co-workers reported a *N*-alkylation of N*H*-1,2,3-triazole **41** using alkylbromide under basic conditions using the phase transfer catalyst Bu_4NBr to afford a mixture of three possible regioisomers: **42** (*N*2), **43** (*N*1) and **44** (*N*3) (Scheme 10). According to the authors, two or three regioisomers were isolated depending on the nature of the triazole substituents, but the regioisomer **42** was formed as the major product in all studied case. The only exception was the alkylation of **41a** (R = H, R^1 = F, R^2 = Br, R^3 = benzyl), where the isomers **42a** and **43b** were obtained in 45:55 ratio. Moreover, the **1a** alkylation provided the highest yield of isomer **44**, where **44a** was obtained in 19% yield. Interestingly, in most cases, the regioisomer *N*3-substituted 1,2,3-triazole **44** was formed in small proportion or was not detected, which was a consequence of steric factors. However, these factors were not valid for a higher proportion of isomer **44** (*N*3) over **43** (*N*1) in the alkylation of **41b** (R = H, R^1 = H, R^2 = CH_3), **41c** (R = Ph, R^1 = H, R^2 = Cl) and **41d** (R = Ph, R^1 = H, R^2 = Cl), which could be explained by a mesomeric effect of the benzene substituents [56].

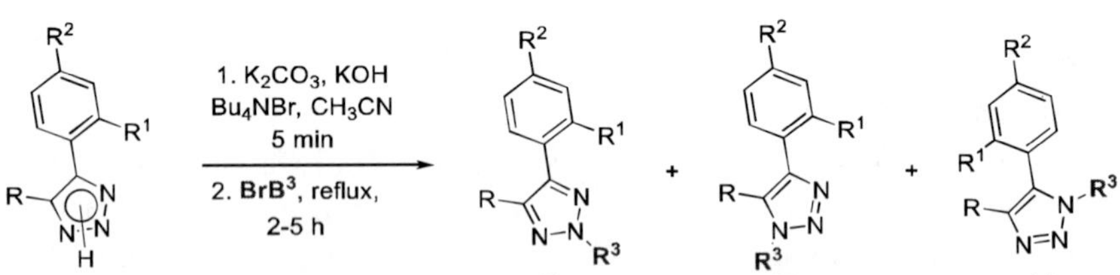

41 42 43 44

Scheme 10. A *N*-alkylation of N*H*-1,2,3-triazole **41** by Jagerovic and co-workers.

Nenajdenko and co-workers described a regioselective *N*-functionalization of 4-Fluoro-5-aryl-1,2,3-N*H*-triazoles **45** with alkyl halides led to the selective formation of *N*2-regioisomers **46** in 76-97% yields (Scheme 11) [57]. When methyl iodide was used, it was possible to observe detectable amounts of *N*1-isomer **48** and *N*3-isomers **47**. The *N*1 position hindered by the aryl group at C-5 of the triazole ring and the decrease of nucleophilicity at the *N*3 position by the fluorine atom explain the high regioselectivity at the N2 position [57].

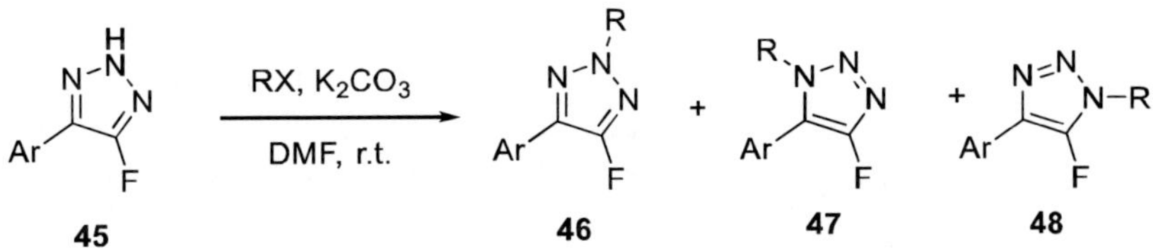

Scheme 11. Regioselective *N*2-functionalization of 4-Fluoro-5-aryl-*NH*-1,2,3-triazoles.

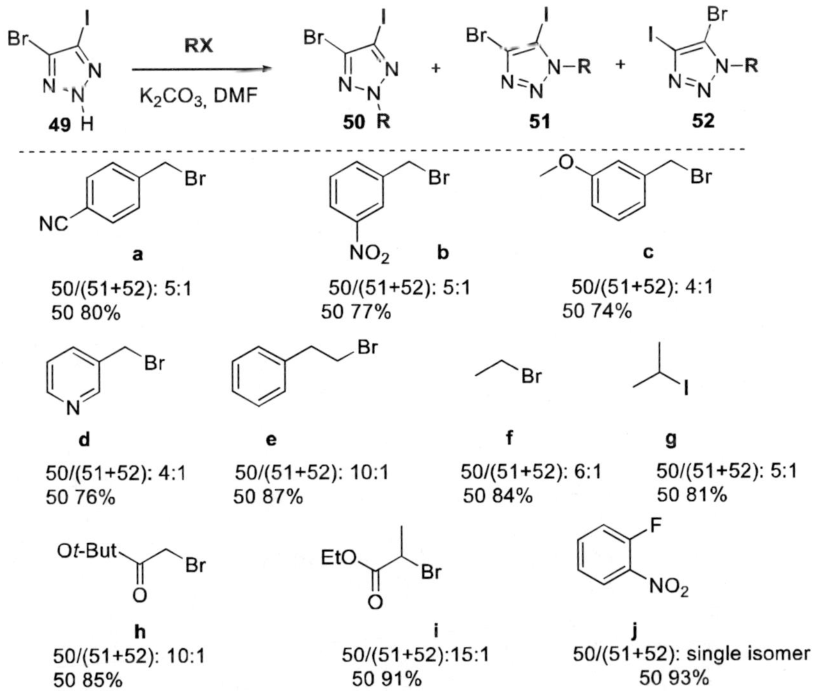

Scheme 12. Regioselective synthesis of N2-alkyl/aryl-1,2,3-triazoles via 4-bromo-5-iodo-1,2,3-triazole.

Regioselective synthesis of N2-alkyl/aryl-1,2,3-triazoles via 4-bromo-5-iodo-1,2,3-triazole (**49**) was performed by Senanayake and co-workers (Scheme 12). For this transformation, several halogenated compounds were used in basic medium. Among the halides, reaction with 2-fluoro-nitrobenzene led to the desired *N*2-arylated product as a single isomer. When bromoacetate and 2-bromoethylbenzene were used, the reactions provided **50**/(**51**+**52**) in a ratio of 10:1, while bromoethane provided **50**/(**51**+**52**) in a ratio of 15:1. Finally, the other alkyl halides showed lower selectivities [58].

Electrochemical dehydrogenative amination of *N,N*-dimethylbenzylamine (**53**) with triazoles and benzotriazoles with excellent regioselectivities was performed. In this methodology, Bu_4NClO_4 was the optimal supporting electrolyte and platinum net was used as cathode and anode (Scheme 13). When triazole was used, the products 54/55 were obtained in 79% overall yield with a ratio of 1:13, while benzotriazoles provide the products **54/55** in 83% yield with a ratio of 1:5. This methodology is a green strategy for organic synthesis [59].

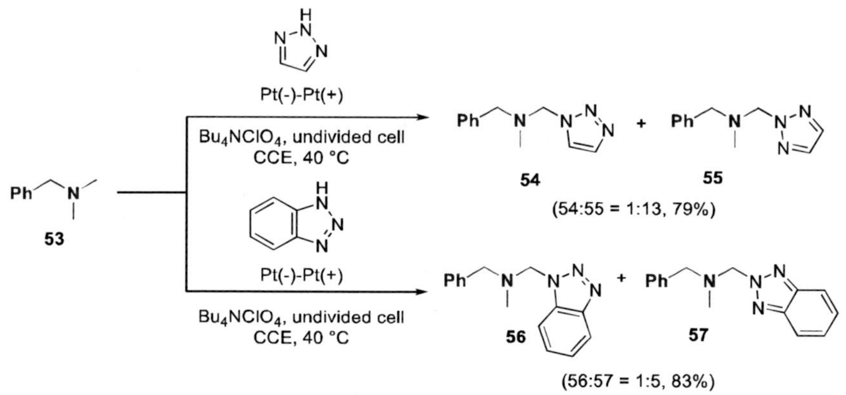

*Reaction conditions: 8 mL of CH_3CN and 2 mL of MeOH, current 10 mA, platinum net anode and platinum net cathode (working area: 2 cm^2), undivided cell, 40 °C, 5 h

Scheme 13. Electrochemical dehydrogenative amination of *N,N*-dimethylbenzylamine with triazoles and benzotriazoles.

A year earlier, Jain's group reported the reaction of 1,2,3-triazoles with *N,N*-dimethyl-benzylamine catalyzed by Cu_2S which led to the synthesis of only *N*1-alkylated products. By using 4,5-unsubstituted and 4-phenyl-substituted 1,2,3-triazoles, only the *N*2-alkylated products **59** were obtained (68-84%) (Scheme 14A). The proposed mechanism consists of the copper-catalyzed decomposition of TBHP to generate a tert-butoxyl radical, and a single electron transfer from amine generates a radical cátion (Scheme 14B). Next, the abstraction of the sp^3 hydrogen of the cation radical to generate imium-type intermediate (**I** or **II**) occurs. Finally, the triazole reacts with **I** to form **III** which, in turn, undergoes decomposition regenerating the catalyst in the catalytic cycle [46].

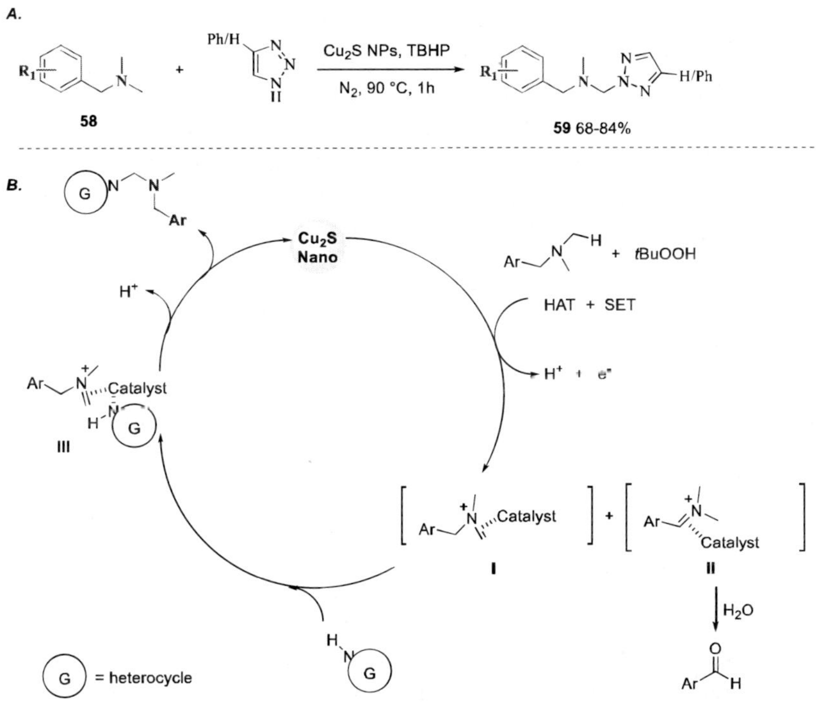

Scheme 14. A. Reaction of 1,2,3-triazoles with *N,N*-dimethylbenzylamine catalyzed by Cu_2S. B. Proposed mechanism.

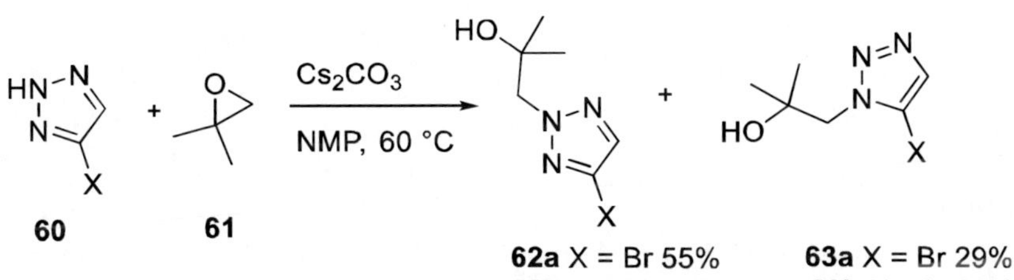

Scheme 15. Reaction between halo-1,2,3-triazoles and epoxide using cesium carbonate as catalyst.

The reaction between bromo- and iodine-1,2,3-triazoles (**60**) with epoxide **61** using catalytic amounts of cesium carbonate provided the *N*2-isomer **62** and *N*1-isomer **63**, with *N*2 being formed in higher yields (Scheme 15) [60].

The N2-selectivity can also be achieved using Brønsted acid to mediate addition of N*H*-1,2,3-triazole **64** to olefins **65**. In this context, *p*-toluenesulfonic acid was successfully used by Chen and co-workers, promoting the synthesis of N2-alkyl 1,2,3- triazole derivatives **66** (Scheme 16A and 16B). A serie of 1,2,3-triazole substrates and olefins were evalueted in this transformation. In general, phenyl triazoles with electron-rich substituents performed better than electron-poor substituents (a-h). Triazole bearing thiophenyl group afford N2-alkyl 1,2,3-triazole in moderate yields with a high *N*2-selectivity (i, 72%, *N*2/*N*1 = 34/1). When several olefins were evaluated, good selectivities were observed (l-o and r), although hindered olefins showed lower selectivity (p-q). The low selectivity, in the last case, may be related to the reversible elimination of the products, due to the stability of the cationic intermediates. When vinyl esters were employed, only N2-substituted 1,2,3-triazole derivatives were obtained in moderate to good yields (s-w) [61]. Based on control experiments, the proposed mechanism starts with the addition of a proton to styrene (Scheme 16C). Then, the cationic intermediate formed is attacked by the electron pair of the nitrogen at the same time that the leaving group tosyl is trapped by the residual amount of water in $CHCl_3$ [61].

A methodology involving Pummerer rearrangement and 1,2,3-triazoles also promoted the synthesis of *N*1/*N*2-isomers in high yields. Specifically, the reaction intermediate formed between DMSO and P_2O_5 decomposes into a thionium ion, which reacts with the triazole forming the respective isomers (Scheme 17). Electronic and steric effects were identified as determinants of regioselectivity. When 4-aryl-1,2,3-triazoles bearing groups at the aryl ring as -CH_3, -OCH_3, F, Cl, Br and NO_2, *N*2-isomer **68** were the major products. However, only phenyl and 4-methoxyphenyl groups gave high regioselectivities (**68/69** being 3.4:1 and 3.0:1, respectively). Furthermore, the introduction of bromine at C5 provided only *N*2 isomers (130°C) as can be seen in the examples involving Ar = Ph and Ar = 4-ClPh. The higher temperature required and the absence of *N*1-isomer can be explained by the fact that bromine is an electron withdrawing group, which acts by decreasing the nucleophilicity of the triazole ring, besides to bromine has a large atomic radius that promotes steric hindace in position N1 [62].

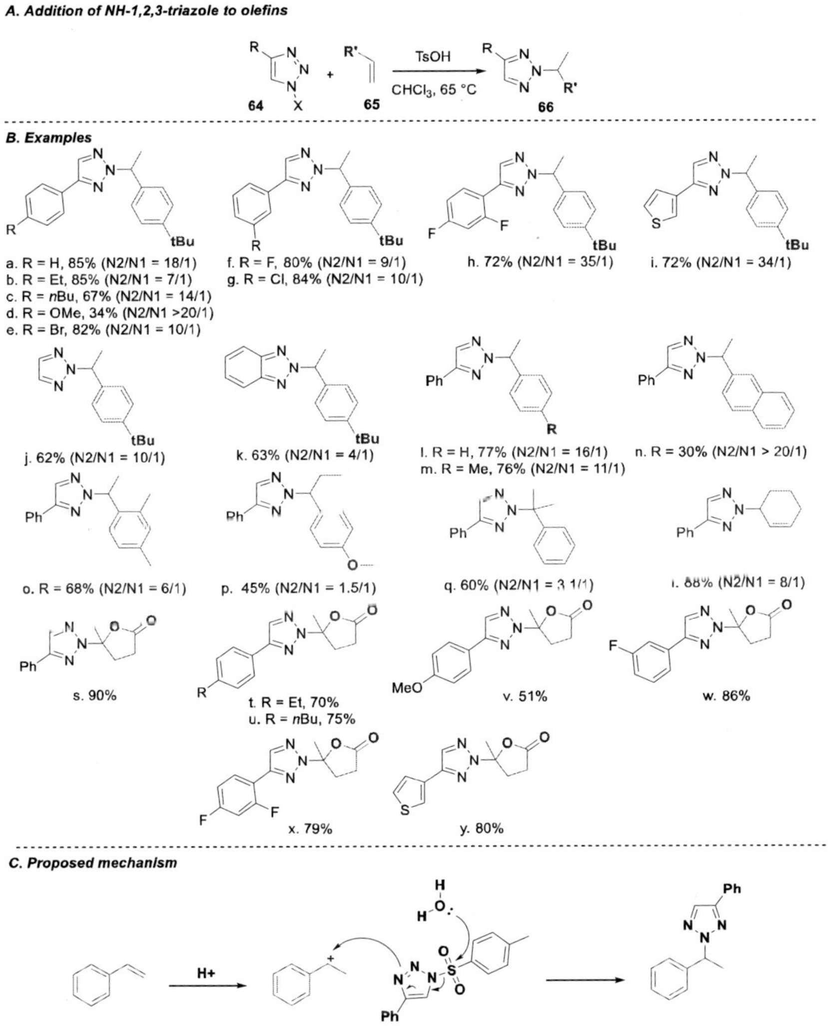

Scheme 16. A. Addition of N*H*-1,2,3-triazole to olefins catalyzed by *p*-toluenesulfonic acid; B. Some examples; C. Resumed mechanism.

Another regioselectivity study was reported by Vereshchagin and co-workers, which studied the *N*-alkylation of 4-nitro-1,2,3-triazoles (**70**) in order to elucidate the orienting role of the nitro group. The reactions were carried out with the sodium salts of the triazole with different alkyl halides in acetone at room temperature. In all cases, a mixture of *N*1 (**71**) and *N*2-isomers (**72**) were obtained with a predominance of **72** (Scheme 18). According to the authors, the absence of *N*3-isomer is due the inductive and shielding effects of the nitro group hindering the neighboring this nitrogen atom [63].

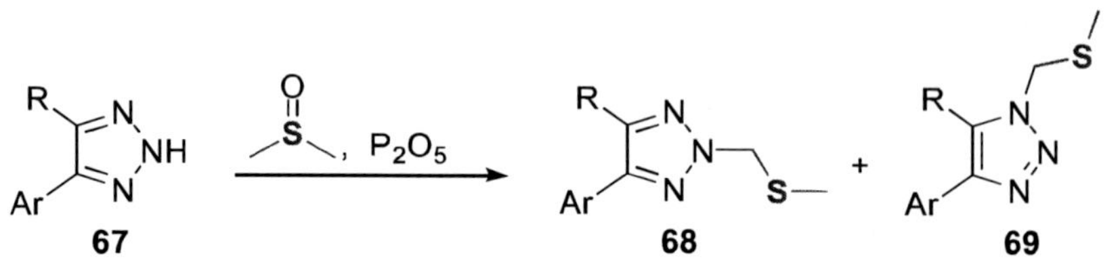

	Yield	ratio
70 °C, R = H, Ar = C_6H_5	95%	68/69 = 3.4:1
70 °C, R = H, Ar = *p*-Cl-C_6H_4	92%	68/69 = 2.3:1
130 °C, R = Br, Ar = C_6H_5	94%	Only **68**
130 °C, R = Br, Ar = *p*-Cl-C_6H_4	91%	Only **68**

Scheme 17. Pummerer rearrangement.

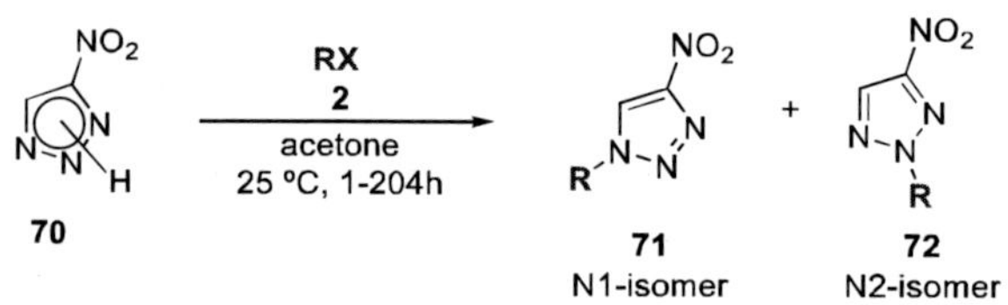

RX	Yield % N1:N2
CH_3I	40:60
C_3H_5Br	40:60
n-C_3H_7Br	45:55
iso-C_3H_7Br	30:70
$C_6H_5OCH_2Br$	33:67

Scheme 18. *N*-alkylation of 4-nitro-1,2,3-triazoles by Vereshchagin and co-workers.

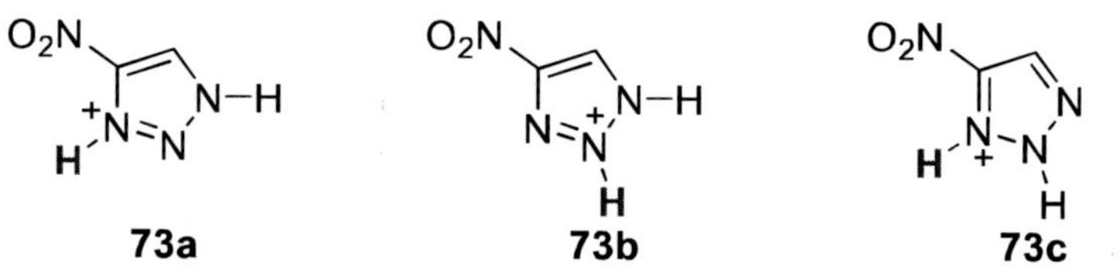

Figure 19. Structures of protonated triazoles **73**.

The treatment of nitrotriazoles with concentrated sulfuric acid produces **73a**, **73b** and **73c** protonated triazoles (Figure 19). Among these, **73a** is the lowest-energy isomer (Quantum chemical calculations: $\Delta E° = -46.8$ kJ mol^{-1} and $\Delta G°_{298,H2O} = -13.5$ kJ mol^{-1}). Furthermore, the N2 atom of **73a** is the free reactive site [64].

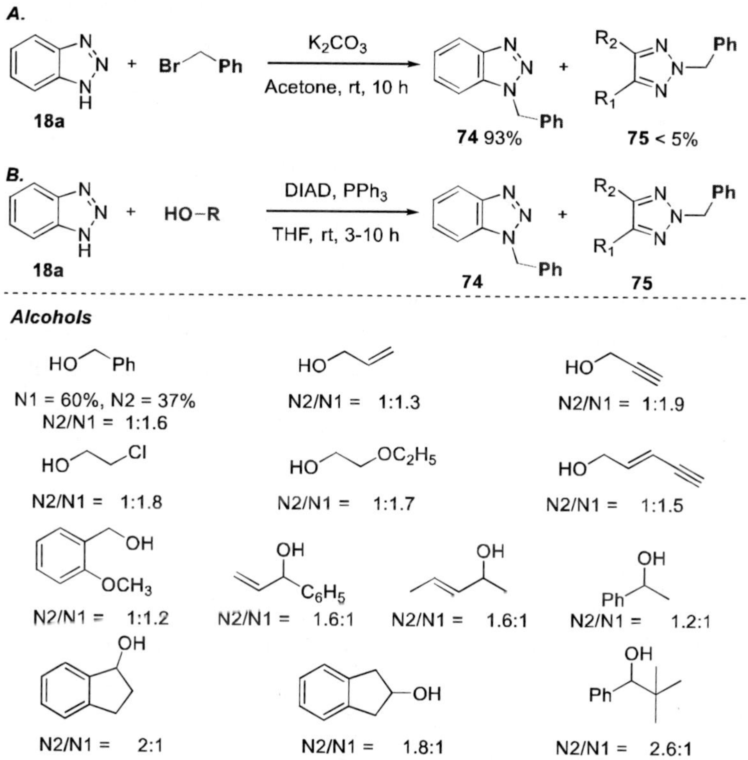

Scheme 19. A. Alkylation reaction of benzotriazole with benzyl bromide; B. Mitsunobu reaction of N*H*-triazoles with alcohols.

In this context, a study of the alkylation reaction of nitrotriazoles with alcohols as alkylating agents showed good *N*2-selectivity when isopropyl and sec-butyl alcohols were employed. However, an increase in the reaction time could led the *N*2-isomers undergo isomerization into *N*1-isomers. For example, when the propan-2-ol was used during 1 hour of reaction a mixture of *N*1-, *N*2- and *N*3-isomers was observed in the rate of 0.7:80.6:18.5 while 144 hours of reaction gave a mixture of isomers in the rate of 16.5:60.0:23.5 (*N*1:*N*2:*N*3). According to the authors, isomerization is greater when cyclohexanol is used (3 h, *N*1:*N*2:*N*3 = 73.4:20.3:6.3). On the other hand, the alkylation of triazole with tert-butyl alcohol gave only *N*1-isomer. The reaction with propanol is slower than with secondary and tertiary alcohols. In this case, n-propanol is rearranged to form the most stable carbocation. In summary, the order of isomerization is propan-2-ol < butan-2-ol < cyclohexanol [64].

Whereas there are methods that use sterically bulky substituents in 1,2,3-triazoles to induce *N*2-selectivity, the Mitsunobu reaction of N*H*-triazoles with alcohols can be an alternative to induce *N*2-selectivity through the modification of the carbon electrophiles. For example, benzotriazole has high N1-preference. This can be seen in the alkylation reaction of benzotriazole with benzyl bromide that gave only trace amounts of *N*2-isomer (*N*1 = 93%, *N*2 = <5%) (Scheme 19A). On the other hand, Mitsuonobu reaction of triazole **18a** with alcohols gave the mixture of **74** (*N*1) and **75** (*N*2), being the *N*2-isomer obtained as major product when secondary alcohols were used (Scheme 19B). It is important to mention that a range of different triazoles showed the same trend under the Mitsunobu condition. According to the authors, these results support the hypothesis that the Mitsunobu reaction promotes the kinetic addition of N2 [65].

3.1.3. Non-Regioselective N-Alkylations Approaches of 1,2,3-Triazois

A series of non-regioselective *N*-alkylations approaches of 1,2,3-triazoless were also published without the methodologic studies. In these cases, the regioselective were not evaluated and just appears as consequence of a protrotopic tautomerism.

A mixture of regioisomers **78** (*N*1) and **79** (*N*2) substituted 1,2,3-triazole was described by Livi and co-workers, which reported the nucleophilic substitution reaction of 4-benzoyl-1,2,3-triazole **76** with of 2-methoxybenzyl chloride **77** (Scheme 20). The reaction was carried out in reflux with acetone and potassium carbonate. The triazole isomers 1-*N*-benzyl-substituted **78** (*t*R = 16.57 min) and 2-*N*-benzyl-substituted **79** (*t*R = 9.95 min) was obtained in 71% total yield, and each isolated isomer in 31% and 24% yield, respectively [66].

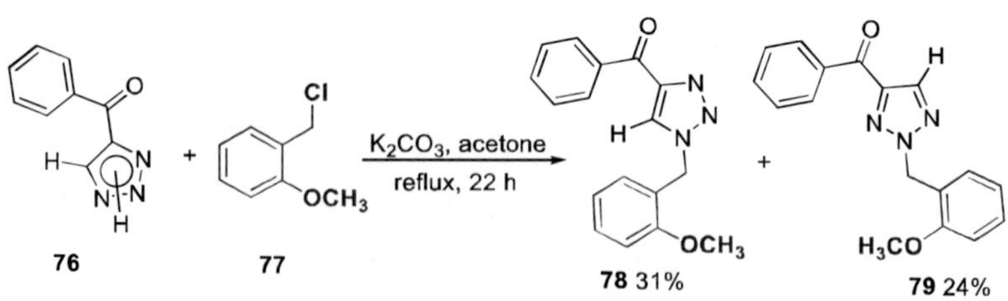

Scheme 20. A *N*-alkylation of N*H*-1,2,3-triazole by Livi and co-workers.

Raić-Malić's group also obtained a mixture of the *N*1- and *N*2-isomers when perfomed a *N*-alkylation of a series of N*H*-1,2,3-triazoles **80** with 2-acetoxymethyl-4-iodobutyl acetate in the presence of potassium carbonate. In all cases, the *N*2-isomer **82** were majorated formed (Scheme 21) [67].

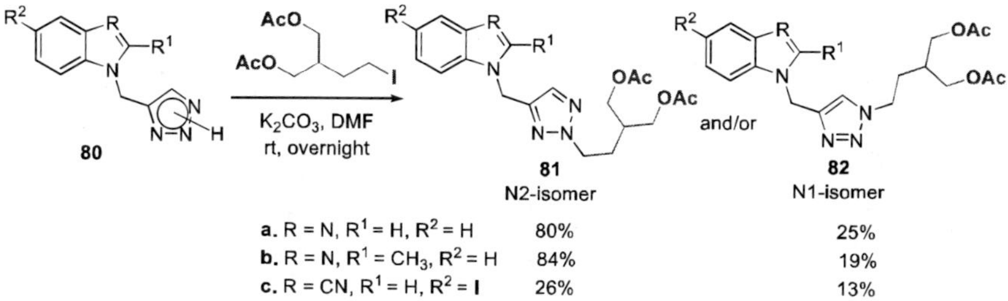

Scheme 21. A *N*-alkylation of N*H*-1,2,3-triazole by Raić-Malić's and co-workers.

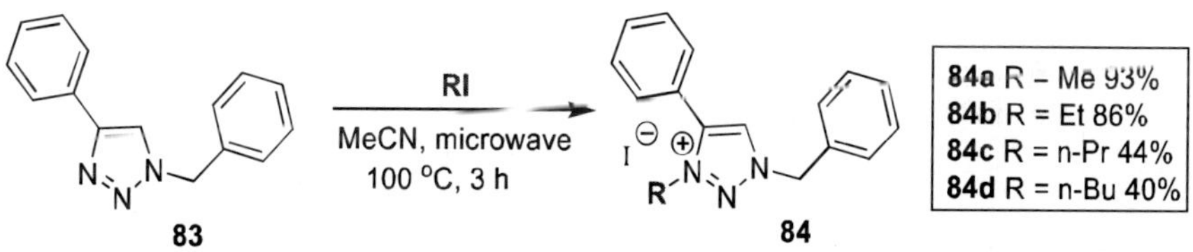

Scheme 22. Alkylation of 1,2,3-triazoles using microwave.

1,2,3-Triazoles can be converted to other heterocycles [68], but it is also possible to introduce another alkyl group into the 1,4-disubstituted 1*H*-1,2,3-triazole compounds. Page and co-workers prepared the 1*H*-1,2,3-cationic triazoles **84** through an alkylation reaction. In this work, the authors unequivocally confirmed, by X-ray crystallography, that the insertion of the methyl group occurred in nitrogen N3 (Scheme 22) [69].

He and Shreeve described the methylation of diammonium 4,4'-bis(5-nitro-1,2,3-2H-triazolate) (**85**) with dimethyl sulfate obtaining four isomers: 1,2'-Dimethyl-5,5'-dinitro-4,4'-bis(1,2,3-triazole) (**86,** traces), 2,2'-dimethyl-5,5'-dinitro-4,4'- bis(1,2,3-triazole) (**87**), 1,3'-dimethyl-5,5'-dinitro-4,4'-bis(1,2,3-triazole) (**88**), and 2,3'- dimethyl-5,5'-dinitro-4,4'-bis(1,2,3-triazole) (**89**) (69% yield, 87:88:89 = 3:1:6) (Scheme 23).

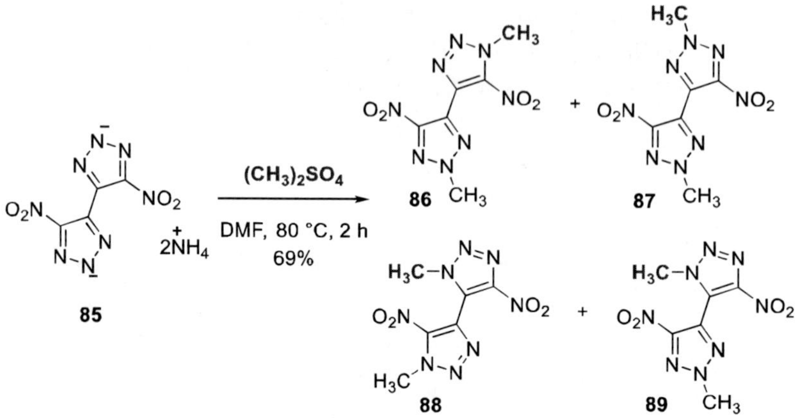

Scheme 23. Methylation of diammonium 4,4'-bis(5-nitro-1,2,3-2*H*-triazolate) **85**.

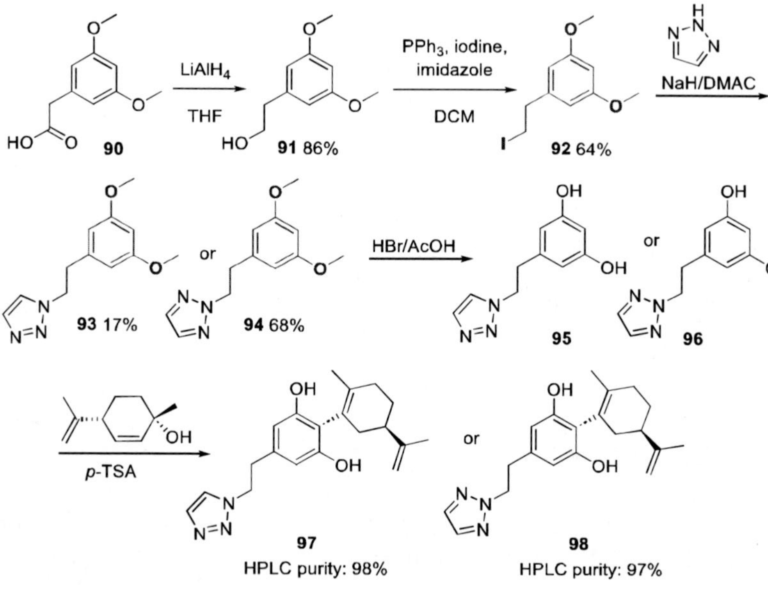

Scheme 24. Synthesis of cannabidiol derivatives **97** and **98**.

In 2017, a study was reported involving a series of cannabidiol derivatives capable of preventing damage to hippocampal neurons induced by ammonium acetate and ethanol at clinically relevant concentrations. Among the compounds evaluated, the N1-1,2,3-triazole **97** demonstrated a similar potency to cannabidiol, while the *N*2-regioisomer **98** was inactive. **97** and **98** isomers were obtained through five synthetic steps (Scheme 24). Initially, a reduction reaction of carbonyl group was carried out to produce alcohol **91**. In turn, **91** was converted into **92** through the reaction with PPh_3, iodine and imidazole. By reacting **92** and 1,2,3-triazole in the presence of NaH and DMAC, a mixture of the **93** and **94** regioisomers was obtained. Due to the difference in polarity between the **93** and **94** isomers, purification through chromatographic column became feasible. Specifically, the **95** isomer (68%) is less polar than the **93** isomer (17%). Subsequently, each isomer was treated separately with HBr/AcOH, leading **95** and **96**. Thereafter, the crude was used directly for the next step without purification, obtaining the cannabidiol derivatives bearing triazoles in their structure (**97** and **98**) [70].

3.2. Arylation Reactions

2-aryl-1,2,3-triazoles are an important class of bioactive compounds (Scheme 25). Thus, *N*2-regioselective arylation reactions have been developed to access such compounds. Generally, the regioselectivity of these reactions is influenced by the same factors discussed in the previous topic. Among the existing approaches to accessing N2-aryl-1,2,3-triazoles, the following stand out: SNAr, Cu(I) catalyzed aryl amidation, Cu(II) mediated boronic acid coupling, use of nanoparticles as a catalyst and diaryl iodonium salts [71].

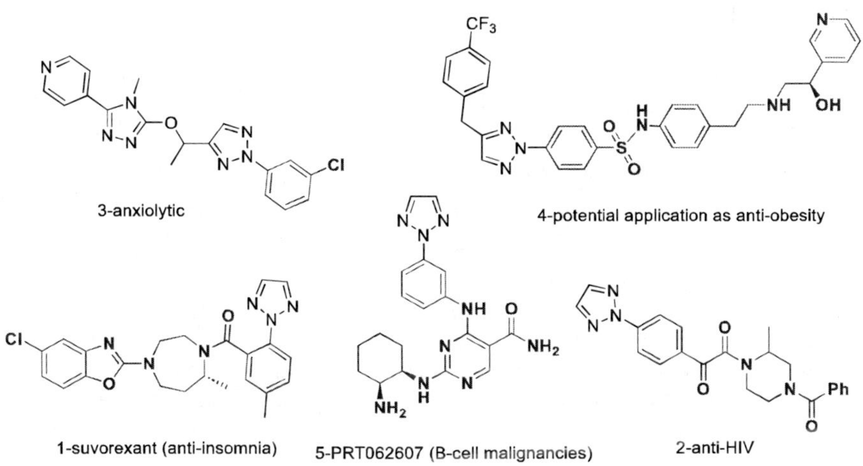

Scheme 25. Structures of bioactive compounds.

N2-Regioselective arylation from C4, C5 disubstituted-1,2,3-N*H*-triazoles by SNAr, Cu(I) catalyzed aryl amidation and Cu(II) mediated boronic acid coupling led to N2-aryl triazoles successfully. The influence of groups C4 and C5 was evaluated in each case. To SNAr, it was observed that **99a** had modest selectivity for *N*1-isomer, **99b** had low selectivity with N-2 as the main product, while **99c** and **99d** had excellent regioselectivity for *N*2-isomer (Scheme 26A). In the reaction of Cu(I) catalyzed aryl amidation, when the same triazoles were evaluated, benzotriazole **99a** gave the *N*1-isomer as the major product (60%) with 8% *N*2-isomer. The selectivity was increased when triazole **99b** was employed, giving *N*2-isomer (68%) and *N*1-isomer (11%). When **99c** and **99d** were used, only *N*2-isomers were isolated with excellent yields. Finally, Cu(II) mediated boronic acid coupling using triazole **99c**, the *N*2-isomer was the only product isolated (Scheme 26C) [72].

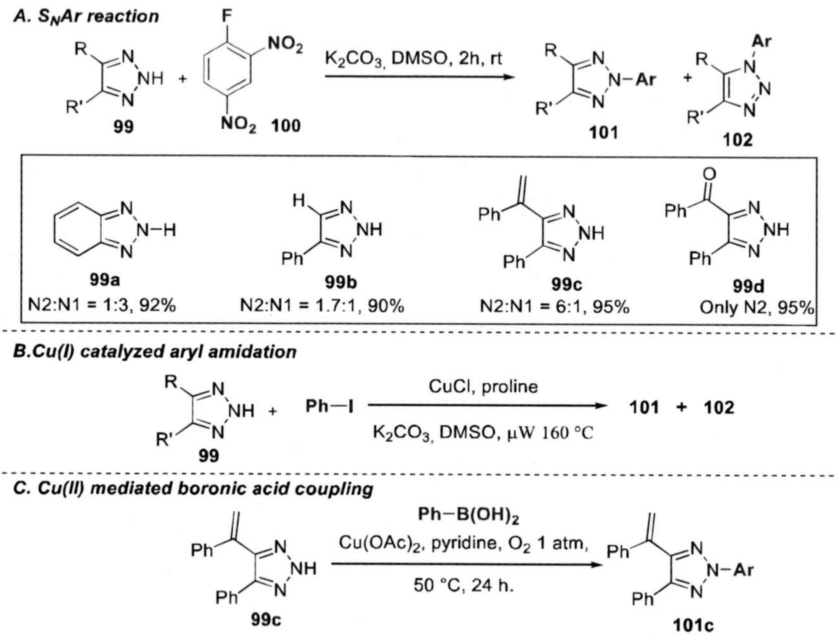

Scheme 26. N2-regioselective arylation from 1,2,3-N*H*-triazoles by A. SNAr; B. Cu(I) catalyzed aryl amidation and C. Cu(II) mediated boronic acid coupling.

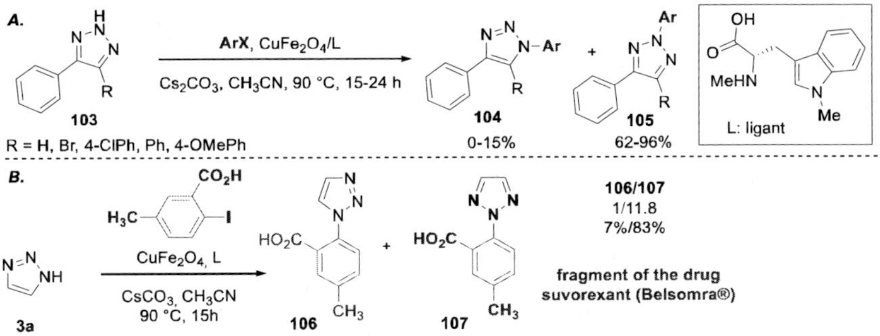

Scheme 27. A. *N*2-regioselective arylation of 1,2,3-triazoles using recyclable $CuFe_2O_4$ nanoparticles as catalyst and a tryptophan-based ligand; B. Synthesis of fragment of the drug suvorexant.

Another example of *N*2-regioselective arylation of 1,2,3-triazoles has reported by Miranda and co-workers. In this work, it was used aryl halides as arylation reagents, recyclable $CuFe_2O_4$ nanoparticles as catalyst and a tryptophan-based ligand that provided N2-arylated-1,2,3 triazoles in good yields and good N2-selectivity (up to 1:16 for N1:N2 when R = H) (Scheme 27A). It's important to mention that when R was Br-, 4-ClPh, Ph, 4-OMePh only *N*2-isomer was detected. This result demonstrated the influence of the steric and electronic effects of the substituents at positions C4 and C5 of the 1,2,3-triazole. Furthermore, when R was Br- (62%) and 4-ClPh (71%) there was a decrease in yields when compared to R = 4-OMe (96%). As an application of this methodology, a fragment of the drug suvorexant (Belsomra®) was achieved in excellent yield, 83% (Scheme 27B) [73].

Diaryl iodonium salts **110** are stable and can be synthesized from inexpensive reagents as arenes and iodoarenes. Due to these advantages, the use of such salts as arylation agents of 1,2,3-triazoles has shown promise. When such salts were used as the arylation reagent of 1,2,3-triazole **109**, N2-aryl-1,2,3-triazoles **111** were obtained in good yields (66–97%) and high regioselectivity (Scheme 28A). The proposed mechanism proceeds through a ligand-exchange, process followed by reductive elimination (Scheme 28B). DFT calculation studies have shown that *N*1-isomer is the kinetic product and *N*2-isomer is the thermodynamic product (Scheme 28B). Thus, greater regioselectivity is achieved at high temperatures [71].

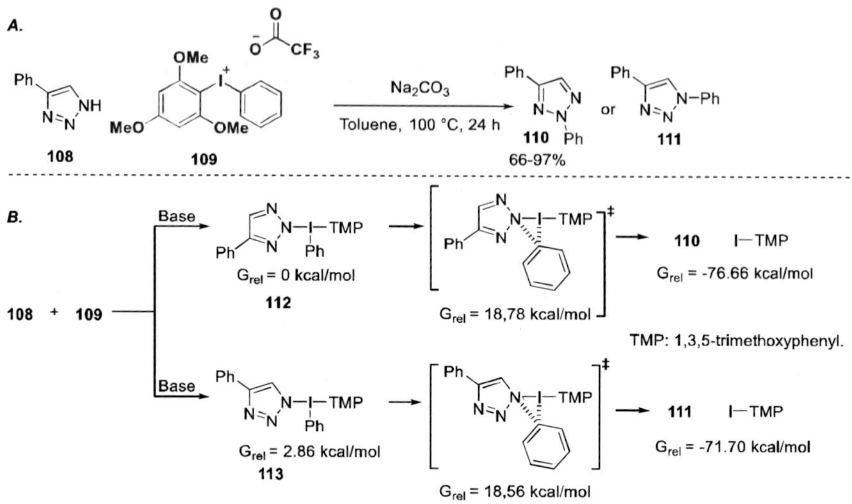

Scheme 28. A. Synthesis of N2-aryl-1,2,3-triazoles using diaryl iodonium salts as arylation agents; B. Results of DFT calculation studies.

Although a study on regioselectivity has not been carried out, the reaction of 1,2,3-triazole **3a** with **114** employing $Fe(acac)_3$ and CuO in presence of base provided a mixture of isomers leading **115** in 52% yield and 116 in 24% yield (Scheme 29). The *N*2-isomer **115** is an intermediate for the synthesis of the spirocompounds **119**, which is a series of compounds that have been investigated as CNS-targeted inverse agonists of the ghrelin receptor [74].

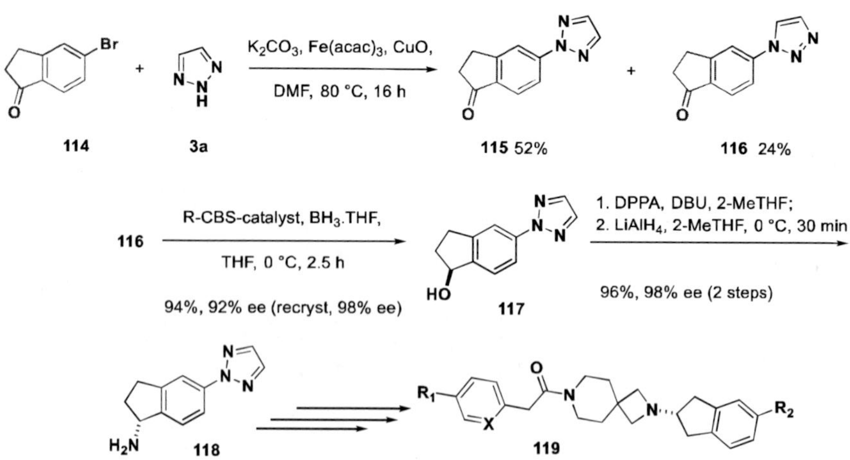

Scheme 29. Synthesis of spirocompounds **119.**

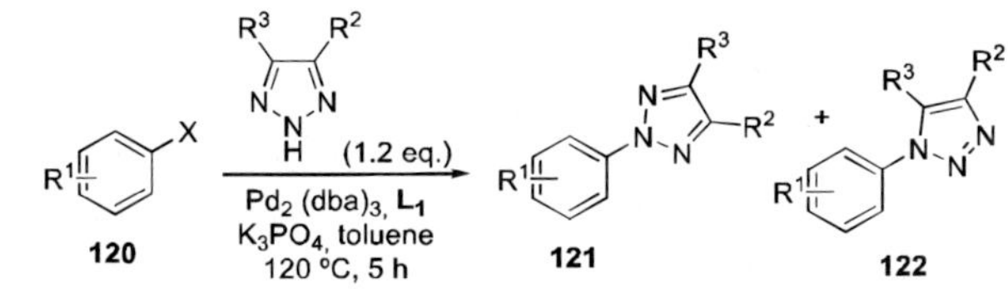

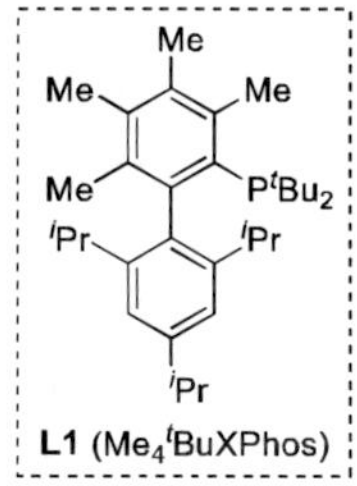

a R^1= 3-CN; R^2= H; R^3= H	89% (97:3)
b R^1= 3-NO_2; R^2= H; R^3= H	87% (98:2)
c R^1= 4-CHO; R^2= H; R^3= H	79% (97:3)
d R^1= 4-Cy; R^2= H; R^3= H	78% (96:4)
e R^1= 2-CN; R^2= H; R^3= H	46% (99:1)
f R^1= 4-OMe; R^2= H; R^3= H	90% (98:2)
g R^1= H; R^2= Ph; R^3= H	90% (97:3)
h R^1= 4-CO_2Me; R^2= 3-Py; R^3= H	85% (96:4)
i R^1= H; R^2= Bu; R^3= H	91% (98:2)
j R^1= 3-OBn; R^2= CH_2NHBoc; R^3= H	91% (99:1)
k R^1= 3-F; R^2= CH_2OBn; R^3= H	94% (99:1)

Scheme 30. Pd-Catalyzed *N*-arylation of 1,2,3-triazoles by the use of a very bulky L1 biarylphosphine ligand.

Ueda and co-workers report that high levels of N2-selectivity can be obtained in the Pd-catalyzed *N*-arylation of 1,2,3-triazoles by the use of a very bulky L1 biarylphosphine ligand (Scheme 30). This method allowed the first highly selective N2 arylation of 4-substituted and 4,5-unsubstituted 1,2,3-triazoles with aryl bromides, chlorides and triflates, providing the alkylated products in high yields (78-94%) and high selectivity (95:5 to 99:1 - *N*2:*N*1). According to the authors, the reaction was significantly less efficient without catalyst preheating, presumably due to the inhibitory effect of 1,2,3-triazole on the in situ formation of the catalytically active Pd(0)-ligand complex. For 1,2,3-triazoles unsubstituted with 4,5, a slight reduction in N2-selectivity was observed for the reactions of aryl chlorides with *p*-electron groups and the yields were decreased when aryl halide containing an ortho substituent was used probably because of the unfavorable steric interaction between the bulky ligand and the ortho substituent. The *N*-arylation of 4-substituted 1,2,3-triazoles was examined by reaction of 4-phenyl-1,2,3-triazole with bromobenzene yielding the *N*2-arylated products **121** in high selectivity (>94%) [75].

Pike and co-workers reported the replacement of the fluorine atom in 3-fluoro-2-nitropyridine (**123**) by the 1,2,3-triazol nucleous using K_2CO_3 in acetonitrile during 50°C (Scheme 31). According to the authors, the high electronegativity of fluorine would favor nucleophilic aromatic substitution increasing selectivity. As a result of the reaction between 3-fluoro-2-

nitropyridine and 1,2,3-triazole, two isomers were afforded in an approximately 4:1 ratio (**124/125**) [76].

Kuang and co-workers reported a copper-promoted one-pot sandmeyer-type reaction for the synthesis of N-aryl-triazoles. They developed an efficient strategy based on the Sandmeyer-type reaction for the synthesis of N1/N2-1,2,3-triazole derivatives by employing functionalized triazoles and readily available arylamines as nitrogen sources and aromatic hydrocarbons (Scheme 32) [77].

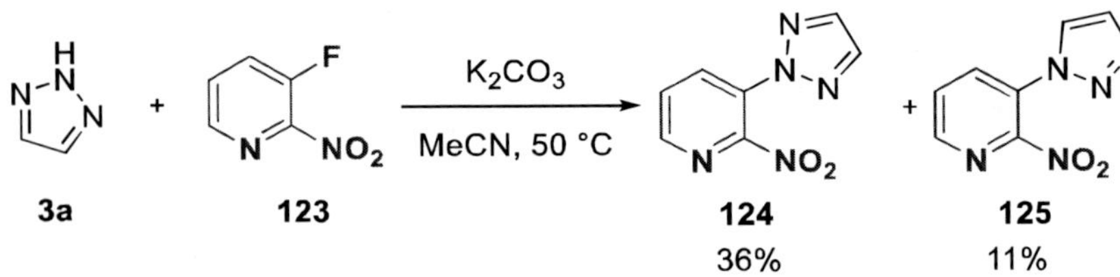

Scheme 31. Reaction between 3-fluoro-2-nitropyridine **123** and 1,2,3-triazole **3a**.

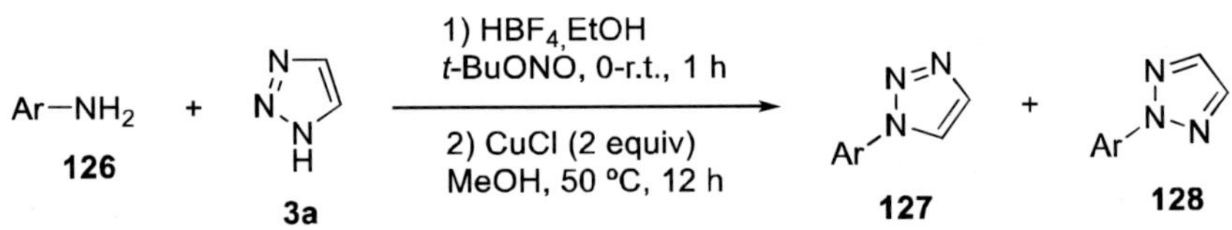

Scheme 32. Copper-promoted one-pot sandmeyer-type reaction for the synthesis of N-aryl-triazoles.

3.3. Amination Reactions

As well as in the alkylation and arylation reactions, the effect of prototropic tautomerism was found in amination reactions of 1,2,3-triazoles containing a nitro group in the structure. Such amination products have application as energetic compounds, due to the high heats of formation. In this topic, 1,2,3-triazole amination methodologies and the influence of prototropic tautomerism on these reactions will be discussed.

Piercey and co-workers reported the *N*-amination of sodium salt of 4-nitro-1,2,3- triazole (**129**) using *o*-tosylhydroxylamine (THA) (Scheme 33). As a result, 2-amino-4-nitro-1,2,3-triazole (**130**) and 1-amino-4-nitro-1,2,3-triazole (**131**) were obtained in 14,3% and 15,4% yield, respectively. According to the authors, **130** undergoes deamination in purification by column chromatography and recrystallization in hot water. The instability of the **130**-isomer justifies the lower yield of this isomer. Specifically, in this study, these compounds were classified as primary explosives with performance similar to that of PETN [78].

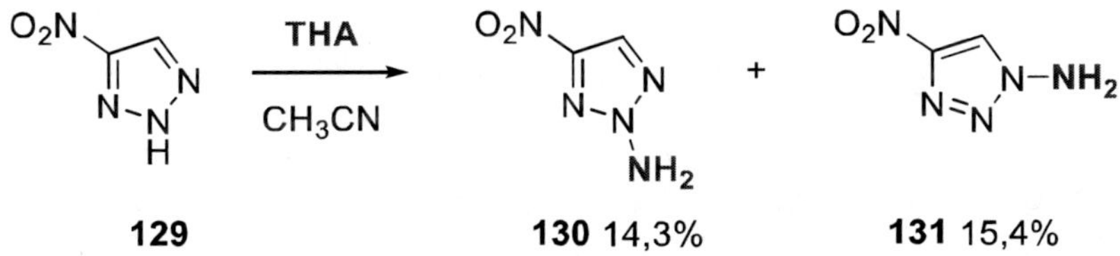

Scheme 33. N-Amination of sodium salt of 4-nitro-1,2,3-triazole using THA.

Similarly, in the work of Shreeve and co-workers, triazoles **130** and **131** were obtained in slightly higher yields through the reaction with hydroxylamine-*O*-sulfonic acid (Scheme 34). Then, an efficient oxidative coupling reaction was performed leading to the obtainment of two regioisomeric energetic molecules, (*E*)-1,2-bis(4-nitro-1*H*-1,2,3-triazol-1-yl)diazene (**134**, 72%) and (*E*)-1,2-bis(4-nitro-2*H*-1,2,3-triazol-2-yl)diazene (**133**, 61%). It is important to mention that, in the last step, the isomers **130** and **131** were treated separately with sodium dichloroisocyanurate and AcOH [79].

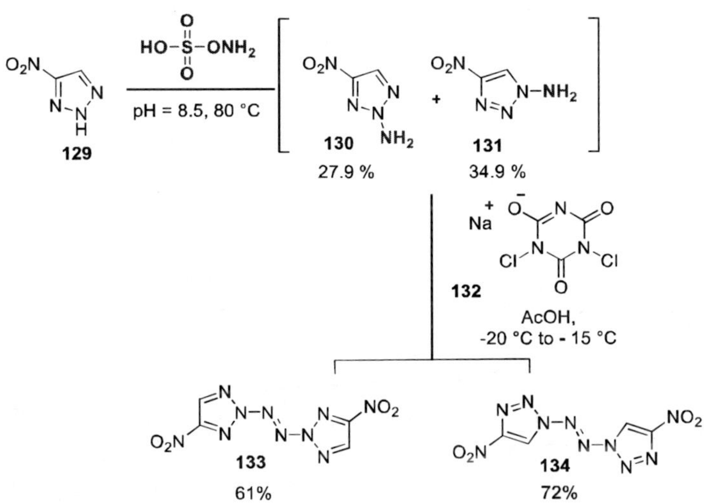

Scheme 34. Synthesis of regioisomeric energetic molecules.

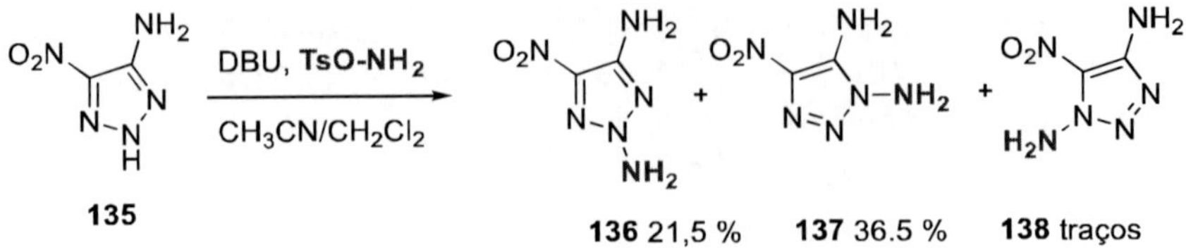

Scheme 35. Amination of 4-amino-5-nitro-1,2,3-triazole **135**.

Amination of 4-amino-5-nitro-1,2,3-triazole **135** employing DBU and TsO-NH_2 resulted in the formation of three isomers, **136**, **137** and **138** (Scheme 35). Compound **137** was the major product with 36.5%, compound **136** was obtained in 21.5% yield, while only a trace of **138** was formed [80].

The amination of diammonium 4,4'-bis(5-nitro-1,2,3-2*H*-triazolate) (**139**) employing THA led to the formation of three isomers: 2,2'-diamino-5,5'-dinitro-4,4'-bis(1,2,3-triazole) (**140**), 3,3'-diamino-5,5'-dinitro-4,4'-bis(1,2,3-triazole) (**141**) and 2,3'-diamino-5,5'-dinitro-4,4'-bis(1,2,3-triazole) (**142**) (41%, 140:141:142 = 2:5:3) (Scheme 36).

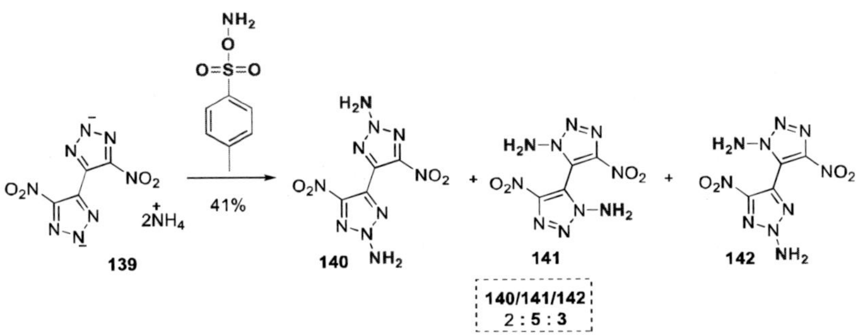

Scheme 36. Amination of diammonium 4,4'-bis(5-nitro-1,2,3-2*H*-triazolate) employing THA.

3.4. Acylation Reactions

Unlike the alkylation, arylation and amination reactions, there are few reports, as far as we know, about the formation of more than one regioisomer in the same methodology. In general, the acylation reaction of 1,2,3-triazoles yields a single isomer. Often, the *N*2 regioisomer is obtained in acylation reactions, although *N*1 and *N*3 isomers are also reported.

As examples of methodologies for obtaining *N*2 regioisomers we can mention the acetylation of **143** with acetic anhydride that occurred exclusively at N2-atom (Scheme 37A). Similarly, when the triazole **99c** was acylated, only regioisomer **145** was obtained in 98% yield (Scheme 37B) [81].

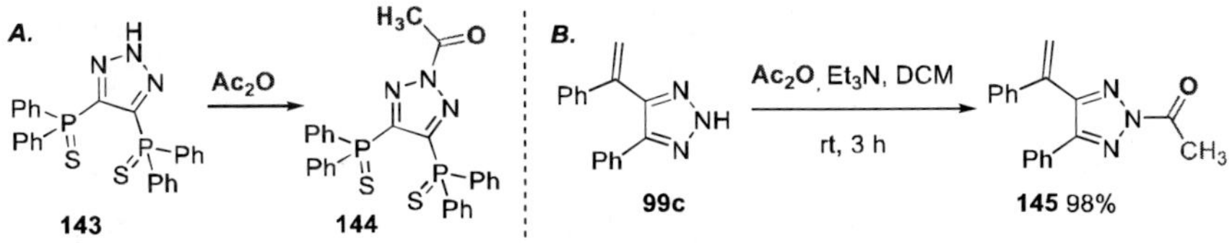

Scheme 37. Acylation reactions for obtaining of *N*2-regioisomers.

In the previous example, both triazoles have substituent groups at C4 and C5 positions, which may provide the steric hindrance needed to promote *N*2-acylation. However, some cases the the steric effects are not the only operant. Analyzing the acetylation reaction of fluorotriazole **146**, it is possible to observe that the N-1 position is impeded by the aryl group and the fluorine atom promotes a decrease in nucleophilicity at the N-3 position (Scheme 38) [57].

Scheme 38. Acetylation reaction of fluorotriazole **146**.

Silva and co-workers reported the synthesis of 1,2,3-triazole **149**, which is a product of the Dimroth rearrangement (Scheme 39). However, the position of hydrogen in the triazole ring could not be confirmed by spectroscopy techniques. For this reason, acetylation of Dimroth rearrangement product was considered to obtain information about which tautomer was preferentially formed. As result, the reaction of 1,2,3-triazole **148** with acetic anhydride, pyridine and DMAP gave only the *N*2-isomer **149** in 58% yield [82].

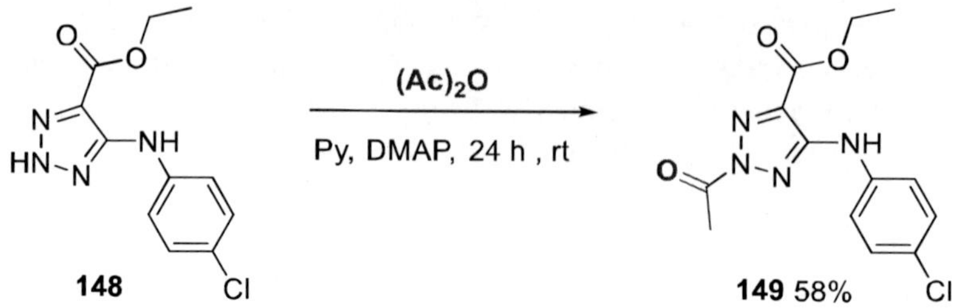

Scheme 39. Acetylation of 1,2,3-triazole **148**.

The reaction of 5-amino-1-aryl-1,2,3-triazole-4-carbothioamides **150** with acetyl chlorides in presence of DBU led to obtaining 2-acyl-5-arylamino-1,2,3-triazole-4-carbothioamides **151** (Scheme 40). In method I, there is the formation of the rearrangement adduct in situ with its subsequent acylation (one-pot regioselective method), while in method II, the Dimroth rearrangement product is synthesized, isolated and subsequently acylated in a next step. These results suggest that the rate-determining step of the reaction is the rearrangement of triazoles, not acylation. The authors believe that regioselectivity is probably due to steric hindrance to the attack of nitrogen atoms in positions 1 and 3 of the 1,2,3-triazole ring by electrophiles [83].

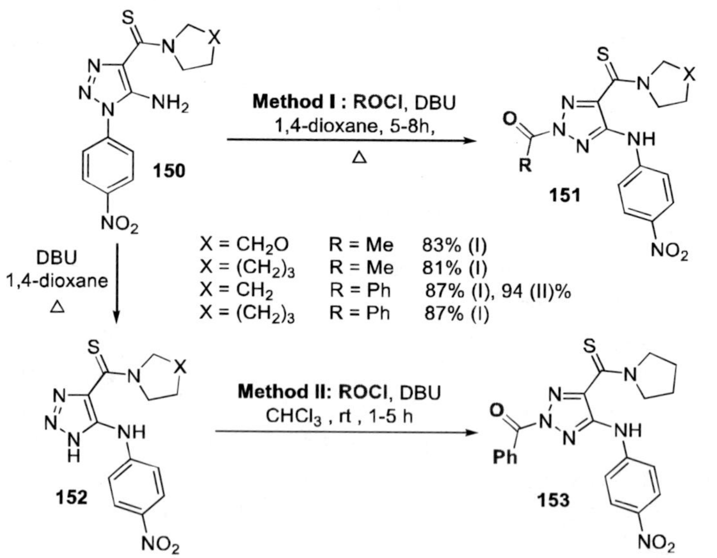

Scheme 40. Acylation reactions of 5-amino-1-aryl-1,2,3-triazole-4-carbothioamides **150**.

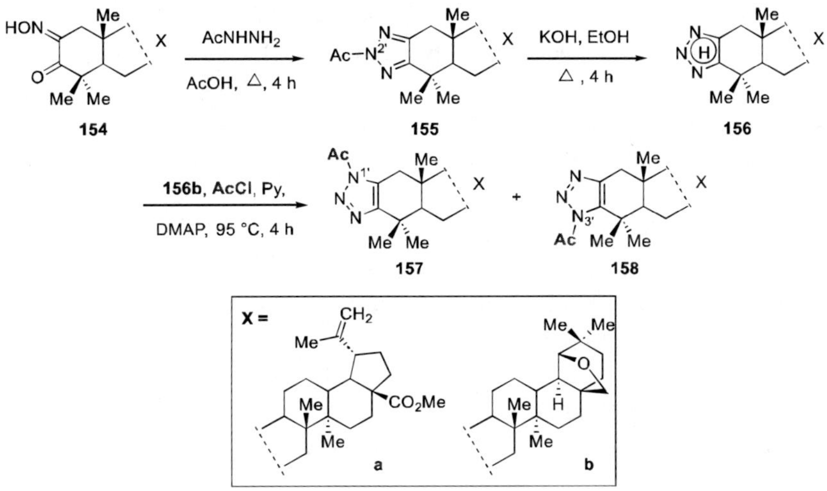

Scheme 41. Acylation reaction described by Grishko and co-workers.

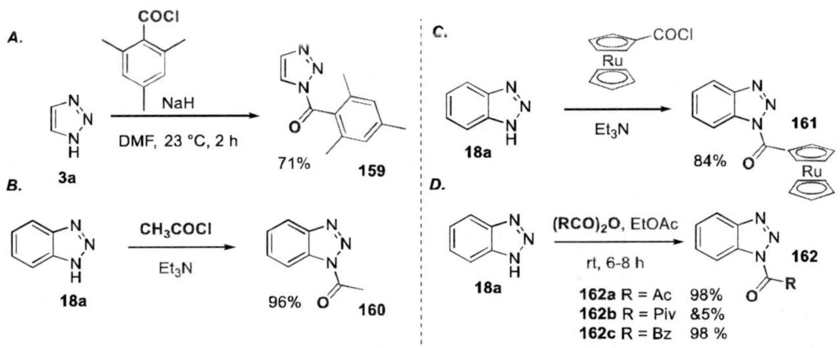

Scheme 42. Acylating methodologies of unsubstituted triazoles and benzotriazoles.

The formation of *N*1 and *N*3 isomers was observed in the work of Grishko and co-workers. The reactions between **154a** and **154b** with acetic acid hydrazide gave 2'-*N*-acetyl-1,2,3-triazoles **155a** and **155b** in 41 and 38% yield, respectively (Scheme 41). Next, by basic hydrolysis of 2'-*N*-acetyl-1,2,3-triazoles, unsubstituted triterpene 1,2,3-triazoles **156a** and **156b** were obatined in both the 1'(3')*H*- and 2'*H*-forms. Subsequently, the acylation of 18αH-oleanane (**156b**) 1,2,3-triazole gave a mixture of **157** and **158** in 48 and 47% yield, respectively [84].

In general, benzotriazoles and unsubstituted triazoles when acylated provide N1-products (Scheme 42). Even using bulkier acylating agents as

2,4,6-trimethylbenzoyl chloride, ruthenocenoyl chloride and pivalic anyhydride, the *N*1-isomer is the only product obtained. Note that in the case involving an unsubstituted triazole, the base used was sodium hydride. Although not mentioned by the authors, it is possible to infer that the use of a strong base also may have contributed to a greater selectivity of the reaction [85, 86, 87].

3.5. Other Reactions: Vinylation, Silylation, Sulfonylation and Oxylation Reactions

Although there are still few reports, the effect of prototropic tautomerism can be observed in some types of reactions such as vinylation, silylation, sulfonylation and oxylation. In some cases, it is possible to evaluate the influence of certain groups on the selectivity of reactions, while in others it is not clear yet.

As an example of a vinylation reaction, we can mention the copper catalyzed vinylation of 1,2,3-triazoles with boronic acids (Scheme 43). NMR analysis of the crude mixture determined that the *N*2-isomer **164** is formed in high selectivity, with bulky 4,5-disubstituted triazoles having a selectivity >30:1, while the monosubstituted triazoles showed selectivity of 5:1. However, disubstituted triazoles were less reactive, requiring a longer reaction time. According to the authors, purification is not a problem, since the *N*2-isomer has a significantly lower polarity compared to *N*1. In this regioselective transformation, the stability and low amount of the catalyst, in addition to the absence of base and the use of triazoles with different functionalities and electronic properties were pointed out as advantages of the method [88].

The silylation is strategically used to obtain sulfonylated and acylated 1,2,3-benzotriazoles (Scheme 44). The nitrogen activated by the silyl group reacts with methanesulfonyl chloride and benzenesulfonyl chloride to give **166a** (60%) and **166b** (59%). Heating **166a** with carboxylic acids in the presence of triethylamine led to acylbenzothiazoles (**167a-e**) in excellent yields (80-95%). It is important to mention that only N1-substituted-benzotriazoles were obtained in both the sulfonylation and the acylation reaction [89].

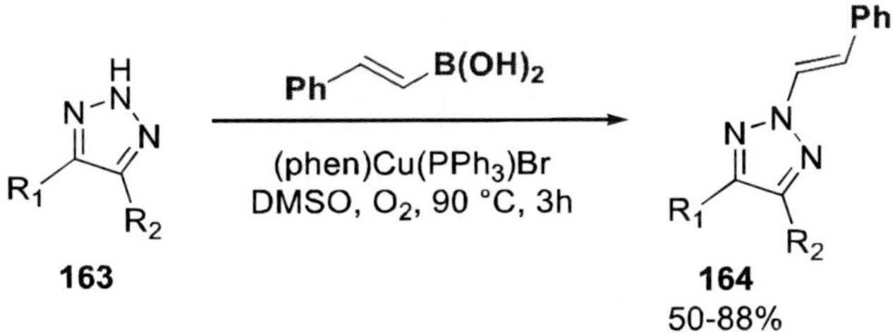

Scheme 43. Copper catalyzed vinylation of 1,2,3-triazoles with boronic acids.

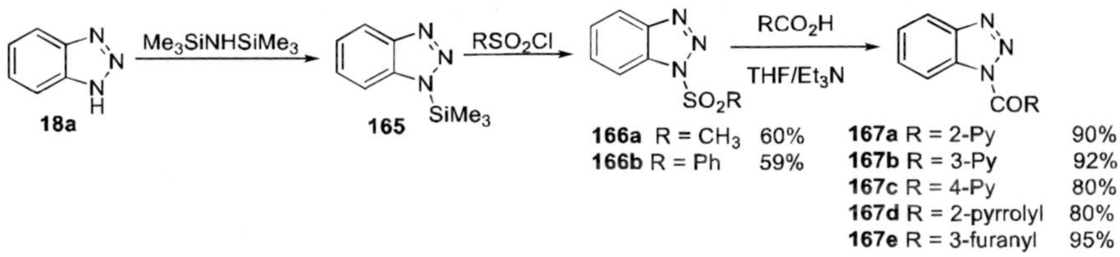

Scheme 44. Synthesis of acylated and sulfonylated benzotriazoles through the silylation step.

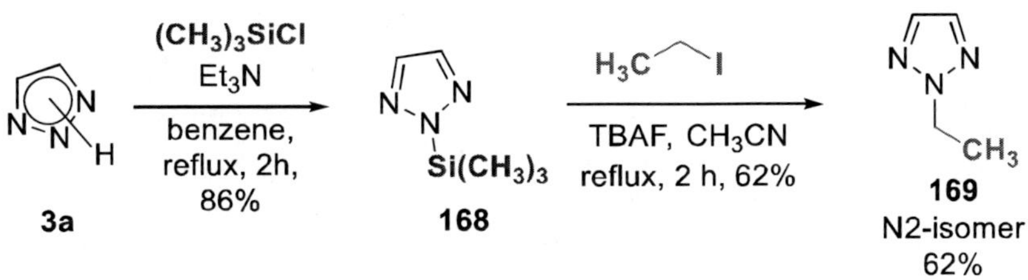

Scheme 45. Regioselective *N*-alkylation of 1,2,3-triazole through a silylation.

However, N2-1,2,3-triazoles can also be obtained through the silylation step. For example, Ohta and coworkers disclosed a regioselective *N*-alkylation of 1,2,3-triazole **3a** through a silylation step. First, the authors synthetized the *N*-trimethylsilyl triazole **168** using chlorotrimethylsilane in benzene for 2 hours. The insertion of sylil group was regioselective and lead the exclusive formation of the *N*2-isormer **168**. Then, to improve the yields of alkylated product, the authors reported the *N*-alkylation step using different alkyl, benzil and phenetyl halides and tetrabutylammonium fluoride or bromide as an additive. The highest yield was achieved using ethyl iodide and tetrabutylammonium fluoride under reflux in acetonitrile for 2 hours (62%) (Scheme 45) [90].

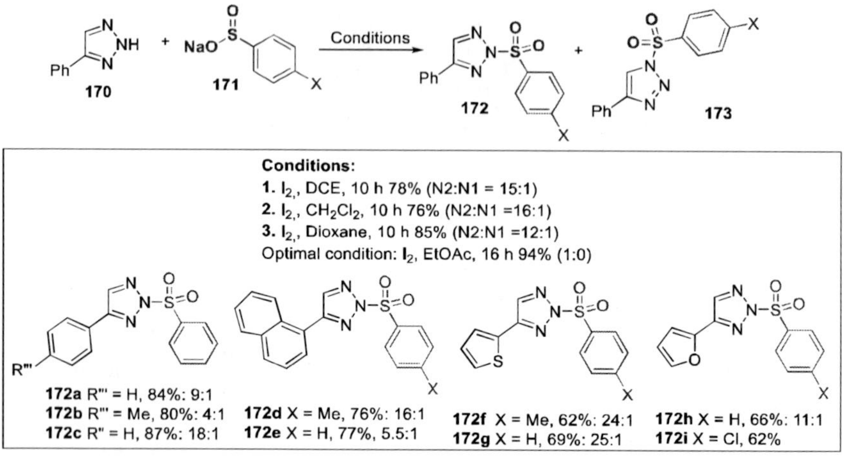

Scheme 46. Sulfonylation of N*H*-1,2,3-triazoles with sodium sulfinates in the presence of molecular iodine.

Nanubolu and co-workers reported a method for sulfonylation of N*H*-1,2,3-triazoles **170** with sodium sulfinates in the presence of molecular iodine (Scheme 46). Specifically, when DCE, DCM and 1,4-dioxane were used as solvents, a mixture of *N*2/*N*1 regioisomers was detected with *N*2-isomer **172** being the major. However, only the *N*2-isomer was obtained when EtOH, EtOAc, CH_3CN and THF was used. For the authors, the influence of the solvent is not clear, but they believe that the rapid interconversion of N1- and N2-triazole may be involved. In some cases, where R, R', R'', R'' and X were evaluated, even using the optimal condition that uses EtOAc as a solvent, it was observed the formation of the *N*1-isomer in a smaller proportion [91].

Murakami and co-workers reported a 1,2-rearrangement of a sulfonyl group occurs on treatment of 1-sulfonyl-1,2,3-triazoles **173a** catalyzed by DMAP to give an equilibrium mixture of 1-sulfonyl- and 2-sulfonyl isomers (9:1) (Scheme 47) [92].

Another example is the reaction of 5-amino-1-aryl-1,2,3-triazole-4-carbothioamides with sulfonyl chlorides that occurs via Dimroth rearrangement with formation of 5-arylamino-2-sulfonyl-1,2,3-triazoles-4-carbothioamides (Scheme 48) [84].

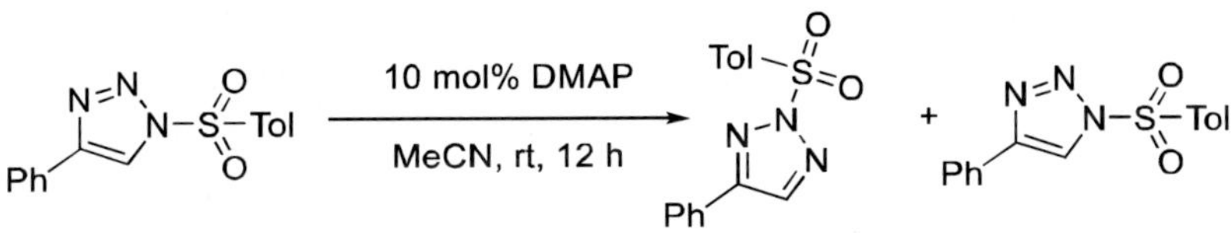

Scheme 47. 1,2-Rearrangement of a sulfonyl group.

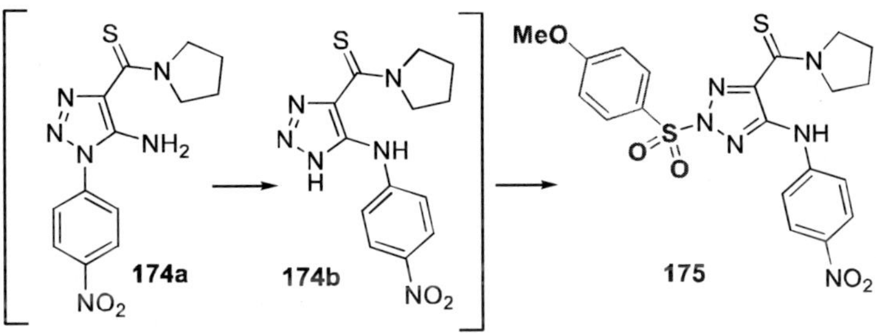

Scheme 48. Sulfonylation reaction of 5-amino-1-aryl 1,2,3-triazole-4-carbothioamides with sulfonyl chlorides.

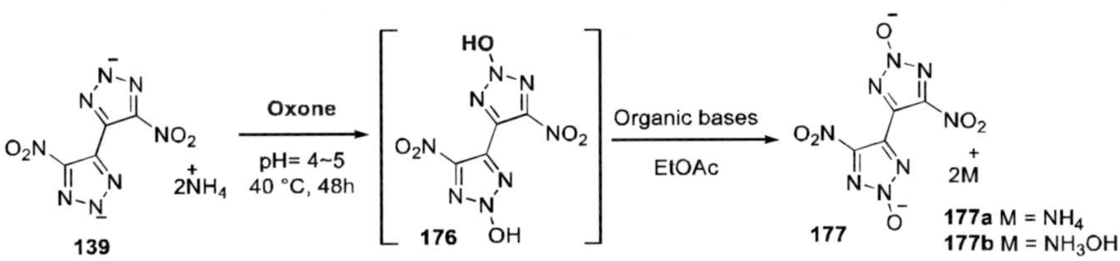

Scheme 49. *N*2-oxidation of 1,2,3-triazoles.

Finally, *N*2-oxidation of 1,2,3-triazoles was also described by He and Shreeve (Scheme 49). Specifically, the *N*-oxidation of diammonium 4,4'-bis(5-nitro-1,2,3-2*H*-triazolate) **139** with oxone gave a single isomer **176**. After treatment with organic bases, **177a** and **177b** were obtained [70].

Final Remarks

There are a large number of drugs containing the triazole nucleus available to treat various systemic diseases. There has been considerable interest in the development of new heterocyclic compounds with novel biological activities. Azole compounds, particularly triazoles, have a broad spectrum and are among the molecules with the most diverse biological activities. This special class of heterocyclic compounds has a broad spectrum of biological activities, such as antimicrobial, cytotoxic, antihistamine, anticonvulsant, analgesic, anti-inflammatory, insecticidal, antimycotic, antimycobacterial, anticancer, antiprotozoal, antimalarial, and antiulcer activity. Many reactions have been developed to produce these compounds, but it is necessary to understand their tautomeric equilibrium and how these tautomers behave in different reactions. This understanding is fundamental for the development of new triazole compounds with new and improved biological activities.

References

[1] Gilchrist, T. L. *Heterocyclic Chemistry*, 3rd ed.; Pearson Education India: Singapore, 2005, pp. 298-307.

[2] Khurshed Bozorova, Jiangyu Zhao, Haji A. Ais (2019). 1,2,3-Triazole-containing hybrids as leads in medicinal chemistry: A recent overview. *Bioorganic & Medicinal Chemistry*, 27, 3511-3531.

[3] Júlio O. F. Melo, Claudio L. Donnici, Rodinei Augusti, Vitor F. Ferreira, Maria Cecília B. V. de Souza, Maria Lourdes G. Ferreira, Anna C. Cunha (2006). Heterociclos 1,2,3-Triazólicos: Histórico, Métodos De Preparação, Aplicações e atividades Farmacológicas. *Química Nova*, 29: 569-579.

[4] Chemistry of 1,2,3-triazoles in Topics in *Heterocyclic Chemistry*, Wim Dehaen and Vasiliy A. Bakulev, (Eds.), Springer, Switzerland 2015.

[5] Speicher, A. *The Chemistry of Heterocycles: Structures, Reactions, Synthesis and Applications*, 2nd ed.; Wiley-VCH & Company: Deutschland, 2003.

[6] Bambas L. L. *Chemistry of heterocyclic compounds: Five member Heterocyclic compounds with nitrogen and sulfur or nitrogen, sulfur and oxygen* (except thiazole). John Wiley & Sons: New Jersey, 2008.

[7] Kharb, R., Sharma, P. C., Shaharyar, M. Pharmacological significance of triazole scaffold; *Journal of Enzyme Inhibition and Medicinal Chemistry*, 2011; 26(1): 1-21.

[8] Vitor F Ferreira, David R da Rocha, Fernando C da Silva, Patrícia G Ferreira, Núbia A Boechat, Jorge L Magalhães. Novel 1H-1,2,3-, 2H-1,2,3-, 1H-1,2,4-and 4H-1,2,4-triazole derivatives: a patent review (2008 - 2011); *Expert Opin. Ther. Patents* (2013) 23(3):319-331.

[9] a) Thirumurugan, P., Matosiuk, D., Jozwiak, K. Click Chemistry for Drug Development and Diverse Chemical-Biology Applications; Chem. Rev. 2013, 113, 4905-4979; b) Rajeev Kharb, Prabodh C. Sharma, Mohammed S. Yar, Pharmacological significance of triazole scaffold. *Journal of Enzyme Inhibition and Medicinal Chemistry* 26: 1-21 (2011); c) Jinjian Wei, Y. Wang, X. L. Wang. (2011) Recent advances of 1, 2, 3-triazole compounds in medicinal chemistry. *Journal of Chinese Pharmaceutical Sciences,* 46: 481-485.

[10] Jalani, H. B., Karagoz, A. € C‚., Tsogoeva, S. B. Synthesis of Substituted 1,2,3-Triazoles via Metal-Free Click Cycloaddition Reactions and Alternative Cyclization Methods; *Synthesis* 2017, 49, 29-41.

[11] Yu L. Angell, Kevin Burgess. Peptidomimetics via copper-catalyzed azide-alkyne cycloadditions; *Chemical Society Reviews* 36, 1674-1689 (2007).

[12] Daniel S. Pedersen, Andrew Abell. 1,2,3-Triazoles in peptidomimetic chemistry; *European Journal of Medicinal Chemistry*, 3, 2399-2411 (2011).

[13] Sandip G. Agalave, Suleman R. Maujan, Vandana S. Pore. Click chemistry: 1,2,3-triazoles as pharmacophores. *Chemistry- An Asian Journal*, 6: 2696-2718 (2006).

[14] a) Evangelia D. Chrysina, Éva Bokor, Kyra-Melinda Alexacou, Maria-Despoina Charavgi, George N. Oikonomakos, Spyros E. Zographos, Demetres D. Leonidas, Nikos G. Oikonomakos, László Somsák. Amide-1,2,3-triazole bioisosterism: the glycogen phosphorylase case. *Tetrahedron: Asymmetry* 20, 733-740 (2009); b) Victoria D. Bock, Dave Speijer, Henk Hiemstra, Jan H. van Maarseveen 1, 2, 3-Triazoles as peptide bond isosteres: synthesis and biological evaluation of cyclotetrapeptide mimics; *Organic & Biomolecular Chemistry*, 5: 971-975 (2007); c) W. Seth Horne, Maneesh K. Yadav, C. David Stout, M. Reza Ghadiri (2004). Heterocyclic peptide backbone modifications in an alpha-helical coiled coil. *Journal of American Chemistry Society*, 126: 15366-15367.

[15] Agalave, S. G., Maujan, S. R., Pore, V. S. Click Chemistry: 1,2,3-Triazoles as Pharmacophores; *Chem. Asian J.* 2011, 6, 2696-2718.

[16] David C. Schröder, Oliver Kracker, Tanja Fröhr, Jerzy Góra, Michal Jewginski, Anke Nieß, Iris Antes, Rafał Latajka, Antoine Marion, Norbert Sewal. 1,4-Disubstituted 1H-1,2,3-Triazole Containing Peptidotriazolamers: A New Class of Peptidomimetics With Interesting Foldamer Properties; *Frontiers in Chemistry*, 7: 155 (2019).

[17] a) Ze-Yi Yan, Yan-Ning Niu, Hai-Long Wei, Lu-Yong Wu, Ya-Bin Zhao, Yong-Min Liang (2006). Combining proline and 'click chemistry': a class of versatile organocatalysts for the highly diastereo- and enantioselective Michael addition in water. *Tetrahedron: Asymmetry*, 17: 3288-3293; Srivari Chandrasekhar, Togapur P. Kumar, Kothapalli Haribabu, Chada R. Reddy; Hydroxyphthalimide allied triazole-pyrrolidine catalyst for asymmetric Michael additions in water. *Tetrahedron: Asymmetry*, 21: 2372-2375 (2010); c) Ya-Bin Zhao, Lin-Wei Zhang, Lu-Yong Wu, Xing Zhong, Rong Li, Jian-Tai Ma; Silica-supported pyrrolidine-triazole, an insoluble, recyclable organocatalyst for the enantioselective Michael addition of ketones to nitroalkenes. *Tetrahedron: Asymmetry*, 19: 1352-1355 (2008); d) Charlotte M. Zammit, MartinWills; Use of triazole-ring formation to attach a Ru/TsDPEN complex for asymmetric transfer hydrogenation to a soluble polymer. *Tetrahedron: Asymmetry*, 24: 844-852 (2013).

[18] a) Yasushi Yoshida, Shinobu Takizawa, Hiroaki Sasai; Design and synthesis of spiro bis(1,2,3-triazolium) salts as chiral ionic liquids; *Tetrahedron: Asymmetry,* 23: 843-851(2012); b) Stijn Raiguel, Joice Thomas, Koen Binnemans, Wim Dehaen; Multi-Gram Scale Synthesis of 1,2,3-Triazolium Ionic Liquids and Assay of Their Resistance towards Bases. *European Journal of Organic Chemistry*, 35: 4850-4856 (2018); c) John M. C. Puguan, Lilian B. Boton, Hern Kim Triazole-based ionene

exhibiting tunable structure and ionic conductivity obtained via cycloaddition reaction: A new polyelectrolyte for electrochromic devices. *Solar Energy Materials and Solar Cells*, 188: 210-218(2018); d) Manh L. Nguyen, Jaeduk Byun, Suwoong Kim, June W. Hyun K. Hur, Tae J. Shin, Byoung-Ki Cho; Ferroelectrically Switching Helical Columnar Assembly Comprising Cisoid Conformers of a 1,2,3-Triazole-based Liquid Crystal. Angewandte Chemie International Edition, 58: 2749-2753 (2019).

[19] a) Luana da S. M. Forezi, Mariana F. C. Cardoso, Daniel T. G. Gonzaga, Fernando de C. da Silva, Vitor F. Ferreira. Alternative Routes to the Click Method for the Synthesis of 1,2,3-Triazoles, an Important Heterocycle; *Medicinal Chemistry*, 18: 1428-1453 (2018); b) Khan M. Faraz, Verma Garima, Akhtar Wasim, Marella Akranth, Alam M. Mumtaz, Akhter Mymoona, Husain Asif, Hasan S. Misbahul, Shaquiquzzaman Mohammad, Haider S. Rashiduddin; Synthetic Trends Followed for the Development of 1,2,3-Triazole Derivatives. *International Journal of Drug Development and Research*, 9: 22-25 (2017).

[20] a) Christian Therrien, Roger C. Levesque. Molecular basis of antibiotic resistance and β-lactamase inhibition by mechanism-based inactivators: Perspectives and future directions. *FEMS Microbiology Reviews,* 24: 251-262 (2000); b) Fahmi Y. Khan, Abdelhaleem Elhiday, Imran F. Khudair, Hind Yousef, Aisha H. Omran, Samer H. Alsamman, Maha Elhamid. Evaluation of the use of piperacillin/tazobactam (Tazocin®) at Hamad General Hospital, Qatar: Are there unjustified prescriptions? *Infection and Drug Resistance,* 5: 17-21(2012); c) Mercedes Delgado-Valverde, Eva Torres, Adoración Valiente-Mendez, Benito Almirante, Silvia Gómez-Zorrilla, Núria Borrell, Juan E. Corzo, Mercedes Gurgui, Manuel Almela, Lara García-Álvarez, María C. Fontecoba-Sánchez, Luis Martínez-Martínez, Rafael Cantón, Julia Praena, Manuel Causse, Belén Gutiérrez-Gutiérrez, Jason A. Roberts, Andras Farkas, Álvaro Pascual, Jesús Rodríguez-Banõ; Impact of the MIC of piperacillin/tazobactam on the outcomefor patients with bacteraemia due to Enterobacteriaceae: the Bacteraemia-MIC project. *Journal of Antimicrobial Chemotherapy*, 71: 521-530 (2016).

[21] a) George G. Zhanel, Phillip Chung, Heather Adam, Sheryl Zelenitsky, Andrew Denisuik, Frank Schweizer, Philippe R. S. Lagacé-Wiens, Ethan Rubinstein, Alfred S. Gin, Andrew Walkty, Daryl J. Hoban, Joseph P. Lynch 3rd, James A. Karlowsky Ceftolozane/tazobactam: a novel cephalosporin/β-lactamase inhibitor combination with activity against multidrug-resistant gram-negative bacilli. *Drugs,* 74: 31-51 (2014); b) Timothy E. Long, Justin T. Williams. Cephalosporins currently in early clinical trials for the treatment of bacterial infections; *Expert Opinion on Investigational Drugs*, 23: 1375-1387 (2014).

[22] C. Caroline Blackwell, Earl H. Freimer, Gregory C. Tuke. *In vitro* evaluation of the new oral cephalosporin cefatrizine: Comparison with other cephalosporins; *Antimicrobial Agents and Chemotherapy* 10, 288-292 (1976).

[23] James W. Wheless, Blanca Vazquez Rufinamide: A Novel Broad-Spectrum Antiepileptic Drug; *Epilepsy Currents*, 10, 1-6 (2010).

[24] a) Shahin Hakimian, Andrea Cheng-Hakimian, Gail D. Anderson, John W. Miller (2007). Rufinamide: a new anti-epileptic medication. *Expert Opinion on*

Pharmacotherapy, 8: 1931-1940; b) Martin J. Brodie, William E. Rosenfeld, Blanca Vazquez, Rajesh Sachdeo, Carlos Perdomo, Allison Mann, Santiago Arroyo. Rufinamide for the adjunctive treatment of partial seizures in adults and adolescents: a randomized placebo-controlled trial. *Epilepsia*, 50: 1899-909 (2009); c) Cecilie J. Landmark, Philip N. Patsalos. Drug interactions involving the new second- and third-generation antiepileptic drugs. *Journal Expert Review of Neurotherapeutics*, 10: 119-140 (2010).

[25] IUPAC. *Compendium of Chemical Terminology*, 2nd ed. (the "Gold Book"). Compiled by A. D. McNaught and A. Wilkinson. Blackwell Scientific Publications, Oxford (1997). Online version (2019-) created by S. J. Chalk. ISBN 0-9678550-9-8. https://doi.org/10.1351/goldbook.

[26] Wofford, D. S., Forkey, D. M., & Russell, J. G. (1982). Nitrogen-15 NMR spectroscopy: prototropic tautomerism of azoles. *The Journal of Organic Chemistry*, 47(26), 5132–5137. doi:10.1021/jo00147a018.

[27] Cox, J. R., Woodcock, I. H. H., Vincent, M. A. Tautomerism of 1,2,3- and 1,2,4-Triazoie h the Gas Phase and in Aqueous Solution: A Combined ab Initio Quantum Mechanics and Free Energy Perturbation Study. *J. Phys. Chem.* 1990, 94, 5499-5501

[28] Elguero, J., Katritzky, A. R., Denisko, O. V. Advances in heterocycle chemistry: Prototropic tautomerism of heterocycles: *Heteroaromatic tautomerism-General overview and methodology*. 1963, vol 76, 311-338.

[29] Larina, L. I. Tautomerism and Structure of Azoles: Nuclear Magnetic Resonance Spectroscopy, *Advances in Heterocyclic Chemistry,* 2018, Volume 124, 233-321

[30] a) M. Begtrup, C. J. Nielsen, L. Nygaard, S. Samdal, C. E. Sjøgren, G. O. Sorensen. The Molecular Structure and Tautomer Equilibrium of Gaseous 1,2,3-Triazole Studied by Microwave Spectroscopy, Electron Diffraction and Ab Initio Calculations, *Acta Chemica Scandinavica* A42 (1988) 500-514.

[31] a) Francisco Tom, Jose-Luis M. Abboud, Jose Laynez, Rafael Notado, Lucia Santos, Sven Ove Nilsson, Javier Cataliin, Rosa Ma. Claramunt, Jose Elguero. Tautomerism and Aromaticity in 1,2,3-Triazoles: The Case of Benzotriazole; J. Am. Chem. Soc. 1989, 1 1 1, 7348-7353; b) catalán, J.; Sánchez-Cabezudo, M.; De Paz, J. L. G., Elguero, J., Taft R. W., Anvia, F.. The tautomerism of 1,2,3-triazole, 3(5)-methylpyrazole and their cations*; J. Comput. Chem.* 1989, 10, 426.

[32] Influence of Lone Pair Repulsion vs Resonance Energy on the Francisco Tomás, Javier Catalán, Pilar Pérez, José Elguero. Relative Stabilities of Molecular Structures: A Theoretical Approach to the Equilibrium between 1H- and 2H-Benzotriazole Tautomers; *J. Org. Chem.* 1994, 59, 2799-2802.

[33] Lunazzi, L., Parisi, F., & Macciantelli, D. (1984). Conformational studies by dynamic nuclear magnetic resonance spectroscopy. Part 27. Kinetics and mechanism of annular tautomerism in isomeric triazoles. *Journal of the Chemical Society, Perkin Transactions* 2, (6), 1025. doi:10.1039/p29840001025

[34] Adrien Albert, Peter J. Taylor. The Tautomerism of 1,2,3-Triazole in Aqueous Solution; *J. Chem. Soc. Perkin Trans.* II (1989) 1903.

[35] Toppet, S., Wouters, G., & Smets, G. (1978). 1H and13C NMR study of the prototropic tautomerism of 4(5)-vinyl-1,2,3-triazole in dimethylformamide as

solvent. *Organic Magnetic Resonance*, 11(11), 578–579. doi:10.1002/mrc.1270111110.

[36] Licht, H. H., Ritter, H., Bircher, H. R., & Bigler, P. (1998). Tautomerism in nitrotriazoles: structure investigation by combined1H,13C and15N NMR spectroscopy. *Magnetic Resonance in Chemistry*, 36(5), 343–350. doi:10.1002/(sici)1097-458x(199805)36:5<343::aid-omr272>3.0.co;2-n.

[37] Nguyen, N. T., Dang, P. H., Vu, N. X. T., Le, T. H., Nguyen, M. T. T. Quinoliniumolate and 2H-1,2,3-Triazole Derivatives from the Stems of Paramignya trimera and Their α-Glucosidase Inhibitory Activities: *In Vitro* and in Silico Studies. *Journal of Natural Products* **2017**, 80, 2151-2155.

[38] Semenov, V. A., Larina, L. I., Demina, M. M. Stereochemistry and tautomerism of silicon-containing 1,2,3-triazole: ab initio and NMR study. *Structural Chemistry* **2020**, 31, 1927–1933.

[39] Abboud, J. L. M., Foces-Foces, C., Notario, R., Trifonov, R. E., Volovodenko, A. P., Ostrovskii, V. A., Elguero, J. (2001). Basicity of N-H- and N-Methyl-1,2,3-triazoles in the Gas Phase, in Solution, and in the Solid State − An Experimental and Theoretical Study. *European Journal of Organic Chemistry*, 2001(16), 3013. doi:10.1002/1099-0690(200108)2001:16<3013::aid-ejoc3013>3.0.co;2-y

[40] Goddard, R., Heinemann, O., Krüger, C. Pyrrole and a Co-crystal of 1H- and 2H-1,2,3-Triazole. *Acta Crystallographica Section C Crystal Structure Communications,* 53(12), 1846-1850. (1997).

[41] Chihiro Maeda, Shigeru Yamaguchi, Chusaku Ikeda, Hiroshi Shinokubo, Atsuhiro Osuka. Dimeric Assemblies from 1,2,3-Triazole-Appended Zn(II) Porphyrins with Control of NH-Tautomerism in 1,2,3-Triazole Organic Letters 2008, 10, 549-552.

[42] a) Thais de Brito da Silva, Kathya Ng Ka Ji, Luana da Silva Magalhães Forezi, Raíssa Maria dos Santos Galvão, Ana Flávia Martins Faria, Murilo Lamim Bello, Jackson Antônio Lamounier Camargos Resende, Vinicius Rangel Campos, Fernando de Carvalho da Silva, Robson Xavier Faria,Vitor Francisco Ferreira; Synthesis and Trypanocidal evaluation of ethyl 5-amino-1-aryl-1H-1,2,3-triazole-4-carboxylates and ethyl 5-(arylamino)-2H-1,2,3-triazole-4-carboxylates; unpublished results, 2020; b) Zi-Ping Cao, Hong-Ru Dong, Guo-Liang Shen, Heng-Shan Dong; Synthesis and Crystal Structure of (5-Methyl-3-phenyl-1H-pyrazol-1-yl)-[5-(p-tolylamino)-2H-1,2,3-triazol-4-yl]methanone. *J. Chem. Crystallogr.* 40, 1078-1081 (2010).

[43] F. Toda, K. Tanaka, J. Elguero, Z. Stein, I. Goldberg. Freezing of Equilibrium of 1,2,3-Triazole and 3(5)-Methylpyrazole by Complex Formation with 1,1-Bis(2,4-dimethylphenyl)but-2-yn-1-ol, and X-Ray Crystal Structure of the Complexes; *Chem. Lett*. 1061 (1988)

[44] P. Muthuraja, M. Sethuram, T. Shanmugavadivu, M. Dhandapani. Single crystal X-ray diffraction and Hirshfeld surface analyses of supramolecular assemblies in certain hydrogen bonded heterocyclic organic crystals; *J. Mol. Struct*. 1122, 146-156 (2016)

[45] Prabakaran, K., Khan, F. R. N., Jin, J. S. ZnO nanoparticles-mediated regioselective synthesis of methyl-*N*-alkylated 1,2,3-triazole-4-carboxylates. *Research on Chemical Intermediates*, 2012, 38, 1101-1109.

[46] Gupta, S., Chandna, N., Singh, A. K., Jain, N. Regioselective Synthesis of N^2-Alkylated-1,2,3-Triazoles and N^1-Alkylated Benzotriazoles: Cu_2S as a Recyclable Nanocatalyst for Oxidative Amination of *N,N*-Dimethylbenzylmamines. *Journal of Organic Chemitry*, 2018, 83 (6), 3226-3235.

[47] Yang, J., Duan, J., Wang, G., Zhou, H., Ma, B., Wu, C., Xiao, J. Visible-Light-Promoted Site-Selective N^1-Alkylation of Benzotriazoles with α-diazoacetates. *Organic Letters,* 2020, 22 (18), 7284-7289.

[48] Wang, J., Li, H., Zu, L. Wang, W. *Enantioselective Organocalalytic Michael Addition Reactions between N-Heterocycles and Nitroolefins*. (Organic Letters, 2006, 8 (7), 1391-1394.

[49] Begtrup, M., Larsen, P. Alkylation, acylation and silylation of azoles. *Acta Chemica Scandinavica*, 1990, 44, 1050-1057.

[50] Tanaka, Y., & Miller, S. I. (1973). Selectivities in 1,2,3-triazolide displacements of halides and additions to alkynes. *Tetrahedron*, 29(21), 3285–3296. doi:10.1016/s0040-4020(01)93481-7.

[51] Bulychev, Y. N., Preobrazhenskaya, M. N., Chernyshev, A. I., & Esipov, S. E. (1988). N-alkylation of 4-chloro-5-cyano-1,2,3-triazole with orthoformic acid derivatives. *Chemistry of Heterocyclic Compounds,* 24(7), 756–759. doi:10. 1007/bf00633170.

[52] Chen, Y. Liu, X., Petersen, J. L., Shi, X. Conformational control in the regioselective synthesis of N-2-substituted-1,2,3-triazoles. *Chem. Commun*., 2008, 3254–3256.

[53] Wu, Y., Feng, D., Gao, M., Wang, Z., Yan, P., Gu, Z., Guan, Q., Zuo, D., Bao, K., Sun, J., Wu, Y., Zhang, W. Design and synthesis of 5-aryl-4-(4-arylpiperazine-1-carbonyl)-2H-1,2,3-triazole derivatives as colchicine binding site inhibitors. *Scientific Reports* **2017**, 7, 17120.

[54] Wang, X. J., Sidhu, K., Zhang, L., Campbell, S., Haddad, N., Reeves, D. C., Krishnamurthy, D., Senanayake, C. H. Bromo-Directed N-2 Alkylation of NH-1,2,3-Triazoles: Efficient Synthesis of Poly-Substituted 1,2,3-Triazoles. *Org. Lett*., Vol. 11, No. 23, 2009, 5490-5493.

[55] Wang, X. J., Zhang, L., Krishnamurthy, D., Senanayake, C. H., Wipf, P. General Solution to the Synthesis of N-2-Substituted 1,2,3-Triazoles. *Org. Lett.,* Vol. 12, No. 20, 2010, 4632-4635.

[56] Oliva, C. G., Jagerovic, N., Goya, P., Alkorta, I., Elguero, J., Cuberes, R., Dordal, A. *N*-Substituted-1,2,3-triazoles: synthesis, characterization and evaluation as cannabinoid ligands. *ARKIVOC* **2010**, 2, 127-147.

[57] Motornov, V. A., Tabolin, A. A., Novikov, R. A., Nelyubina, Y. V., Ioffe, S. L., Smolyar, I. V., Nenajdenko, V. G. Synthesis and Regioselectivite N-2 Funcionalization of 4-Fluoro-5-aryl-1,2,3-N*H*triazoles. *European Journal of Organic Chemistry*, 2017, 46, 6851-6860.

[58] Zhang, L., Li, Z., Wang, X. J., Yee, N., Senanayake, C. H. Regioselective Synthesis of Polysubstituted N22-Alkyl/Aryl-1,2,3-Triazoles via 4-Bromo-5-5iodo-1,2,3-triazole. *Synlett,* 2012, 23(7), 1052-1056.

[59] Lian, F., Sun, C., Xu, K., Zeng, C. Electrochemical Dehydrogenative Imidation of N-Methyl_Substituted Benzylamines with Phthalimides for the Direct Synthesis of Phthalimide-Protected gem-Diamines. *Organic Letters*, 2019, 21(1), 156-159.

[60] Sakurada, I. Facile synthesis of bromo- and iodo-1,2,3-triazoles. *Tetrahedron Letters,* 2017, 58(32), 3188-3190.

[61] Shi, J., Zhu, L., Wen, J., Chen, Z. Brönsted acid catalyzed addition of N^1-*p*-methyl toluenesulfonyl triazole to olefins for the preparation of N^2-alkyl 1,2,3-triazoles with high N^2-selectivity. *Chinese Journal of Catalysis,* 2016, 37(8), 1222-1226.

[62] Zhang, C., Zheng, L., Yan, Q., Hu, Q., Jia, F., Chen, Y. A Direct P_2O_5-Mediated Synthesis of Diverse Sulfur-Containing Triazoles via Alkylation of NH-1,2,3-triazoles with Dimethyl Sulfoxide. *Chemistry Select*, 2018, 3(37), 10277-10280.

[63] Vereshchagin, L. I., Kuznetsova, N. I., Kirillova, L. P., Shcherbakov, V. V., Sukhanov, G. T., & Gareev, G. A. (1986). Reactions of 4-nitro-1,2,3-triazole with alkylating agents and compounds with activated multiple bonds. *Chemistry of Heterocyclic Compounds*, 22(7), 745–748. doi:10.1007/bf00522733.

[64] Sakovich, G. V., Sukhanov, G. T., Filippova, Y. V., Sukhanova A. G., Bosov, K. k. Alkylation of 4(5)-nitro-1,2,3-triazole with alcohols ins strongly acidc media. *Russian Chemical Bulletin*, 2013, 62(1), 111-116.

[65] Yan, W., Liao, T., Tuguldur, O., Zhong, C., Petersen, J. L., Shi, X. Mitsunobu, Reaction of 1,2,3-NH-Triazoles: A Regio- and Stereoselective Approach to Functionalized Derivatives. *Chemistry- An Asina Journal*, 2011, 6(10), 2720-2724.

[66] Calderone, V., Giorgi, I., Livi, O., Martinotti, E., Mantuano, E., Martelli, A., Nardi, A. Benzoyl and/or benzyl substituted 1,2,3-triazoles as potassium channel activators. *European Journal of Medicinal Chemistry* **2005**, 40, 521-528.

[67] Maracić, S., Kraljević, T. K., Paljetak, H. C., Perić, M., Matijašić, M., Verbanac, D., Cetina, M., Raić-Malić, S. 1,2,3-Triazole pharmacophore-based benzofused nitrogen/sulfur heterocycles with potential anti-Moraxella catarrhalis activity. *Bioorganic & Medicinal Chemistry* 2015, Volume 23, Issue 23, Pages 7448-7463.

[68] Miura, T., Yamauchi, M., Murakami, M., Facile synthesis of pyrroloindoles via a rhodium(II)-catalyzed annulation of 3-benzylidene-indolin-2-ones and α-imino carbenes. *Chem. Commun*. 2009, 1470.

[69] Page, P. C. B., Stephenson, G. R., Harveya, J., Slawinb. A. M. Z. Ruthenium-Free Preparation of 1,5-Disubstituted Triazoles by Alkylative Debenzylation of 1,4-Disubstituted Triazoles. *Synlett,* 2016, 27, 2500-2504.

[70] Kinney, W. A., McDonnell, M. E., Zhong, H. M., Liu, C., Yang, L., Ling, W., Qian, T., Chen, Y., Cai, Z., Petkanas, D., Brenneman, D. E. *ACS Medicinal Chemistry Letters*, 2016, 7(4), 424-428.

[71] Roshandel, S., Lunn, M. J., Rasul, G., Ravinson, D. S. M., Suri, S. C., Prakash, G. K. S. Catalyst-Free Regioselective N^2 Arylation of 1,2,3-Triazoles Using Diaryl Iodonium Salts. *Organic Letters*, 2019, 21(16), 6255–6258

[72] Liu, Y., Yan, W., Chen, Y., Petersen, J. L., Shi, X. Efficient Synthesis of *N*-2-Aryl-1,2,3-Triazole Fluorophores via Post-Triazole Arylation. *Organic Letters,* 2008, 10(23), 5389-5392.

[73] Lopes, A. B., Wagner, P., Kümmerle, A. E., Bihel, F., Bourguignon, J. J., Schmitt M., Miranda, L. S. M. Development of a L-Tryptophan-Based Ligant for Regioselective Copper Catalyzes N2-Arylation of 1,2,3-Triazoles. *Chemistry Select,* 2017, 2(22), 6544-6548.

[74] McClure, K. F., Jackson, M., Cameron, K., O., Kung, D. W., Perry, D. A., Orr, S. T. M., Zhang, Y., Kohrt, J., Tu, M., Gao, H., Fernando, D., Jones, R., Erasga, N., Wang, G., Polivkova, J., Jiao, W., Swartz, R., Ueno, H., Bhattacharya, S., Stock, I. A., Varma, S., Bagdasarian, V., Perez, S., Kelly-Sullivan, D., Wang, R., Kong, J., Cornelius, P., Michael, L., Lee, E., Janssen, A., Steyn, S. J., Lapham, K., Goosen, *T. Bioorganic & Medicinal Chemistry Letters*, 2013, 23(19), 5410-5414.

[75] Ueda, S., Su, M. Buchwald, S. Highly N^2-Selective Palladium-Caralyzed Arylation of 1,2,3-Triazoles. *Angewandte Chemie Internation* Edition English, 2011, 50(38), 8944-8947.

[76] Culshaw, J. D., Eden, J. M., Ford, S. J., Hayter, B., Perkins, D. R., Pike, K. G. Synthesis of 3-Substituted 2-Aminopyridines via Displacement of 3-Fluoro-2-nitropyridine. *Synlett,* 2012, 23(12), 1816-1820.

[77] Cui, M., Wang, R., Yang, Q., Kuang, C. Copper-Promoted One-Pot Sandmeyer-Type Reaction for the Synthesis of N-Aryltriazoles *The Journal of Organic Chemistry,* 2022, 87 (15), 9654-9662.

[78] Wozniak, D. R., Salfer, B., Zeller, M., Byrd, E. F. C., Piercey, D. G. Sensitive Energetics from the *N*-Amination, of 4-Nitro-1,2,3-Triazole. *Chemistry Open,* 2020, 9(8), 806-811.

[79] Lai, Q., Fei, T., Yin, P., Sheeve, J. M. 1,2,3-Triazole with linear and branched catenated nitrogen chains – The role of regiochemistry in energetic materials. *Chemical Engeneering Journal*, 2021, 410, 128148.

[80] He, C., Yin, P., Mitchell L. A., Parrish, D. A., Sheeve, J. M. Energetic aminated-azole assemblies from intramolecular and intermolecular N-H..O and N-H. .N hydrogen bonds. *Chemical Communications*, 2016, 52, 8123-8126.

[81] Chen, Y., Liu, Y., Petersen, J. L., Shi, X. Conformational control in the regioselective synthesis of N-2-substituted-1,2,3-triazoles. *Chemical Communications,* 2008, 3254-3256.

[82] Silva, T. B., Ji, K. N. K., Pauli, F. P., Galvão, R. M. S., Faria, A. F. M., Bello, M. L., Resende, J. A. L. C., Campos, V. R., Forezi, L. S. M., Silva, F. C., Faria, R. X., Ferreira, V. F. Synthesis and *in vitro* and *in silico* studies of 1*H*- and 2*H*-1,2,3-triazoles as antichagasic agents. *Bioorganic Chemistry*, 2021, 105250.

[83] Dianova, L. N., Beryozkina, T. V., Filimonov, V. O., Fan, Z., Slepukhin, P. A., Bakulev, V. A. The tandem Dimroth rearrangement and sulfonylation/acylation as regioselective method for the synthesis of 5-arylamino-2-sulfonyland 2-acyl-5-arylamino-1,2,3-triazole-4-carbothioamides. *Chemistry of Heterocycles Compounds*, 2020, 56, 1341-1347.

[84] Galaiko, N. V., Nazarov, A. V., Tolmacheva, L. A., Slepukhin, P. A., Vikharev, Y. B., Maiorova, O. A., Grishko, V. V. *Synthesis of Triterpene A-Condensed Azoles,* 2014, 50(1), 65-75.

[85] Takahashi, Y., Ikeda, H., Kanase, Y., Makino, K., Tabata, H., Oshitari, T., Inagaki, S., Otani, Y., Natsugari, H., Takahashi, H., Ohwada, T. Elucidation of the *E*-Amide Preference of *N*-Acyl Azoles. *The Journal of Organic Chemitry*, 2017, 82(21), 11370-11382.

[86] Erasmus, E., Swarts, J. C. Intramolecular communication and electrochemical observation of the 17-electron ruthenocenium cation in fluorinated ruthenocene-

containing β-diketones; polymorphism of $C_{10}H_{21}$ and $C_{10}F_{21}$ derivatives. *New Journal of Chemistry*, 2013, 37, 2862-2873.

[87] Temperini, A., Terlizzi, R., Testaferri, L., Tiecco, M. Additive-Free Chemoselective Acylation of Amines. *Synthetic Communications*, 2009, 40(2), 295-302.

[88] Motornov, V., Latyshev, G. V., Kotovshchikov, Y. N., Lukashev, N. V., Beletskaya, I. P. Copper(I)-Catalyzed Regioselective Chan-Lam *N*2-Vinylation of 1,2,3-Triazoles and Tetrazoles. *Advanced Synthesis & Catalysis,* 2019, 361(14), 3306-3311.

[89] Katritzky, A. R., Shobana, N., Pernak, J., Afridi, A. S., Fan, W. Q. Sulfonyl derivatives of benzotriazole: Part 1. A novel approach to the activation of carboxylic acids. *Tetrahedron,* 1992, 48(37), 7817-7822.

[90] Ohta, S., Kawasaki, I., Uemura, T., Yamashita, M., Yoshioka, T., & Yamaguchi, S. (1997). Alkylation and Acylation of the 1, 2, 3-Triazole Ring. *Chemical & Pharmaceutical Bulletin*, 45(7), 1140–1145.

[91] Reddy, R. J., Shankar, A., Waheed, M., Nanubolu, J. B. Metal-free, highly regioselective sulfonylation of *NH*-1,2,3-triazoles with sodium sulfinates and thiosulfonates. *Tetrahedron Letters*, 2018, 59, 2014-2017.

[92] Murakami, M., Yamauchi, M., Miura, T. Preparation of 2-Sulfonyl-1,2,3-triazoles by Base-Promoted 1,2-Rearrangement of a Sulfonyl Group. *Heterocycles,* 2010, 80(1), 177.

Chapter 6

Application of 1,2,3-Triazoles in Biosurfactants

Thorsten Heidelberg*

Chemistry Department, Faculty of Science, Universiti Malaya, Kuala Lumpur, Malaysia

Abstract

Non-ionic surfactants are indispensable industrial products with wide application spectrum. Environmental concerns and performance advantages are constantly driving the market towards utilization of renewable resources. Particularly carbohydrates have gained significant relevance. Immiscibility of the precursors for surfactant antipodes cause challenges for the synthesis of sugar-based surfactants. One possibility to address these is the use of triazole-linkers, which can be easily created by click chemistry coupling. The flexibility of the coupling approach enables a customized design of the surfactants for specific target applications. Of particular relevance is the tuning of the relative surface areas for the surfactant antipodes, which determines the geometry of surfactant assemblies.

Besides this generic structure-based optimization of surfactant properties, the triazole-based coupling provides unique opportunities for life-science applications, particularly for drug delivery. These cover bio-conjugation of surfactant assemblies, particularly vesicles, for targeted drug delivery, as well as stimulated changes of surfactant assembly geometry, giving rise to potential induced drug delivery. An example, which also may find application in environmental chemistry or diagnostic sensors, are macrocyclic ethers incorporating triazoles as chelating ligands for the coordination of metal cations.

* Corresponding Author's Email: heidelberg@um.edu.my.

In: Triazoles and Their Uses
Editor: Karl Roland
ISBN: 979-8-88697-800-1

Keywords: click-chemistry, sugar-based surfactants, renewable resources, packing theory, conjugation of surfactant assemblies

Introduction

Surfactants

Surfactant are indispensable industrial products with wide application spectrum. This ranges from cleaning products over emulsifiers, templating and surface modification reagents up to supramolecular assemblies. Moreover, the application diversity is reflected in a broad and heterogeneous distribution of application fields, catering washing, food, lubrication, oil recovery as well as personal care and even medical products. Effectively, surfactants find applications in many industrial sectors and have significant impact on our daily life.

The key feature of a surfactant is a molecular design comprising of incompatible domains, also termed as antipodes. The structural design of a typical surfactant comprises of a hydrophilic and a hydrophobic domain, as illustrated in Figure 1A. Unlike for compound mixtures, the antipodes cannot be separated macroscopically, owing to the chemical bond connecting them. The incompatible domains, therefore, separate microscopically, giving rise to micro-structured molecular assemblies, as shown in Figure 1B. These form the basis for practically all surfactant applications, catering either surface activity or assembly phenomena.

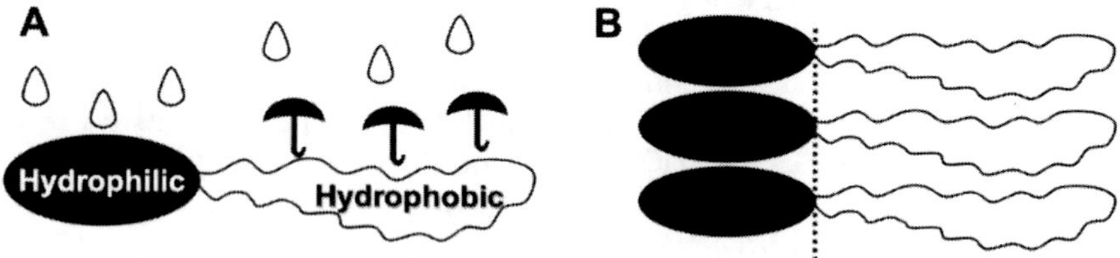

Figure 1. Molecular structure (A) and assembly of surfactants (B).

Surfactants are typically produced from oleochemical resources by attaching a hydrophilic group as antipode to the hydrocarbon-based hydrophobic domain. Depending on the nature of the hydrophilic group, surfactants are classified as either anionic, cationic or non-ionic. The presence or absence of a charge has significant impact on the surface activity and the assembly behavior of the surfactant. This reflects interactions with counterions, affecting the antipodal design of the surfactant. A simple example is given in Figure 2, where low pH can practically remove the hydrophilic antipode, thereby turning a surfactant into a non-surface-active compound.

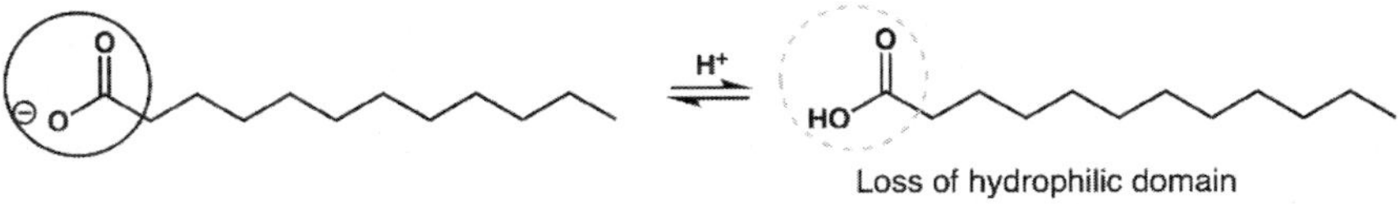

Figure 2. pH-induced loss of interfacial activity for ionic surfactants.

In view of low ion interactions, non-ionic surfactants exhibit lowest sensitivity of their surface and assembly behavior towards external perturbation, like pH variations. The corresponding high stability of their surfactant properties favors their use for particular delicate applications, like medical formulations. Unfortunately, classical non-ionic surfactants, which are typically obtained by addition of ethylene oxide oligomers to hydrocarbon precursors with a hydroxyl group, are poorly biodegradable, reflecting the ether linkages within the molecule. Biodegradability of particularly the chemical linkage between the surfactant antipodes is, however, a critical aspect for environmental concerns. This reflects the ability of surfactants to solubilize hydrophobic compounds, e.g., agricultural chemicals, and subsequently transfer them into ground water reservoirs. Therefore, so-called biosurfactants consistently gain growing industrial interest. The term 'biosurfactant' here does not reflect a surfactant of biological origin, but a surfactant obtained from renewable resources with bio-degradable chemical bond between the surfactant antipodes. Carbohydrates are particularly interesting precursors for the hydrophilic surfactant domain, owing to easy access of resources.

Sugar-Based Surfactants

Carbohydrates, also termed as 'sugars,' provide highly hydrophilic head groups without ion requirement. This reflects the large number of hydroxyl groups, which mediate effective interactions with water. Sugar-based surfactants are typically non-ionic. However, ionic charges can be added to increase the hydrophilicity of the head group. Common approaches cover the oxidation to poly-hydroxy carboxylic acids and the sulfation of hydroxyl groups. The alternative introduction of phosphoric esters, on the other hand, is avoided due to waste-water related eutrophication. Alternatively to the introduction of negative charges, the hydrophilicity of carbohydrate head groups can be increased by ethylene oxide oligomers. Although this approach generates poorly bio-degradable ether bonds, the latter do not affect the biodegradation of the surfactant properties, since the chemical bond that connects the hydrophobic domain to the carbohydrate head group is readily biodegradable.

Sugar-based surfactants differ significantly in terms of their chemical stability, reflecting the chemical connection between the surfactant antipodes. Possible linkages cover esters, amides and glycosides, as displayed in Figure 3. All these bonds are susceptible to enzymatic hydrolysis, which forms the basis for their biodegradation. Lowest in stability and production price is the ester. Sugar esters can be prepared by enzymatic syntheses using most economic resources. (Gumel et al., 2011) While they are favorable emulsifiers in food, their potential for high value products, particularly in medical formulations, is affected by the easy hydrolysis. More stable are amides, typically obtained by reductive amination of carbohydrate resources followed by acylation with biodiesel. (Hildreth, 1982) However, strong hydrogen bond-mediated intramolecular interactions of the surfactants cause application constraints. Their typical high Krafft points require use at higher temperature, (Maunier et al., 1997) thereby making he surfactant energetically unfavorable.

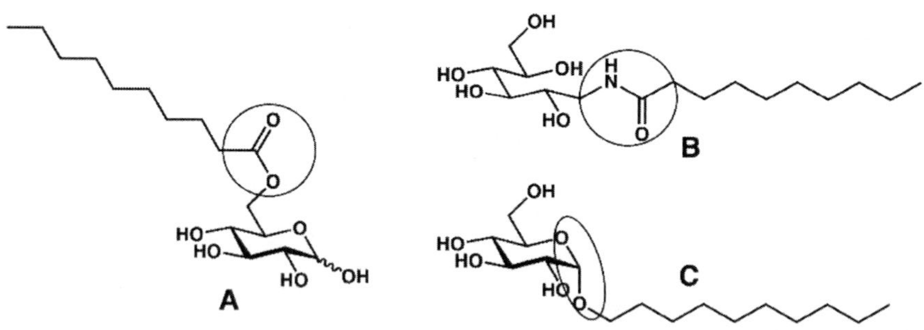

Figure 3. Glucose-based C_{10}-surfactants; A: ester, B: amide, C: glycoside.

Most promising in terms of chemical stability is a glycosidic connection of the surfactant antipodes. Glycoside-based surfactants have gained a reasonable share on the surfactant market in so-called alkyl polyglucosides (APGs). These surfactants are produced by acid catalyzed condensation of glucose, as most economic reducing carbohydrate resource, with fatty alcohols. (van Rybinski et al., 1998) However, the production requires specialized reactors and vacuum requirements due to the high boiling points of fatty alcohols. Moreover, the reaction must be performed at high temperature in presence of homogenizers to overcome the immiscibility of the surfactant precursors. The incompatible solubility profile of the surfactant antipodes, which forms the basis for the surfactant's properties, constitutes an obstacle for the synthesis of surfactants. Only very few solvents enable the formation of a homogeneous phase for effective reaction of the surfactant antipode precursors, unless protection groups are applied, (Vill et al., 1989) rendering the synthesis economically non-viable. Unlike sugar-amide-based surfactants, glycosides cannot be prepared in low molecular weight alcohols, as these solvents constitute more reactive competitors for the glycosylation. Therefore, a recent strategy for the preparation of glycosidic surfactants aims for the separation of the glycosylation and the coupling of surfactant antipodes. (Sani et al., 2012)

Discussion

Alkyl Triazole Glycosides (ATGs)

The efficiency of the copper-catalyzed Huisgen reaction (Kolb et al., 2001) provides a promising opportunity for the coupling of surfactant antipodes. The most important feature hereby is the solvent compatibility, which enables the use of low molecular weight alcohols as reaction medium. The required functional group, either terminal alkyne or azide, can be introduced to the carbohydrate surfactant precursor via glycosylation of a low molecular weight spacer, while the complementary reaction center is introduced at the precursor of the hydrophobic domain. The surfactant synthesis (Sani et al., 2012) is shown in Figure 4. Slow biodegradability of the triazole linkage between the surfactant antipodes requests for an alternative biodegradation linkage between the surfactant antipodes. This can be found in the glycosidic connection between the triazole and the carbohydrate. Initial biodegradation studies of triazole-linked surfactants indicate no environmental concern. (Oldham et al., 2013).

The triazole linkage between the surfactant antipodes enables previously unknown opportunities for an application-specific design of glycosidic surfactants. This particularly reflects the possibility to tune the interfacial surface areas of the surfactant antipodes. The resulting molecular design determines the assembly behavior of the surfactant, as previously described in the packing theory. (Israelchvili et al., 1976) The surface ratio of the surfactant antipodes defines the molecular shape of the surfactant, which is subsequently reflected in the geometry of the supramolecular assembly, as illustrated in Figure 5.

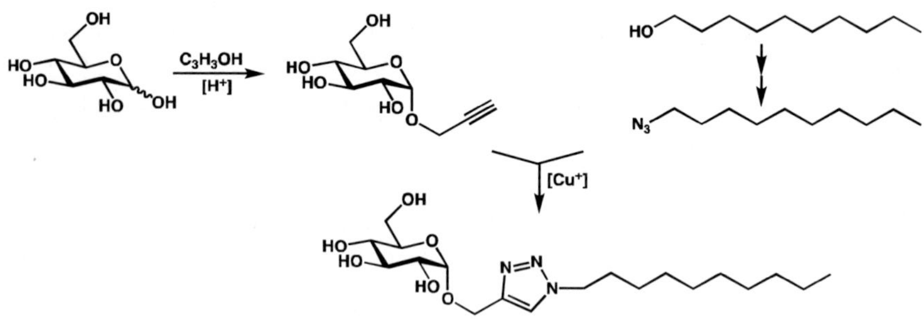

Figure 4. Synthetic approach towards ATGs.

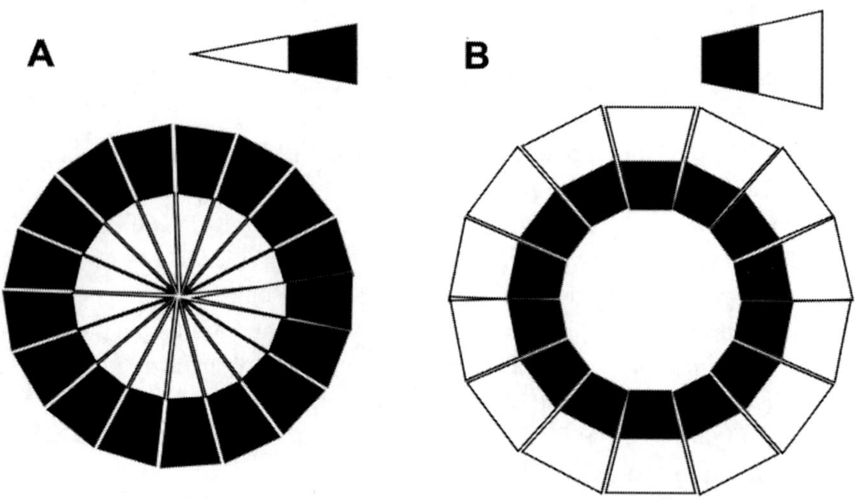

Figure 5. Surfactant shape and assembly geometry. A: conical shape with dominant hydrophilic domain for micellar assemblies B: wedge shaped surfactant with dominant hydrophobic domain for water-in oil emulsions.

The geometry of a surfactant assembly is of major importance for its application. The molecular-induced curvature of a surfactant assembly defines the preferred shape for an interface. A surfactant with preferred assembly curvature matching the microscopic interface of an application is more effective to stabilize a macroscopic 2-phase system, e.g., an emulsion, than another surfactant, which has to adopt to an unfavorable assembly curvature. Conformational flexibility of a carbon chain, as typically applied for the hydrophobic domain, ensures a surfactant's ability to adopt to different interface geometries. However, an induced chain coiling affects the enthalpy of the surfactant assembly, leading to only metastable macroscopic systems, that rearrange over time. Therefore, the tuning of the surface areas of the surfactant antipodes enables an enhanced control over the surface activity and, subsequently, the stability of an application's interface.

The coupling of surfactant antipodes by the Huisgen reaction enlarges the structural diversity of the surfactant. This reflects reduced constraints in terms of chemical reactivity for surfactant precursors on the one hand, as well as the possibility of multiple couplings on the other. As a result, the surface area of both surfactant antipodes can be tuned, leading to either convex, planar or concave assembly geometries based on an aqueous medium. All of these assembly geometries can be related to specific applications: For cleaning applications, requesting for high solubility of the surfactant in water, micellar assemblies, with significant convex curvature, ensure effective capacity for the encapsulation of oil-based dirt. Higher capacity for emulsions, however can be achieved with a planar, or lamellar surfactant assembly. This reflects larger droplet sizes with reduced surface curvature. On the other hand, a lamellar assembly of the surfactant increases the chance for macroscopic aggregates with reduced surfactant solubility. Finally, oil-based emulsions (water in oil) favor a concave surfactant assembly.

A wide range of ATGs, varying in molecular design as well as chain length of the hydrophobic domain, has been synthesized. Figure 6 provides an overview on the structures, while the surfactant properties as summarized in Table 1. In order to compare the properties of ATGs with common APGs, selected data for alkyl glucosides, as model compounds for ATGs without oligomerization of the carbohydrate, are included in Table 1 as well. In order to indicate the surfactant design in terms of packing theory matters, ATGs are classified based on their number of hydrophilic head groups (H_n) and their hydrophobic chains (T_n) connected via the triazole linker.

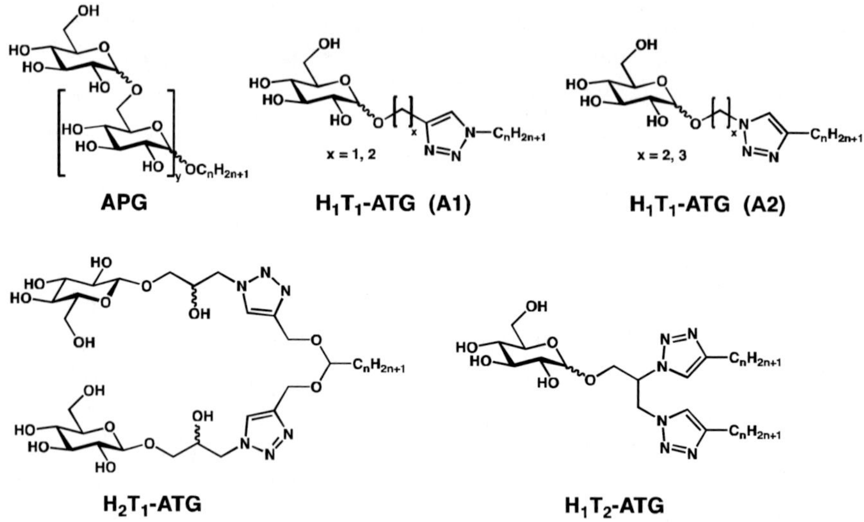

Figure 6. Structures of ATGs and APG for correlation with Table 1.

Table 1. Surfactant properties of ATGs and APG analogs

	Sugar	Hydrophobic			T_K	CMC	γ_{CMC}	a
Type	**Head**	**Domain**	**Chain**		**[°C]**	**[mmol L^{-1}]**	**[mN m^{-1}]**	**[Å^2]**
APG [1]	β-Glc	straight C_8	C_8	y=0		25	30	42
H_1T_1-ATG, (A1)[2]	β-Glc	Guerbet C_8	C_8	x=1	<10	8.0		
H_1T_1-ATG, (A1)[2]	α/β-Glc	Guerbet C_8	C_8	x=1	<10	8.0		
H_1T_1-ATG, (A1)[2]	β-Glc	straight C_8	C_8	x=1	<10	1.3	29	
H_1T_1-ATG, (A1)[2]	α-Glc	straight C_8	C_8	x=1	<10	1.1	29	
H_1T_1-ATG, (A1)[2]	α/β-Glc	straight C_8	C_8	x=1	<10	1.2		
H_1T_1-ATG, (A1)[3]	α-Glc	straight C_8	C_8	x=2	<10	2.8	29	65
H_1T_1-ATG, (A2)[3]	α-Glc	straight C_8	C_8	x=2	<10	3.6	29	41
H_1T_1-ATG, (A2)[3]	α-Glc	straight C_8	C_8	x=3	<10	1.2	29	40
H_1T_1-ATG, (A1)[3]	β-Glc	straight C_9	C_9	x=1	<10	1.9	29	34
APG [1]	β-Glc	straight C_{10}	C_{10}	y=0	<10	2.2	28	47
H_1T_1-ATG, (A1)[2]	β-Glc	straight C_{10}	C_{10}	x=1	<10	0.77	29	
H_1T_1-ATG, (A1)[2]	α/β-Glc	straight C_{10}	C_{10}	x=1	<10	0.82		
APG [1]	β-Glc	straight C_{12}	C_{12}	y=0		0.19	28	36
H_1T_1-ATG, (A1)[2]	α-Glc	straight C_{12}	C_{12}	x=1	<10	0.14		
H_1T_1-ATG, (A1)[2]	β-Glc	straight C_{12}	C_{12}	x=1	<10	0.11	29	
H_1T_1-ATG, (A1)[2]	α/β–Glc	straight C_{12}	C_{12}	x=1	<10	0.16		
H_1T_1-ATG, (A1)[2]	β-Glc	Guerbet C_{12}	C_{12}	x=1	<10	0.58		
H_1T_1-ATG, (A1)[2]	α/β-Glc	Guerbet C_{12}	C_{12}	x=1	<10	0.50		
H_1T_1-ATG, (A1)[4]	β-Glc	straight C_{11}	C_{12}	x=1	<10	0.57	34.7	53
H_1T_1-ATG, (A1)[2]	β-Glc	straight C_{14}	C_{14}	x=1	42	0.065	29	

	Sugar	Hydrophobic			T_K	CMC	γ_{CMC}	a
Type	**Head**	**Domain**	**Chain**		**[°C]**	**[mmol L^{-1}]**	**[mN m^{-1}]**	**[$Å^2$]**
H_2T_1-ATG [4]	β-Glc	straight C_{12}	C_{14-15}			0.15	39.8	53
H_1T_1-ATG, (A1) [2]	β-Glc	straight C_{16}	C_{16}	x=1	53	0.0045	29	
H_1T_2-ATG [5]	β-Glc	Straight C_8	C_{18-19}		52-53	0.0010	25.6	49
H_1T_1-ATG (A1) [2]	β-Glc	straight C_{18}	C_{18}	x=1	64			

[1] (Kozo et al., 1961), [2] (Sani et al., 2012), [3] (Ng et al., 2017), [4] (Ali et al., 2014), [5] (Ng et al., 2021).

ATGs with moderate convex assembly geometry can be obtained by coupling of mono-functional alkyl glycosides with simple monofunctional hydrocarbon precursors. (Sani et al., 2012) These surfactants can adopt to both planar and convex interface and match the assembly behavior of APGs. The molecular design can apply either glycosides with azide- or alkyne-functionality. (Ng et al., 2017) The choice of the functional group does not affect the surfactant properties significantly. However, in view of a minimization of the linker between the two surfactant domains the alkyne is best attached to the sugar. Propargyl glycosides, therefore, represent the most practical surfactant precursors. Owing to failed crystallization attempts, anomeric mixtures with α-dominance, as obtained from Fischer glycosylation of propargyl alcohol, are most economic. In terms of the surfactant properties, the anomeric configuration of an ATG is practically irrelevant. (Sani et al., 2012) Moreover, diastereomeric mixtures are common for APGs as well.

Increasing chain length of the hydrophobic domain reduces the critical micelle concentration (CMC) of ATGs. (Sani et al., 2012) This is a common feature of surfactants. However, the decrease does not match the pattern of APGs, which follow the common surfactant trend of a decrease by one magnitude upon increase of two carbon atoms. (Tanford, 1980) In general, simple straight chain ATGs exhibit lower CMCs compared to the corresponding APGs. This reflects a hydrophobic effect of the triazole linker. However, the effect reduces upon increasing chain length of the hydrophobic domain. This feature may be explained by the different nature of interactions of the triazole and the hydrophobic domain. Particularly domain crystallization effects of increasing hydrocarbon domains are not enhanced by the triazole linker.

Unusual Surfactant Designs Based on ATGs

While the hydrophobic effect of the triazole linker disfavors applications for simple straight chain surfactants, ATGs gain application potential upon use of branched hydrocarbon domains. (Sani et al., 2012) This reflects synthetic challenges for corresponding APG analogs, owing to dominant eliminations under Fisher glycosylation conditions. The branched hydrophobic domain moves the surfactant design from a convex assembly towards a more balanced planar geometry. Potential applications cover the already indicated emulsion of larger amounts for both, oil-in-water and water-in-oil. Moreover, a lamellar assembly behavior provides potental for life science applications, particularly vesicular drug delivery systems.

A similar design can be obtained by a double click strategy, leading to a biantennary ATG comprising two triazole linkers. (Ng et al., 2021) The synthetic approach is displayed in Figure 7. Owing to the larger surface area of the triazoles, compared to a hydrocarbon chain, the surfactant assemblies are moving towards a convex geometry. The application, hence, focuses on water-in oil applications at preferably lower water loading. The presence of the triazoles enhances interactions of the surfactant with aromatic oil contents. (Ng et al., 2021) This feature can enhance the stability of an emulsion, thereby providing beneficial contribution of the triazole to the surfactant performance. On the other hand, close proximity of triazoles in a surfactant gives rise to a chelating coordination effect, (You et al., 2014) which can affect the removal of the copper catalyst from the surfactant after the Huisgen coppling. (Ng et al., 2021).

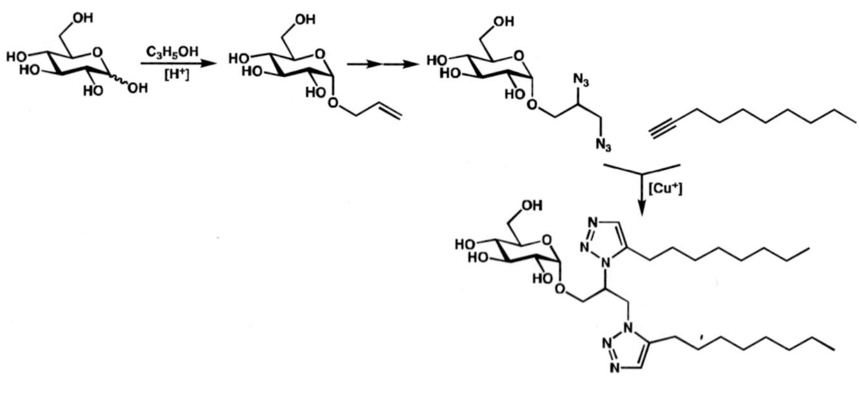

Figure 7. Synthesis of biantennary APG by double click coupling.

A different surfactant design, aiming for increased water solubility, applied a double click coupling with two hydrophilic domains. (Ali et al., 2014) The parallel alignment of two sugars substantially increases the surface area of the hydrophilic domain, leading to exclusively micellar assemblies with potentially smaller size compared to the classical single straight chain ATG. The synthetic scheme is shown in Figure 8. Unlike in previous examples, the carbohydrate was not propargylated but carried an azide. The reason for the reverse functional group alignment was the more convenient synthesis for the hydrophobic domain precursor. A potential disadvantage of the azido-alkyl glucoside is remaining surfactant activity after the biodegradation of the glycoside. This reflects the two hydroxyl groups at the linker. However, the acetalic linkage between the hydrophobic domain and the triazole linker prevents remaining surfactant activity after waste water treatment. In fact, the acetal at the hydrocarbon domain breaks more easily than the glycoside.

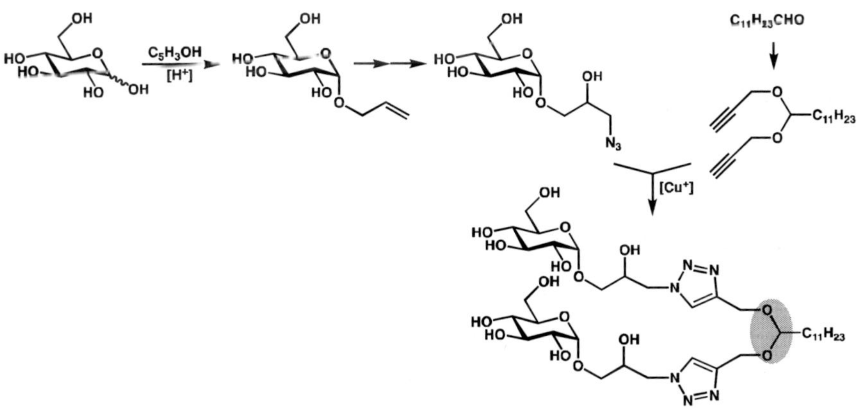

Figure 8. Synthesis of double headed ATG for cleaning applications.

Application Potential of 1,2,3-Triazoles for Vesicular Drug Delivery

The previously indicated chelating effect of a bis-triazole may be considered for a cation-stimulated rearrangement of a surfactant assembly, based on repulsive ionic interactions. Interesting is such a feature, if the surfactant assembly refers to a drug delivery system. A rearrangement of the assembly geometry may then lead to the release of the encapsulated drug. Practical surfactant assemblies for drug delivery purpose are particularly vesicles.

These resemble miniature cell-like assemblies, which are formed by a spherical surfactant bilayer. The conversion of a vesicular assembly into micelles reflects a release of an encapsulated compound, owing to the dynamic nature of micelle formation and cleavage, whereas vesicles exhibit a significantly higher assembly stability.

However, the coordination efficiency of a simple bis-triazole is not expected to cause sufficient charges to destabilize a vesicular surfactant assembly. On the other hand, the critical assembly concentration (CAC) of a sugar-based surfactant with anellated crown ether increases significanty in presence of suitable counter ions, (Sabah et al., 2011) thereby providing a potential cation-stimulated release. Unfortunately, the synthesis of crown ethers is tedious. On the other hand, the copper catalyzed azide alkyne cycloaddition (CuAAC) provides an easy possibility to convert a simple alkyl glycoside derivative into a crown-ether anellated system. (Salman et al., 2015) The reaction sequence in shown in Figure 9. Although the increase of the CAC upon cation complexation is so far insufficient for an effective release mechanism, the concept may be further developed to incorporate elements of a coordination induced increase of the surface area for the hydrophilic domain. A resulting change in assembly geometry from a lamellar phase to micelles may enable a more effective stimulated release.

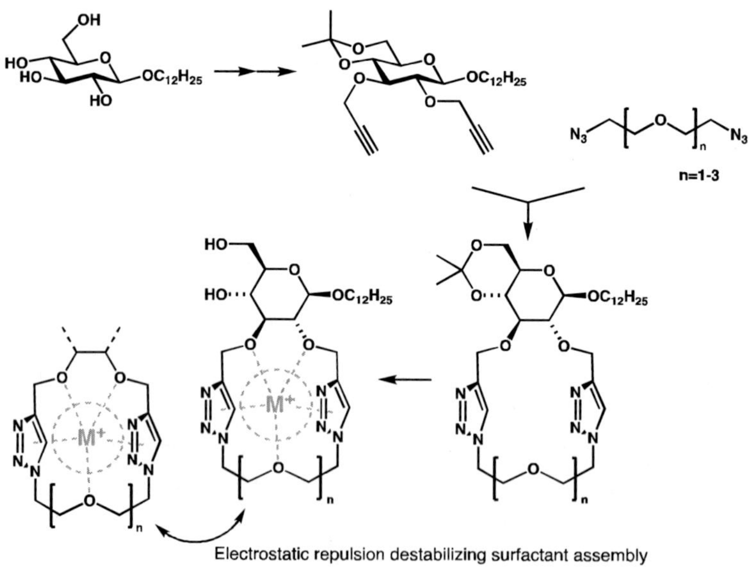

Figure 9. Synthesis of macrocyclic ligands via CuAAC for potential cation triggered drug delivery.

The efficiency of the Huisgen coupling enables powerful bioconjugations, leading to advanced materials for life science applications, particularly in medicinal context. (Kolb et al., 2001) Its application on sugar-based surfactants enables a potential modular targeted drug delivery system. The basis of such an approach can be a functionalized biosurfactant, which enables a biofunctionalization via CuAAC, as shown in Figure 10. (Goh et al., 2019) The biantennary sugar-based surfactant ensures strong intermolecular binding with a base surfactant, reflecting a simple biantennary sugar-based surfactant. Unlike for ionic surfactants, the cohesive interactions are not limited to the hydrophobic domain, but include hydrogen bonding between surfactant head groups as well.

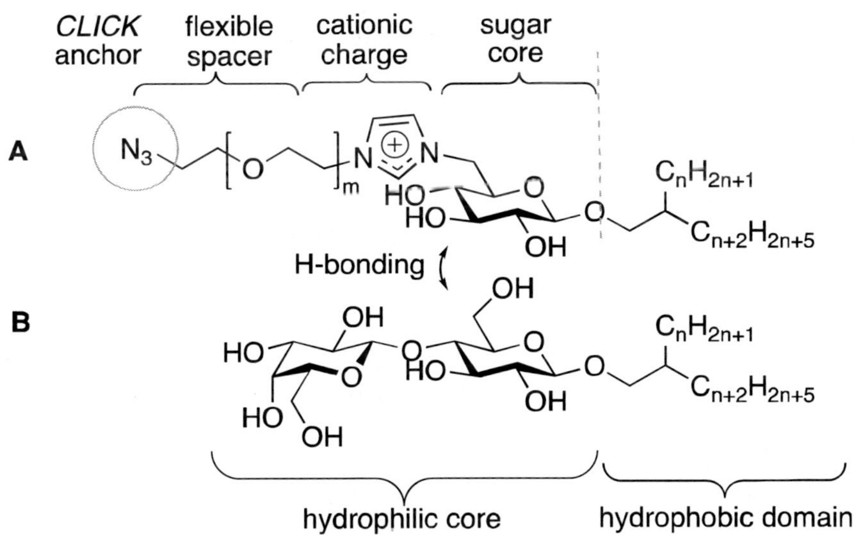

Figure 10. Functionalized sugar-based surfactant (A) for biofunctionalization of vesicles with base surfactant (B).

The distribution of the bio-anchor on a vesicular assembly is random, owing to repulsive interactions. Dynamic light scattering experiments with complementary zeta potential measurements indicate homogeneously small sized vesicles with narrow surface charge distribution. (Goh et al., 2019) A drug loaded vesicular carrier could subsequently selectively biofunctionalized at the exterior membrane, thereby avoiding loss of receptor antigen, as visualized in Figure 11. A water flexible spacer ensures the accessibility of the azide anchor on the surfactant for the biofunctionalization. Upon coupling with a propargyl functionalized antigen, the vesicle turns into a targeted drug delivery system. Application of different receptor antigens enable a modular system as flexible response for different target cells, which will be killed upon the delivery of a 'generic' cell-toxic. Unwanted side effects will be minimized based on the encapsulation of the drug and its receptor specific delivery.

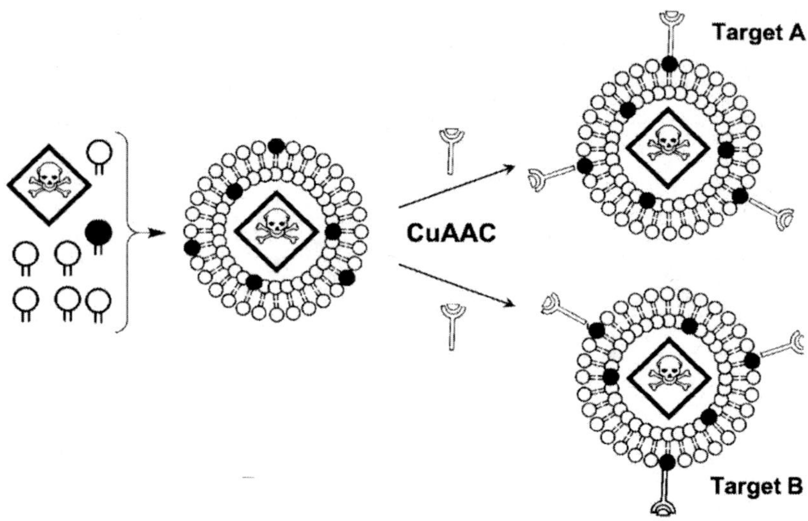

Figure 11. Modular vesicular drug delivery system.

Conclusion

A 1,2,3-triazole linkage between surfactant antipodes enables a valuable tuning of the assembly behavior in sugar-based surfactants. While simple analogs of APGs suffer from a hydrophobic participation of the triazole linker without providing effective advantages, the click coupling approach is particularly interesting to create surfactants with either multiple hydrophobic chains or parallel arranged hydrophilic head groups. The resulting surfactants are optimized for an explicit assembly geometry and should be considered as application-specific rather than general surfactants. This, however, does not limit the application potential to only one application field. Since it resembles the assembly type, various fields may apply. In general, the triazole is solely considered as linking group. However, it can contribute actively to an emulsifying performance via π-stagging interaction with aromatic oil contents. Multiple triazole linkages can lead to chelating behavior with potential negative implications on the removal of the catalyst from the surfactant.

Besides providing an effective connection strategy for surfactant antipodes, a triazole on a surfactant may enable potential drug delivery applications. For this, the triazole can either mediate an effective bioconjugation of a surfactant assembly for targeted drug delivery, or enable an effective integration of a macrocyclic ligand into a sugar-based surfactant. The latter provides potential for the development of a cation stimulated release mechanism in drug delivery.

Disclaimer

None

References

Ali, T. H., Hussen, R. S. D., Heidelberg, T. New Y-shaped surfactants from renewable resources, *Colloid Surf. B* (2014) 123:981-985.

Goh, E.W., Heidelberg, T., Hussen, R. S. D., Salman, A. A. Imidazolium-linked azido-functionalized Guerbet glycosides: Multifunctional surfactants for biofunctionalization of vesicles, *ASC Omega* 4 (2019) 4: 17039-17047.

Gumel, A. M., Annuar, M. S. M., Heidelberg, T., Chisti, Y. Lipase mediated synthesis of fatty acid esters, *Proc. Biochem.* (2011) 46:2079-2090.

Hildreth, J. E., N-D-Gluco-N-methylalkanamide compounds, a new class of non-ionic detergents for membrane biochemistry. *Biochem. J.* (1982) 207:363–366.

Israelachvili, J. N., Mitchell, B. D.; Ninham, J. W. Theory of self-assembly of hydrocarbon amphiphiles into micelles and bilayers. *J. Chem. Soc., Faraday Trans. 2* (1976) 1525–1568.

Kolb, H. C., Finn, M. G., Sharpless, K. B. Click chemistry: diverse chemical function from a few good reactions. *Angew. Chem., Int. Ed.* (2001) 40:2004−2021.

Kozo, S., Tokio, Y., Ryohei, H. The surface tension and the critical micelle concentration in aqueous solution of β-D-alkyl glucosides and their mixtures, *Bull. Chem. Soc. Jp.* (1961) 34:237-241.

Maunier, V., Boullanger, P., Lafont, D., Chevalier, Y. Synthesis and surface-active properties of amphiphilic 6-aminocarbonyl derivatives of D-glucose. *Carbohydr. Res.* (1997) 299:49–57.

Ng, S. H., Heidelberg, T., Salman, A.A. Spacer effect on triazole-linked sugar-based surfactants, *J. Dispers. Sci. Technol.* (2017) 38:105-109.

Ng, S. H., Mazlee, M. T. F., Heidelberg, T. Biantennary alkyl triazole glycosides by double-click-coupling for water-in-oil-emulsi!cation, *J. Surfact. Deterg.* (2021) 24:473-482.

Oldham, E. D., Seelam, S., Lema, C., Aguilera, R. J., Fiegel, J., Rankin, S. E., Knutson, B. L., Lehmler, H.-J. Synthesis, surface properties, and biocompatibility of 1,2,3-triazole-containing alkyl β-D-xylopyranoside surfactants, *Carbohydr. Res.* (2013) 379:68-77.

Sabah, K., Heidelberg T., Hashim, R. Novel crown ethers on glucose based glycolipids, *Carbohydr. Res.* (2011) 346:891-896.

Salman, A. A., Tabandeh, M., Heidelberg, T., Hussen, R. S. D. Functional glycolipid-crown-ethers by click chemistry, *Carbohydr. Res.* (2015) 406:41-45.

Sani, F. A., Heidelberg, T., Hashim, R., Farhanullah. Alkyl triazole glycosides (ATGs)—A new class of bio-related surfactants, *Colloid Surf. B* (2012) 97:196-200.

Tanford, C. The hydrophobic effect: Formation of micelles and biological membranes, Somerset/Wiley, New York, 1980.

van Rybinski, W., Hill, K. Alkyl Polyglycosides-Properties and Applications of a new Class of Surfactants, *Angew. Chem. Int. Ed.* (1998) 37:1328–1345.

Vill, V., Boecker, T., Thiem, J., Fischer, F. Studies on liquid-crystalline glycosides, *Liq. Cryst.* (1989) 6:349-356.

You, X., Wei, Z. A multidentae ligand based on two triazole groups from 'click chemistry' and its copper(II) and iron(II) complexes: Synthesis, structure, and electrochemistry. *Transit. Met. Chem.* (2014) 39:675–680.

Chapter 7

Triazoles as Potential Corrosion Inhibitors: Recent Developments and Future Perspectives

Sowmya Ramkumar[1]
D. Nalini[1,*]
D. S. Chauhan[2]
and M. A. Quraishi[3]

[1]Department of Chemistry, PSGR Krishnammal College for Women, Coimbatore, Tamil Nadu, India
[2]Modern National Chemicals, Second Industrial City, Dammam, Saudi Arabia
[3]Interdisciplinary Research Center for Advanced Materials, King Fahd University of Petroleum and Minerals, Dhahran, Saudi Arabia

Abstract

Industries are continuously searching for solutions to reduce the losses caused by the massive issue of corrosion, which consumes company earnings. Prevention is the best defense against corrosion. Customized solutions using various corrosion inhibitors are one of the best methods of corrosion prevention. Organic compounds rich in heteroatoms and electron-rich cores, serve as barriers that prevent corrosion by forming an adsorbed layer. Triazoles stand out as a significant five-member nitrogen heterocycle among nitrogen-containing heterocyclic compounds with outstanding uses. Five-membered triazole come in two different structural varieties: 1,2,3-triazole and 1,2,4-triazole. Both 1,2,3- and 1,2,4-triazoles may support a wide range of substituents (electrophiles and nucleophiles) around the core structures due to their

* Corresponding Autor's Email: nalini@psgrkcw.ac.in

In: Triazoles and Their Uses
Editor: Karl Roland
ISBN: 979-8-88697-800-1

structural features, which allows for the development of numerous unique compounds. The excellent corrosion protection action of triazole derivatives can be attributed to the presence of electron-withdrawing and electron-donating substituents that allow an improved protection performance compared to that of the parent triazole molecule. Computational studies have provided evidence that the triazole- based molecules undergo donor-acceptor type interactions with metallic substrates. Triazoles are used in a variety of commercial applications, including physiologically active compounds, agrochemicals, corrosion inhibitors, dyes, and optical brighteners. This paper deals with the review on recent developments on triazole as corrosion inhibitors and their application as corrosion inhibitors and their recent trends in the field of corrosion of different metals in a variety of corrosive environments.

Keywords: heterocycles, triazoles, corrosion inhibitors

Introduction

Strategic Impact of Corrosion Damage

The great expanded industrial applications and importance of metals based on their properties have been recognized by many investigators. However, despite its importance, the problem associated with corrosion, still remains a cause of concern to all. Huge losses of natural and artificial resources have been noted annually all over the globe as a result of corrosion (Hackerman 1993; Bineet Kumar Singh, 2010). In the petroleum and gas industries, more than half of the reported failures of pipelines are caused by failure due to corrosion and subsequent rupture of the pipe wall, resulting in large losses of products, environmental pollution and ecological disaster. The three main reasons for the importance of corrosion are: economics, safety and conservation. To reduce the economic impact of corrosion, corrosion engineers, with the support of corrosion scientists, aim to reduce material losses, as well as the accompanying economic losses, that result from the corrosion of piping, tanks, metal components of machines, ships, bridges, marine structures, and so on. Loss due to corrosion is a waste not only of the metal, but also of the energy, the water, and the human effort that was used to produce and fabricate the metal structures (Parameswari 2010). In addition, rebuilding corroded equipment requires further investment of all these resources - metal, energy, water, and human.

"The cost of corrosion works out too much higher than any of the calamities the nation has faced over the years" – NACE International.

The economic consequences of corrosion affect technology. A great deal of the development of new technology is held back by corrosion problems. To highlight the damage of corrosion to the environment, impact on people and waste of resources, *April 22nd* is observed as the *International Corrosion Awareness Day builds upon Earth Day.*

Owing to the need of the hour, numerous researchers work in the field corrosion science and engineering as evidenced from the tremendous increase in the number of publications related to the field of corrosion. In many industries, the need to use materials safely is a primary consideration. Physical requirements can be satisfied easily, but corrosion effects seriously complicate the selection of suitable materials. Generally, the costs and problems associated with corrosion resistant materials have made the use of corrosion inhibitors as a practical and economic alternative.

Corrosion Inhibitors

A corrosion inhibitor is a chemical that, when introduced to an environment in small concentrations, effectively slows the pace at which a metal exposed to that environment corrodes (Fouad, 1999; Sastri, 2001; Yu, 2010; Sowmya, 2019; Sounthari, 2013; Nalini, 2014; Raja, 2016). The usage of corrosion inhibitors depends on the mechanism used and the environment in which they are placed. Studies have shown that the efficiency of inhibition can be qualitatively related to the amount of adsorbed inhibitor on the metal surface. The adsorption of inhibitors is governed by the residual charge on the surface of the metal and by the nature and chemical structure of the inhibitor. Inhibitors contain a polar group which are the functional group based on nitrogen, sulfur, or oxygen that has a pair of unshared electrons which are available to bond with metal surfaces. They lay down an organic film on the entire metal surface forming a protective layer and thereby isolating the metal from the corrosive environment. Thus, such organic moieties have the property of retarding either the anodic or cathodic or both the corrosion reaction and can take up different roles in the inhibition mechanism (Merimi, 2022; Ansari, 2016; Nalini, 2013; Sowmya, 2015).

Triazoles as Corrosion Inhibitors

Pickling is a metal surface treatment used to remove impurities, such as stains, inorganic contaminants, rust or scale from ferrous metals, copper, precious metals and aluminum alloys. A solution called pickle liquor, which contains strong acids, is used to remove the surface impurities. It is commonly used to descale or clean steel in various steelmaking processes. A wide range of applications are seen for the acid inhibitors in the industrial field as a component in pre-treatment composition and in cleaning solutions for industrial equipment and also in acidification of oil wells.

Triazoles stand out among the heterocyclic nitrogen-containing molecules for their exceptional pharmacological uses. Five-membered triazoles come in two different structural varieties: 1,2,3-triazole and 1,2,4-triazole. Both 1,2,3- and 1,2,4-triazoles potentially support a wide range of substituents (electrophiles and nucleophiles) around the core structures due to their structural properties, making it possible for the development of numerous novel active molecules. The therapeutic properties of triazoles and their derivatives include antibacterial, antiviral, antitubercular, anticancer, anticonvulsant, analgesic, antioxidant, anti-inflammatory, and antidepressant actions. Because of this bioactive nature of triazoles, they can be considered as a better environmentally friendly inhibitor system for metal corrosion. Triazole derivatives which were initially explored as biologically active molecules and useful pharmaceutical substances, have now been recognized as corrosion inhibitors for steels, copper, aluminium, and their alloys in corrosive conditions. Numerous researchers (Yu. I. Kuznetsov, 2021; Zarrouk, 2012; Khiati, 2011; Al Maofari, 2014; Chebabe, 2014; Rajkumar, 2014; Rao, 2014; Kumar, 2012; El Attar, 2015; Pour-Ali, 2016; Hrimla, 2021) have been working on the application of may triazole systems as corrosion inhibitors for varied metals in different environment. Various metehos and methodologies are employed by the reserahcers widely for finding the efficacy of the considered triazoles system.

Triazoles have the core structure of a five-membered heterocyclic ring with the chemical formula $C_2H_3N_3$ containing two carbon and three nitrogen atoms. Two significant isomers (Figure 1), 1,2,3-triazole (v-triazole) and 1,2,4-triazole(s-triazole), were formed in the five-membered ring when nitrogen atoms were arranged in a maximum of two different positions. Both triazoles' are planar with all the atoms in sp^2 hybridization. Both forms have six pi (Π) electrons accessible, which are delocalized around the ring to give them their aromatic properties. Furthermore, triazoles are energy-rich heterocycles due to the presence of 3 N atoms.

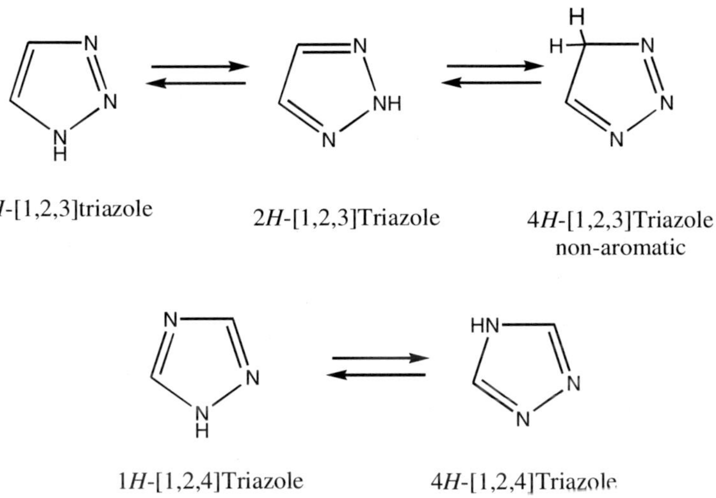

Figure 1. Isomeric Structures of Triazoles.

Literature Survey on Triazoles as Corrosion Inhibitors

Triazole compounds are well-known for their biological effects in addition to their ability to inhibit corrosion on mild steel. Due to their chemical stability against basic and acid hydrolysis, as well as in oxidative and reductive conditions, triazoles are best suited for corrosion inhibtion applications.

Triazole structures are also comparatively resistant to deterioration. Triazole derivatives, which are aromatic, electron-rich systems, can bind to metal surface by weak interactions such cation-, anion-, coordination-, hydrogen-, van der Waals-, or hydrophobic effect. As a result, they exhibit a variety of corrosion inhibitory properties. A collective survey on the literatures of past studies on triazoles (Dhaif, 2019; Sampath, 2020; Negrón Silva, 2014; González Olvera, 2016; Zhang, 2010; Elkhotfi, 2017; as corrosion inhibitors and their study practise helps in the medley of a new methodology in the outset of the research from which new areas of research can be inferred. There are numerous parameters determining the selection of an effective acid inhibitor and study methodology like, effect of structure of the organic inhibitor, substituent effect on the performance of the inhibitor, effect of concentration, time and temperature on the inhibition performance, Corrosion environment and metal employed, latest advancements in corrosion studies, new methodologies, etc. The ability of many triazole derivatives to inhibit corrosion has been studied, and research has shown that triazole derivatives are effective inhibitors for varied metals under different conditions. Moreover, triazole derivatives have been regarded as environmentally acceptable substances among the various nitrogen-containing species tested as anti-corrosion agents. In this review, the use of substituted triazoles as corrosion inhibitors for steel and other metals is investigated by evaluating different triazole derivatives with respect to their capacity to protect against damage in various acidic conditions and at varied doses of triazole inhibitors, as determined by electrochemical measurements.

1,2,3- Tiazoles as Acid Corrosion Inhibitors

By using a wide range of approaches, 1,2,3-triazole derivatives have been investigated for their ability to suppress metal corrosion in a variety of environments. Investigation of the protective coating and explanation of the mechanistic features of corrosion inhibition have been carried out using electrochemical techniques in conjunction with surface examination methods. The 1,2,3-triazole derivative was principally studied for corrosion inhibition in hydrochloric acid, sulphuric acid, phosphoric acid, and nitric acid, which provided a good simulation of actual industrial settings where metals are exposed to various acids.

This section focus is on previously published studies that have examined the use of substituted 1,2,3-triazoles as corrosion inhibitors. The following discussion covers the types of metals, corrosive media, corrosion inhibitor dosages, and their corresponding values of inhibition efficiency.

1,2,3-triazole-5-carboxylate (MHTC & OxTDC), subsiuted 1,2,3-triazole methanol, 1,4-disubstituted 1,2,3-triazole, pyrimidine dione based 1,2,3 – triazoles were studied for their corrosion resistance property against acid corrosion of mildsteel and the results of which are tabulated in Table 1. The results of Table 1 are discussed as follows.

According to an investigation by Rouifi et al., 2019 new 4-amino-1-((8-hydroxyquinolin-5-yl)methyl)-1,2,3-triazole-5-carboxylate (MHTC) exhibits corrosion inhibition and adsorption behaviour for carbon steel exposed to 1 M HCl. To examine the corrosion performance capability of MHTC, Tafel curves, electrochemical impendence spectroscopy (EIS) measurements, and mass loss (ML) estimates were made. The inhibitory efficacy of MHTC improves with concentration; at 1mM, it provides the highest level of 91% inhibition efficiency. According to polarisation research, MHTC functions as a mixed corrosion inhibitor. The Langmuir isotherm was followed during MHTC's adsorption on the surface of CS. The surface of the carbon steel was examined using SEM and UV-visible methods. the DFT approach was used to do a quantum chemical analysis of MHTC's reactivity. Using Monte Carlo (MC) simulations, the adsorption model of the MHTC molecule on the CS surface was examined. The results showed that the MHTC adsorbs on the Fe(110) surface in a flat orientation.

Rahmani et al., 2019 investigated the potential of 4,5-diethyl 1-[(4-ethyl-2-phenyl-4,5-dihydro-1,3-oxazol-4-yl)methyl], a new triazole derivative. 4,5-dihydro-1H-1,2,3-triazole-4,5-dicarboxylate (OxTDC) was examined using gravimetric and electrochemical methods as an anticorrosion agent for mild steel (MS) in a 1 mol/L HCl solution. To identify the ideal concentration and temperature range for the optimum corrosion-inhibiting action, the OxTDC was tested at a range of concentrations (0.001-1 mM) and four temperatures (308, 318, 328, and 338 K). At 298 K, the synthetic inhibitor's 94.8% inhibition efficiency was visible. At 1 mM OxTDC, the corrosion current density dropped from 1071.9 A cm2 (blank) to 67.1 A cm2. This derivative functions as a mixed-type inhibitor, and in an acidic environment, the adsorption mode corresponds to the Langmuir isotherm. Additionally, the outcomes of molecular simulations (DFT combined with molecular dynamics) demonstrated the adsorption of the OxTDC compound on the surface of the iron. The electronic interactions of OxTDC molecules with metal atoms were made clearer by quantum mechanics calculations.

As corrosion inhibitors for mild steel in 1M HCl, Qi Ma et al., 2017 synthesised and researched certain 1,2,3 triazoles, including (1-(pyridin-4-ylmethyl)-1H-1,2,3-triazole-4-yl)methanol (PTM) and (1-(benzyl-1H-1,2,3-triazole-4-yl)methanol (BTM). In the concentration range of 0.21 mM, it is discovered that PTM performs as a more potent inhibitor than BTM. The triazole derivatives can adsorb on the mild steel surface by either taking electrons from iron surfaces or sharing the lone pair electrons of N atoms with iron atoms, according to computational chemistry studies. The pyridine segment should be in charge of PTM's increased inhibitory efficiency because of its robust contact with the mild steel surface in aqueous systems.

Table 1. Assesment on the corrosion inhibitor studies on some 1,2,3,- triazole

S No	Name of the Inhibitor System	Chemical Strcture	Metal / Medium	Reference
1	4-amino-1-((8-hydroxyquinolin-5-yl)methyl)-1,2,3-triazole-5-carboxylate (MHTC)	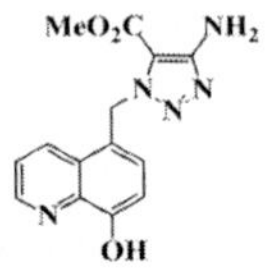 (MHTC)	Mild steel / 1M HCl	Rouifi, F. Benhiba et al., 2019
2	4,5-dihydro-1H-1,2,3-triazole-4,5-dicarboxylate (OxTDC)	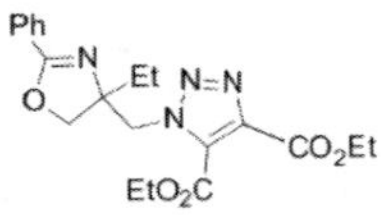	Mild steel / 1 M HCl	Rahmani et al., 2019
3	(1-(pyridin-4-ylmethyl)-1H-1,2,3-triazole-4-yl)methanol (PTM)	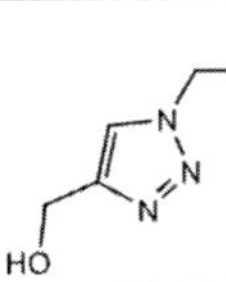 PTM	Mild steel / 1 M HCl	Qi Ma et al., 2017
4	(1-(benzyl-1H-1,2,3-triazole-4-yl)methanol (BTM)	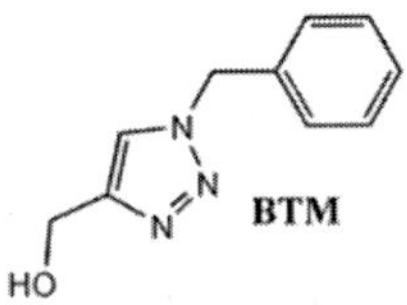 BTM	Mild steel / 1 M HCl	Qi Ma et al., 2017
5	1,3-Bis((1-benzyl-1H -1,2,3-triazol-4-yl)methyl) pyrimidine-2,4(1H,3H)-dione (BBTMPD),	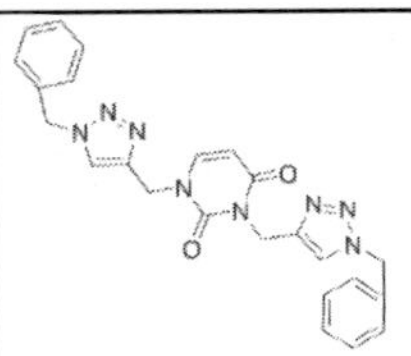	pipeline steel grade API 5L X52/1 M HCl	Espinoza-Vázquez et al., 2014
6	1,3-Bis((1-(4-chlorobenzyl)- 1H-1,2,3-triazol-4-yl)methyl) pyrimidine-2,4(1H,3H)-dione (BClBTMPD),	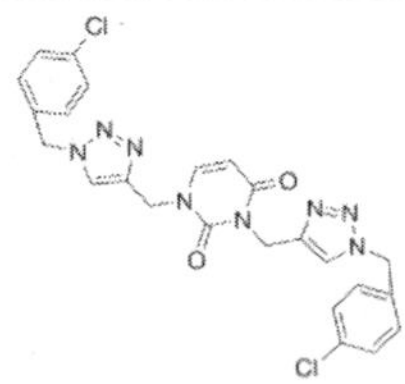	pipeline steel grade API 5L X52/1 M HCl	Espinoza-Vázquez et al., 2014
7	1,3-Bis((1-(4-bromobenzyl)- 1H-1,2,3-triazol-4-yl)methyl) pyrimidine-2,4(1H,3H)-dione (BBrBTMPD),	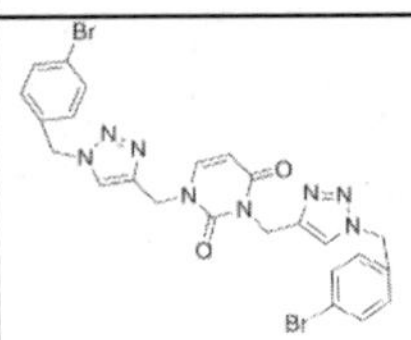	pipeline steel grade API 5L X52/1 M HCl	Espinoza-Vázquez et al., 2014

S No	Name of the Inhibitor System	Chemical Strcture	Metal / Medium	Reference
8	1,3-Bis((1-benzyl-1H-1,2,3-triazol-4-yl)methyl)- 5-methylpyrimidine2,4(1H,3H)-dione (BBTMMPD),	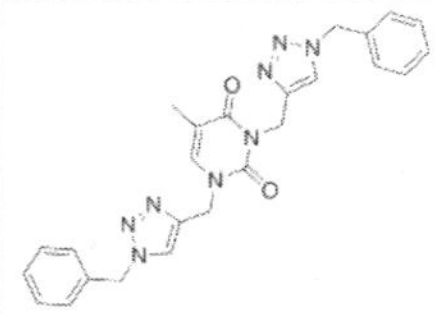	pipeline steel grade API 5L X52/1 M HCl	Espinoza-Vázquez et al., 2014
9	1,3-Bis((1-(4-bromobenzyl)- 1H-1,2,3-triazol-4-yl) methyl)-5-methylpyrimidine2,4(1H,3H)-dione (BClBTMMPD),	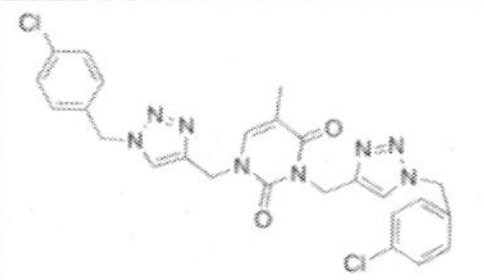	pipeline steel grade API 5L X52/1 M HCl	Espinoza-Vázquez et al., 2014
10	1-Benzyl-4-((benzyloxy)methyl)-1H-1,2,3-triazole (Comound 3-7)	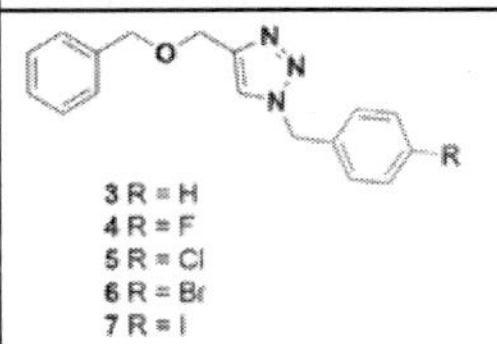	API 5L X52 steel / 1M HCl	Espinoza Vázquez et at, 2021
11	1,3-Bis((1-(4-chlorobenzyl)- 1H-1,2,3-triazol-4-yl) methyl)-5 methylpyrimidine 2,4(1H,3H)-dione (BBrBTMMPD),	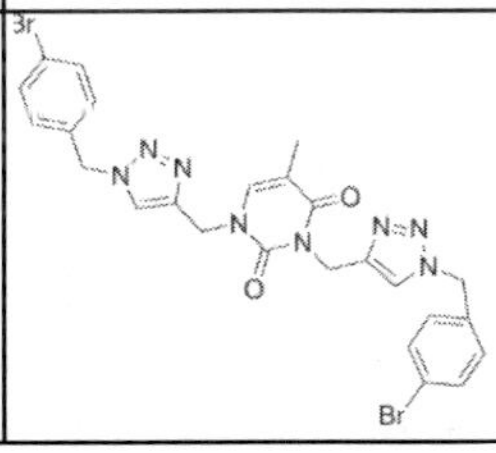	pipeline steel grade API 5L X52/1 M HCl	Espinoza-Vázquez et al., 2014

Espinoza-Vázquez et al., 2014 used electrochemical impedance spectroscopy (EIS) to measure the corrosion inhibition efficiency (IE) of six 1,2,3-triazole benzyl derivatives of uracil and thymine compounds in the system: pipeline steel grade API 5L X52/1 M HCl. It was determined how the inhibitor concentration, 5-200 ppm, affected the IE of all of these drugs. The obtained results demonstrated that, except for one of the di-alkyl derivatives of uracil, which could only achieve an IE of 84% at 200 ppm, the IE increased with increasing inhibitor concentration in all cases, and that the remaining compounds displayed IEs greater than 90% at relatively low concentrations. The 1,2,3-triazole benzyl derivative of uracil, which produced 96% IE with just 5 ppm, was the most notable inhibitor. It was discovered that these inhibitors always adhere to the Langmuir isotherm-described physisorption mechanism. It has also been demonstrated that these substances have an impact on both the anodic and cathodic processes, qualifying them as mixed-type inhibitors that can obstruct both anodic and cathodic sites on API 5L X52 steel surfaces submerged in HCl. For the two most effective inhibitors, the impact of the steel immersion time on their IE in the corroding media was also evaluated. It is demonstrated that even after 350 exposure hours to this extremely corrosive medium, both inhibitors may still offer sufficient protection (IE > 88%) to the steel substrate, despite the fact that their respective IEs decreased almost linearly as a function of time.

Espinoza Vázquez et al., 2021 describes how new triazoles generated from benzyl alcohol were created. These compounds, which were given the numbers 1 through 7, are intended to be employed as corrosion inhibitors for API 5L X52 steel in HCl 1M media. The findings revealed that benzyl alcohol was outperformed by all derivatives in terms of their ability to inhibit. With 94.2% inhibition efficacy at a concentration of 50 ppm, compound 6 had the best inhibitory capacity of all derivatives according to electrochemical impedance spectroscopy and polarisation curves. Additionally, it was discovered that compounds 4, 5, and 6 functioned as mixed inhibitors, while compounds 3, 6, and 7 were inhibitors with anodic behaviour. All of the synthetized derivatives of benzoyl alcohol fit the Langmuir adsorption isotherm. The thermodynamic parameter values observed led to the combined adsorption phenomena being proposed. Assuming the cluster approach, dispersion-corrected DFT calculations obtained free binding energies that closely matched empirically determined standard adsorption free energies. The electrostatic interactions and chemical bonds between the metal surface and the corrosion inhibitor were used to explain the mixed physisorption-chemisorption phenomenon. The highest charge transference to the metal cluster was finally accomplished by compound 6.

1,2,4- Triazoles as Acid Corrosion Inhibitors

Triazole derivatives have a particular affinity for metal surfaces with moving water molecules on their surface. Triazole derivatives are amphoteric in nature and form salts with acids and bases. Additionally, they have a lot of p-electron abundance and unshared electron pairs on the nitrogen atom, which when combined with the metal's d-orbitals, can provide a protective coating. As a result, numerous earlier publications concentrated on the utilisation of 1,2,4-triazole derivatives as corrosion inhibitors extensively which are discussed in the current section.

(E)-4-(4-(dimethylamino)benzylideneamino)-5-(pyridin-4-yl)-2H-1,2,4-triazole-3(4H)-thione (DPT) was a novel triazole derivative that Phadke Swathi, 2022 took into consideration for their corrosion inhibitve protection.

The inhibitory effectiveness (IE) of DPT for the corrosion of C1018 carbon steel was assessed using electrochemical experimental techniques. According to the studies, DPT is extremely effective, with an IE value of 90% at 250 ppm. Additionally, the addition of DPT caused a drop in both the anodic and cathodic current densities, which is a sign of mixed-type inhibition. Additionally, the Langmuir model for DPT-metal adsorption was deduced from the studies. The surface morphological analyses of C1018 carbon steel with and without DPT revealed a significant metal-DPT adsorption. The structure of DPT and its ability to suppress corrosion were correlated using DFFT and Monte-Carlo simulation techniques. The theoretical calculations and the high IE values of DPT found in experimental experiments coincided, and this compound is a potent inhibitor of corrosion of C1018 steel.

As part of their ongoing research into the development of N- and S-containing corrosion inhibitors for acidic solutions, Quraishi and Sardar, 2002 have examined the corrosion-inhibiting properties of four triazoles on mild steel in acidic solutions: 3-hydroxyphenyl-4-phenyl-5-mercapto-1,2,4-triazole (HPMT); 3-aminophenyl-4-phenyl-5-mercapto 1,2,4-triazole (APMT); 3,4-diphenyl-5-mercapto-1,2,4-triazole (DPMT); and 3-cinnamyl-4-phenyl-5-mercapto-1,2,4,triazole (CPMT). By using polarisation research, it was discovered that all inhibitors were mixed types, and they all work to prevent corrosion by adhering to the metal surface in accordance with the Langmuir adsorption isotherm.

Ahmed Al-Amiery et al., synthesised and characterised 4-Amino-3-(2-bromo-5-methoxyphenyl)-1H-1,2,4-triazole-5(4H)-thione (ATH) using spectroscopic techniques and elemental analysis. The weight loss technique, scanning electron microscopy, and the adsorption isotherm were used to investigate ATH's corrosion inhibition of mild steel in corrosive environments. ATH has a higher inhibition efficiency against mild steel corrosion. Langmuir isotherm model fits adsorption data well

In one of the studies, Chauhan et al., 2019 modified chitosan chemically utilising 4-amino-5-methyl-1,2,4-triazole-3-thiol, leading to the first time the inhibitor triazole modified chitosan was created. Utilizing FTIR and NMR, the newly created biopolymer (CS-AMT) was described before being tested as an inhibitor of carbon steel corrosion in 1M hydrochloric acid. Utilizing the weight loss method, electrochemical measurements, and surface analysis, corrosion testing and evaluation were extensively carried out. At 200 mg/L of inhibitor concentration, adhering to the Langmuir isotherm, a maximum corrosion inhibition efficacy of >95% was attained and demonstrated physical and chemical adsorption. The inhibitor was shown to have adhered to the surface of carbon steel in electrochemical research using impedance analysis, and potentiodynamic polarisation revealed a mixed kind of inhibitor behaviour with a cathodic predominance. A thorough theoretical investigation was carried out utilising DFT calculations, Fukui indices analysis, and molecular dynamics (MD) simulation to gain a better understanding of the interaction of inhibitor molecules with the metal surface. The experimental results were supported by the DFT research, which revealed a smaller energy gap for CS-AMT, and the MD simulations, which revealed an increased binding energy for CS-AMT relative to the parent chitosan and triazole moieties.

The mechanism and effectiveness of the inhibition of (Z) -4 - ((2-bromobenzylidene) amino) were examined using electrochemical measurements and the weight loss method. (Z) -4 - ((3-bromobenzylidene) amino) and (Z) -5-methyl-2-4-dihydro-3H-1,2,4-triazole-3-thione 5-methyl-2-4-dihydro-3H-1,2,4-triazole-3-thione was developed by Merimi et al., 2019 to prevent mild steel from corroding in HCl. The results demonstrate that the two triazole compounds are effective inhibitors, and their effectiveness follows the pattern (2i)> (21). According to the Langmuir model, both derivatives are mixed inhibitors and adsorb on the metal surface. An optical microscope investigation reveals that the surface morphology of mild steel has improved in the presence of the investigated inhibitor. The potential of the zero charge (Epzc) at the solution/metal interface was used to investigate the inhibitory mechanism. Dynamic simulation studies proposed the idea of water molecules gradually replacing other molecules on the surface of the iron surface for corrosion protection studies. Calculated and discussed parameters related to kinetics and thermodynamics. It has been demonstrated that there is a strong correlation between theoretical and experimental results when the quantum chemical parameters are estimated using the GAUSSIAN09W suite.

Rugmini Ammal, Projila, and Abraham Joseph, 2018 have determined the effectiveness of 4-(4-hydroxy-3-methoxy benzyledene amino)-4-H-1,2,4-triazole-3, 5-dimethanol, HMATD, as a corrosion inhibitor for mild steel in 0.5M HCl. It has been investigated how several conditions, including temperature and inhibitor concentration, affect efficiency. By lowering the charge transfer through the metal solution interface, the electrochemical impedance spectroscopic studies demonstrated the inhibitory action of HMATD. The mixed type behaviour of HMATD is indicated by the polarisation curves. By adhering to the metal surface and inhibiting the active sites there, the inhibitor molecule blocks activity and complies with the Langmuir adsorption isotherm. Additionally, a number of kinetic and thermodynamic characteristics have been determined.

Three newly synthesised triazole Schiff bases (3-bromo-4-fluoro-benzylidene)-[1,2,4]triazol-4-yl-amine (BFBT), (4-trifluoromethyl-benzylidene)-[1,2,4]triazol-4-yl-amine (TMBT) and (2-fluoro-4-nitro-benzylidene)-[1,2,4]triazol-4-yl-amine (FNBT) were investigated by Turuvekere et al., 2015 as corrosion inhibitors on mild steel in 0.5M hydrochloric acid by chemical (weight loss) and electrochemical (potentiodynamic polarisation and electrochemical impedance) techniques. The inhibition efficiency increases with an increase in inhibitor concentration and decreases with an increase in temperature of the medium. Various activation and adsorption thermodynamic parameters were evaluated. Adsorption of all the three inhibitors follows the Langmuir isotherm. Electrochemical impedance studies show that charge transfer resistance increases with concentration of inhibitors. Polarisation studies prove that inhibitors are of the mixed type. Surface morphology was examined using SEM and FTIR studies. Calculations in quantum chemistry support the findings of experiments. Greater than TMBT and FNBT due to higher E_{HOMO}, lower E_{LUMO}, smaller orbital gap (E), and larger dipole moment, BFBT exhibits the highest level of inhibitory efficiency. All of the investigated approaches exhibit strong correlations with one another.

The corrosion inhibitory effect of two 3,5-bis(disubstituted)-4-amino-1,2,4-triazole derivatives, namely 3,5-bis(4-tolyl)-4-amino-1,2,4-triazole (K1) and 3,5-bis(3,4-dimethoxyphenyl)-4-amino-1,2,4-triazole (K2), for mild steel in 2M H_3PO_4 has been studied using electrochemical impedance spectroscopy (EIS), Tafel polarization curves and weight loss measurements (Belghiti et al., 2016). It was found that the inhibition efficiency of the two investigated inhibitors increases with increase in concentration of inhibitors and decreases with increase in temperature. K1 and K2 show corrosion inhibition efficiency of 80.00 and 86.35%, respectively at 10−3M and 308K. Impedance experimental data revealed a frequency distribution of the capacitance, simulated as constant phase element. Potentiostatic polarization study showed that K1 and K2 are mixed-type inhibitors in 2M H_3PO_4. The results obtained from electrochemical and weight loss studies were in reasonable agreement. The adsorption of K1 and K2 on steel surface obeys Langmuir's adsorption isotherm. The correlation between inhibition efficiency and molecular structure of the inhibitors is investigated by determination of chemical indexes, which were performed using density functional theory (DFT) at (B3LYP/6-31G) (d,p) level. Monte Carlo simulations were further performed to simulate the adsorption of K1 and K2 on Fe (110) surface in the presence of water and the results show that K2 is more effective as corrosion inhibitor for mild steel in 2M H_3PO_4 medium than K1.

Hazazi, Fawzy, and Awad, 2014 employed potentiodynamic polarisation and electrochemical impedance spectroscopy to study the synergistic effect of 4-amino-5-methyl-4H-1,2,4-triazole-3-thiol (AMTT) and halides on the corrosion of mild steel in 0.5 M H_2SO_4 solution (EIS). The protective effect of AMTT was significantly enhanced by the halides, according to experimental results. With increasing AMTT concentration, both protection efficiency (%P) and degree of surface covering rise, and the influence of halides on the protection efficiency follows the trend: Cl- Br- I, which shows that the halide's radii and electronegativity are crucial to the adsorption process. On the other hand, because iodide is so powerful at inhibiting, the synergism occurred in the opposite sequence. AMTT's adsorption are in agreement with the Temkin adsorption isotherm.

Tourabi et al., 2013 investigated the corrosion inhibition of carbon steel in hydrochloric acid solution at 30°C by a new 4-amino-1,2,4-triazole derivative, namely 3,5-bis(2-thienylmethyl)-4-amino-1,2,4-triazole (2-TMAT), using electrochemical impedance spectroscopy (EIS) and polarisation techniques. The experimental results revealed that this organic compound inhibited corrosion well and that the inhibition efficiency increased with inhibitor concentration. Two-time constants determined by charge transfer and inhibitor adsorption, respectively, can be easily outlined. Potentiodynamic polarisation revealed that 2-TMAT is a hybrid inhibitor. The Langmuir isotherm was followed by the adsorption of 2-TMAT on the carbon steel surface in 1M HCl solution, with a very high negative value of the standard Gibbs free energy of adsorption Gads° (chemisorption). The mechanism of corrosion inhibition of carbon steel in 1M HCl medium by 3,5-bis(2-thienylmethyl)-4-amino-1,2,4-triazole was studied using X-ray photoelectron spectroscopy (XPS) (2-TMAT).

Hamdy H. Hassan, 2007 conducted a comparison study on 5-amino-1,2,4-triazole (5-ATA), 5-amino-3-mercapto-1,2,4-triazole (5-AMT), 5-amino-3-methylthio-1,2,4-triazole (5-AMeTT), and 1-ameTT as mild steel corrosion inhibitors in 0.1M HCl solution at 20°C. The metal corrosion behaviour was studied using potentiodynamic polarisation and electrochemical impedance spectroscopy (EIS) techniques in the absence and presence of distinct concentrations of these inhibitors under the influence of various experimental conditions. Open circuit potential (OCP) measurements as a function of time until reaching steady-state potentials (Est) were also established. The studies revealed that 5-AMT was the most effective inhibitor, with inhibition efficiency (IE%) values reaching up to 96% at a concentration of 103M. Polarization curves revealed that the four compounds investigated act as mixed inhibitors. In the absence and presence of the investigated inhibitors, the potential of zero charge (PZC) of mild steel was determined in 0.1M HCl. The chemical structure of the four inhibitors tested was discussed. The results of OCP versus time, polarisation, and impedance measurements agree well.

Three fatty acid 1,2,4-triazoles: 3-undecane-4-aryl-5-mercapto-1,2,4-triazole (triazole 1), 3-heptadeca-8-ene)-4-aryl-5-mercapto-1,2,4-triazole (triazole 2), and 3-deca-9-ene)-4-aryl-5-mercapto-1,2-4-triazole (triazole 3) were synthesised by Quarashi ad Jamal 2002 and their corrosion inhibitive property was ascertained by electrochemical techniques. Electrochemical polarisation studies at room temperature revealed that all triazoles are mixed-type inhibitors, inhibiting both anodic and cathodic reactions. Temkin's adsorption isotherm was followed by these compounds' adsorption onto mild steel from 15% HCl. They have also studies The influence of some more fatty acid triazoles namely, 4-Phenyl-5-undecyl 4H- [1,2,4] triazole-3-thiol (PUTT), 5- Heptadec-8-enyl-4-phenyl-4H- [1,2,4] triazole-3-thiol (HPTT), and 5-Dec-9-enyl-4-phenyl-4H- [1,2,4] triazole-3- thiol (DPTT) on the corrosion of mild steel in both HCl and H_2SO_4 medium by both electrochemical and gravimetric methods. To investigate the mechanism of corrosion inhibition, the activation energy and free energy of adsorption of each triazole were calculated. Potentiodynamic polarisation studies were performed at room temperature, and all of the compounds were mixed type inhibitors that inhibit mild steel corrosion by blocking the metal's active sites.

3-undecane-4-aryl-5-mercapto-1,2,4-triazole (triazole 1), 3(heptadeca-8-ene)-4-aryl-5-mercapto-1,2,4-triazole (triazole 2), and 3(deca-9-ene)-4-aryl-5-mercapto-1,2-4-triazole (triazole 3) were synthesized by Quraishi and Danish Jamal, 2000 and their ability to prevent corrosion in 15% hydrochloric acid was assessed using electrochemical and weight-loss methods. All of the triazoles are mixed-type inhibitors, meaning that they obstruct both cathodic and anodic processes, according to electrochemical polarisation tests conducted at room temperature. These chemicals were adsorbed onto mild steel from 15% HCl in accordance with Temkin's adsorption isotherm.

Table 2. 1,2,4- triazole s corrosion inhbitors

S No	Name of the Inhibitor System	Chemical Framework	IE%	Corrosive Environment	Type of Inhibitor	Reference
1	3,5-Bis(4-methyltiophenyl)-4H-1,2,4-triazole & derivatives	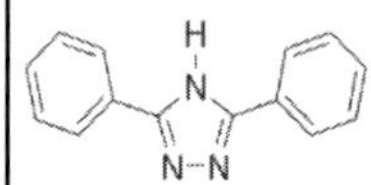	80 – 93%	Mild steel in HCl / H_2SO_4 / H_3PO_4	Mixed Type / Anodic	Bentiss et al., 2000, 2003, 2007; El Mehdi et al., 2003; Arshad et al., 2017; Wang 2012
2	5-Amino-1,2,4-triazole & derivatives	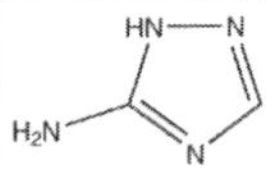	25 – 90%	Mild steel / Carbon steel in HCl	Mixed Type	Hassan et al., 2007; Mert et al., 2011; Tourabi et al., 2014
3	3-Benzylidene amino-1, 2, 4-triazole phosphonate & derivatives	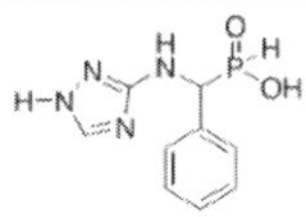	40-60%	Mild steel in neutral aqueous media	Mixed-type	Ramesh & Rajeswari, 2004
4	8-Bromo-5-morpholino-3- (4-propylphenyl)-[1,2,4] triazolo[4,3-c]pyrimidine & derivatives	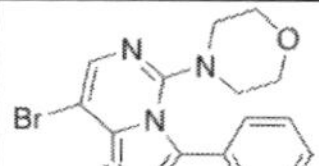	55-93%	Mild steel in HCl	Mixed-type inhibitor	Gurudatt et al., 2015
5	(3-Phenylallylidene) amino-5-(pyridine-4-yl)- 4H-1,2,4-triazole-3-thiol & derivatives	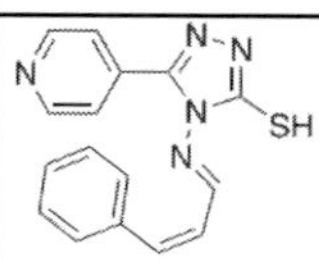	40-95%	Mild steel in HCl	Mixed-type inhibitor	Ansari et al., 2014
6	(3-Bromo-4-fluorobenzylidene)-[1,2,4] triazol-4-yl-amine & derivatives	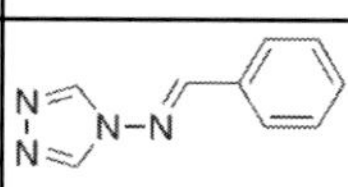	70-90%	Mild steel in HCl	Mixed-type inhibitor	Chaitra et al., 2015
7	1--(4-Fluorophenyl)-2,2-dimethylpropan-1-one O-(2-(4H-1,2,4-triazol4-yl)propan-2-yl) oxime & derivatives	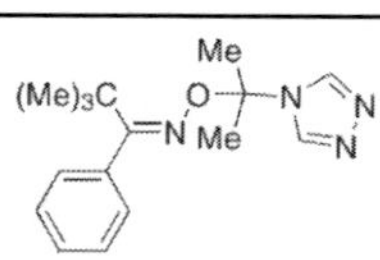	20-95%	Mild steel in HCl	Mixed-type inhibitor	Guo et al., 2014
8	4-amino-1,2,4-triazole & derivatives	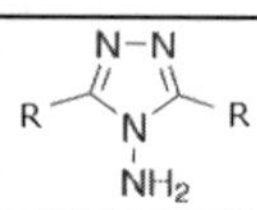	35-95%	Mild steel in a 2.5M H_2SO_4	Mixed-type inhibitor / Chemisorption	Musa et al., 2010

Through experimental research, Bentiss and colleagues have found that 3,5-bis(4-methyltiophenyl)-4H-1,2,4-triazole (4-MTHT) is a highly effective acid corrosion inhibitor. Additionally, several researchers have looked at related 1,2,4-triazole systems, which were compiled in Table 2.

Triazole Derivatives as Inhibitors for Other Metals / Alloys in Varied Environment

Although copper and its alloys have many uses in contemporary industries, they are vulnerable to corrosion from acids and potent alkaline solutions, particularly when these substances are combined with oxygen or other oxidants. Numerous 1,2,4-triazole compounds have been used successfully to prevent copper from corroding. The industries of transportation, construction, electrical engineering, home appliances, containers, aeroplanes, and aerospace have all made extensive use of aluminium and its alloys. Additionally, due to the creation of a resistant oxide layer, aluminium exhibits good corrosion resistance. However, due to this film's amphoteric susceptibility, the metal rapidly dissolves in acidic or basic solutions, particularly those that contain chloride. According to a review of the literature, both alkaline and acidic media frequently use different inhibitors to stop this.

2,5-Bis(2-thienyl) -4-amino 2,5-bis-1,2,4-triazole (2-TAT) (3-thienyl) -4-amino (Tang et al., 2008) was investigated as copper corrosion inhibitors in 0.5M H_2SO_4. The effect of molecular structure on the performance of three triazole derivatives namely benzotriazole (BZT), 5-methyl-1H-benzotriazole (5MBZT), and 3-amino-5-methylthio-1H-1,2,4, triazole (3AMT) as corrosion inhibitors for C1020 steel and pure copper in 2% hydrochloric acid (HCl), which simulate descaling process in a typical desalination plant was investigated by Nuha et al., 2022. 4-amino-4H-1,2,4-triazole-3thiol (ATT), 4-amino-5-methyl-4H-1,2,4-triazole-3thiol (AMTT), and 4-amino-5-ethyl-4H-1,2,4-triazole-3thiol (AETT) were synthesised and the inhibition effect of three 1,2,4-triazole derivatives on copper deterioration in 0.5M HCl solution wasinvestigated (Sudheer and Quraishi, 2013). Two very basic molecules, 1H-1,2,3-triazole and 1,2,4-triazole that were investigated by Stanley et al., 2017 as copper corrosion inhibitors in NaCl solution using a variety of electrochemical and analytical methods. Using Electrochemical, gravimetric and surface analytical techniques, Chauhan et al., 2019 studied the compound 4-amino-5-methyl-4H-1,2,4-triazole-3-thiol (AMTT), 5-(phenyl)-4H-1,2,4-triazole-3-thiol was studied by Sherif et al., 2007 and 3-amino-1,2,4-triazole by Sherif et al., 2007 as a novel corrosion inhibitor for copper in 3.5% NaCl solution. Corrosion tests and DFT simulations were used to assess the effectiveness of imidazole, 1,2,4-triazole, and their 1-methyl derivatives as copper corrosion inhibitors in NaCl solutions. 3-amino-1,2,4-triazole (Xu QJ et al., 2009), bis-(4-amino-5-mercapto-1,2,4-triazol-3-yl)-butane (Khiati et al., 2011), 3,40 -bi-1,2,4-triazole (Al Maofari et al., 2014), 3-methyl 4-amino 1,2,4 triazole (Chieb et al., 2011) and 3-methyl 1,2,4-triazole 5-thione (Rahmouni et al., 2007) were studied for their corrosion resistance property againgt coppe corrosion in 3% NaCl. 1,2,4-Triazole derivatives like, 3-anisalidene amino 1,2,4-triazole phosphonate (Ramesh et al., 2004), 1,2,4-Triazole derivatives such as 1,2,4-triazole, 3-amino-1,2,4-triazole and 2-4-diamino1,2,4-triazole in 1 M H2SO4 (Soumoue et al., 2004); 3,5-diamino-1,2,4-triazole in 2 M HNO3 (Zarrouk et al., 2013);, 5-amino-3-mercapto 1,2,4 triazole, 3,5-bis(2-thienyl)-4- amino-1,2,4-triazole (Chen et al., 2009) were investigated as copper corrosion inhibitors in varied acidic conditions.

Schiff bases of 4-amino-4H-1, 2, 4-triazole-3,5-dimethanol was considered as aluminium corrosion inhibitor in 1N HNO_3 (Sam John et al., 2011). Similarly, several researchers have studied varied 1,2,4- triazoles as inhibitors for aluminum and its alloy in varied environments which are summarized as follows (Table 3).

Spectroscopic and electrochemical techniques have been used to explore the corrosion inhibitive analyses. It is discovered that their main mode of adsorption on the surface of the Al alloy obeys the Langmuir adsorption isotherm. For the purpose of determining the relationship between the inhibitor's inhibitive effect and molecular structure, a quantum chemical technique was employed to determine some of the molecule's electronic properties. In some of the research, the evolution of the Volta potential distribution and the surface topography during corrosion tests were examined using scanning Kelvin probe force microscopy and atomic force microscopy. The free energy, enthalpy, and entropy of adsorption, as well as the activation parameters, were estimated as thermodynamic parameters. Molecular dynamics and density functional theory have been used to examine the behaviours of molecules adhering to metal surfaces.

Due to zinc's numerous industrial uses and economic significance, its ability to prevent corrosion has garnered a lot of interest. Both alkaline and acidic solutions can cause zinc to corrode, with the corrosion being more severe at pH levels below 6 and above 12.5. The influence of pH, potential, inhibitor concentration, zinc ion concentration, and exposure period on the suppression of zinc corrosion by various triazole derivatives was examined. 5-

[4-(methylthio)benzyl]-4H-1,2,4-triazole-3- thiol (Shylesha et al., 2011) was found to be an excellent corrosion resistant inhibitor for zinc in acidic environment.

Table 3. Survey on 1,2,4- triazole as corrosion inhibitor for aluminum and its alloys

S No	Name of the Inhibitor System	Triazole moity	Metal	Corrosive Environment	Reference
1	4-amino-4H-1, 2, 4-triazole-3,5-dimethanol	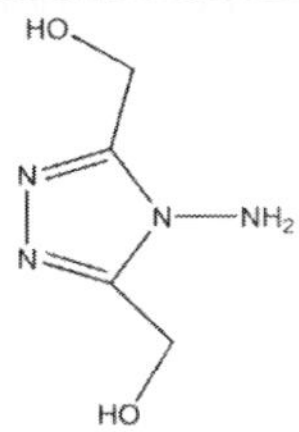	Aluminium	1N HNO_3	Sam John et al., 2011
2	3-amino-1,2,4-triazole-5-thiol	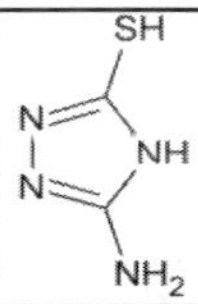	Aluminium	sea water	Sherif 2013
3	3-(4-benzylidene-) amino-1,2,4-triazole	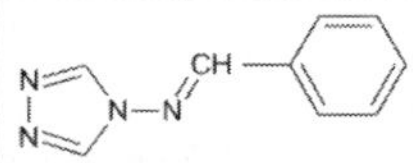	5052Al alloy	HCl	Zhu & Yang, 2013
4	3-amino-1,2,4-triazole-5-thiol	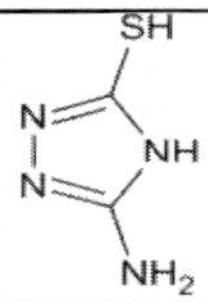	Al 2024-T3 alloy	Aqueous sulphate medium	Qafsaoui & Takenouti, 2010
5	3-ethyl-4-amino-5- mercapto-1,2,4-triazole	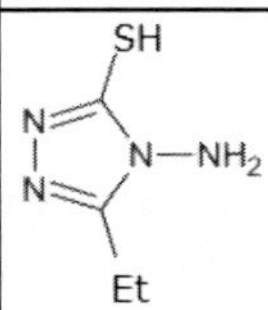	6061 Al-15 vol. pct. SiC(p) composite and its base alloy	Sulphuric acid	Pinto et al., 2016
6	3-amino-1,2,4-triazole	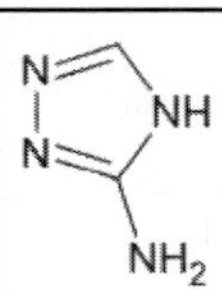	AA2024 aluminium alloy	Neutral chloride solution	Zheludkevich eta, 2005
7	3-ethyl-4-amino-5-mercapto-1,2,4-triazole (EAMT)	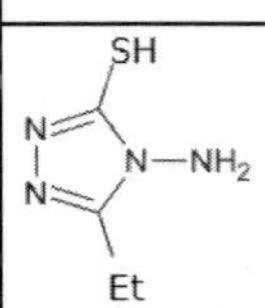	6061-Al alloy	Aqueous sodium hydroxide	Reena Kumari et al., 2011

Table 3. (Continued)

S No	Name of the Inhibitor System	Triazole moity	Metal	Corrosive Environment	Reference
8	4-amino-5-(4-nitrophenyl)-4H-1,2,4-triazole-3-thiol	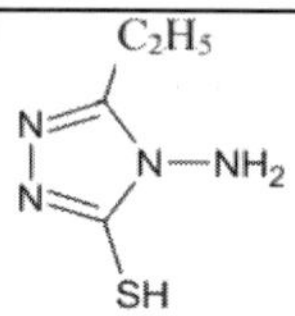	6061 Al-15 vol. pct. SiC(p) composite	Sodium hydroxide	Reena Kumari at el 2014
9	4-amino-4H-1,2,4-triazole-3,5-dimethanol	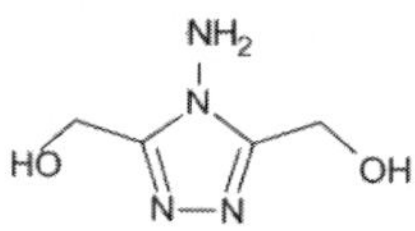	Pure aluminium metal	Nitric acid	John et al., 2011

Salima et al., 2019 developed six brand-new triazole derivatives from benzoic acid that had been replaced with thiocarbohydrazide. Utilizing the weight loss technique, synthetic triazole derivatives were tested and assessed as low-carbon steel corrosion inhibitors. These compounds were discovered to have a promising inhibitory effect on mild steel corrosion in acidic solution. At various temperatures and inhibitor doses, the most effective inhibitor was analysed. Over 96% of inhibition was effective.

An oxo-triazole derivative (DTP) was taken into consideration by Zhihua Tao et al., 2009 Its inhibitory effect on the corrosion of mild steel in sulphuric acid was examined using weight loss, potentiodynamic polarisation, EIS, and SEM. The findings showed that DTP was a superior inhibitor, and the weight loss experiment's and the electrochemical experiment's inhibition efficiencies were in good accord.

Ksenija and Norman, 2005 studied into the impact of a triazole (TRA), benzotriazole (BZT), 1-hydroxybenzotriazole (HBT), 1H-benzotriazol-1-methanol (BTM), and N-(1H benzotriazole-1-ylmethyl)-formamide (FMB) on the corrosion of iron in 1M HCl and $HClO_4$. The inhibitory impact of each of these substances in both acids last upto 16 hrs.

The corrosion of API 5 L X52 steel in 1 M HCl media was studied (Espinoza et al., 2019) in order to assess the newly designed 1,2,3-triazole-containing theophylline derivatives' abilities to suppress corrosion.

Dry Lab Studies on Triazoles as Corrosion Inhibitors

Despite being generally expensive and time-consuming, experimental methods are valuable for illuminating the inhibitory mechanism. The molecular structure has been studied as one of the key elements influencing the adsorption of organic molecules on the metal surface using a variety of quantum chemistry approaches and molecular modelling techniques. Theoretical investigations have found a correlation between the researched triazole inhibitors' molecular and electronic structures and their effectiveness as inhibitors. The inhibition property of inhibitors was characterised using molecular orbital simulations, which will be useful to understand the mechanism underlying corrosion inhibition and to mimic the inhibitor's way of adsorption on the metal surface. Several literatures have been noted for the use of computational tools in elucidating the corrosion inhibiton mechanism of different triazole derivatives (Ansari, 2018; Sowmya, 2019).

The molecular dynamics simulation method and quantum chemical calculations were used to investigate the adsorption behaviour of three triazole derivatives (Khaled 2008) at the Fe surface. The results showed that these compounds significantly inhibited the corrosion of mild steel in 1M HCl solution. The adsorption structures of triazole derivatives were optimised using molecular simulation studies. In the presence of solvent effects, the iron/inhibitor/solvent interfaces were simulated, and the charges on the inhibitor molecules as well as their structural parameters were calculated.

The property and structural parameters of four triazole derivatives (Mohamed K Awad et al., 2010) was investigated using density functional theory (DFT) at the B3LYP/6-31G++ (dp) basis set level, ab initio calculations using the HF/6-31G++ (dp), and semi-empirical PM3 methods. (5-amino 1,2,4 triazole (5-ATA), 5-amino-3-mercapto 1,2,4 triazole (5-AMT), 5-amino-3-methyl thio 1,2,4 triazole (5-AMeTT), 1-amino-3-methyl thio 1,2,4 triazole (1-AMeTT)) were studied to see if there was a link between molecular structure and inhibition efficiency (%IE). The highest occupied molecular orbital energy (E_{HOMO}), the lowest unoccupied molecular orbital energy (E_{LUMO}), the energy gap (ΔE), the dipole moment, the sum of total negative charges (TNC), the molar volume (MV), electronegativity, hardness, softness and the fraction of electrons transferred from the inhibitor molecule to the metal surface (N) have all been calculated. The adsorption energies of the inhibitors with the copper (110) surface were also investigated using the molecular dynamic (MD) method. There was a strong correlation between the theoretical data and the experimental results.

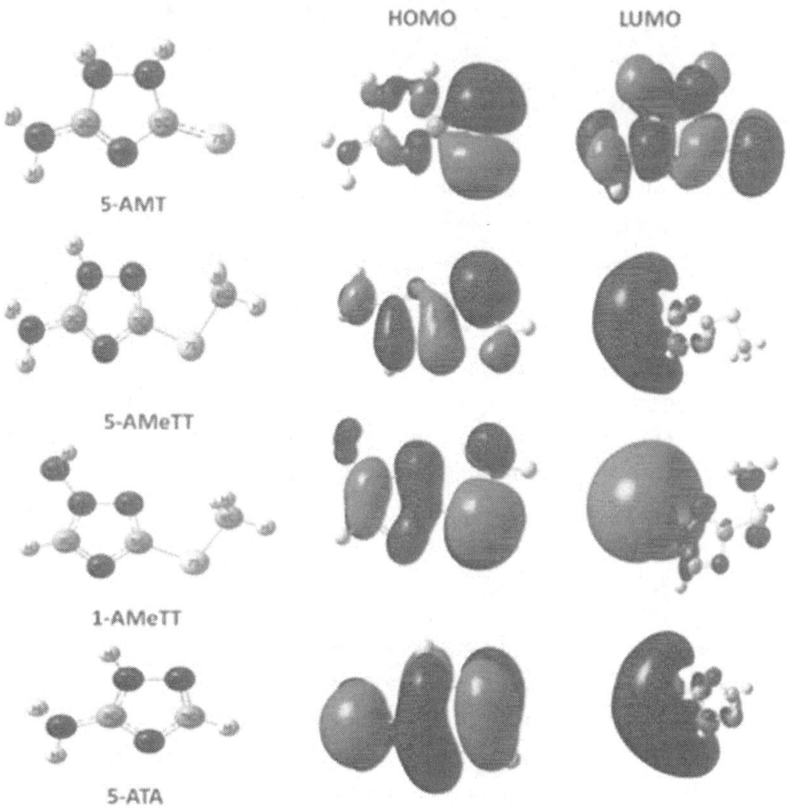

Figure 2. The optimized molecular structures, HOMO and LUMO of the neutral inhibitor molecules using DFT/B3LYP/6-31G++(d p).

El Sayed et al., 2006 used semi-empirical SCF molecular orbital methods to investigate the corrosion inhibition efficiencies of some triazole, oxadiazole, and thiadiazole derivatives for steel in the presence of acidic medium. Geometric structures, total negative charge on the molecule (TNC), molecular orbital energy, core-core repulsion (CCR), dipole moment (μ) and linear solvation energy terms, molecular volume (Vi), and dipolar-polarization were all found to be associated with corrosion inhibition efficiency. To calculate corrosion inhibition efficiency, four equations were proposed. The agreement between the calculated and experimental results was found to be satisfactory; the standard deviations between the calculated and experimental results ranged from 0.03 to 4.18.

For several tiazole compounds in 1M HCl, (Ramzi 2021), derived a potential relationship between experimental inhibition efficiencies (IE_{exp}) of corrosion inhibitors and some quantum characteristics. Efficiency of corrosion inhibition was connected with highest occupied molecular energy level (E_{HOMO}), lowest unoccupied molecular energy level (E_{LUMO}), dipole moment (μ), adsorption energy (E_{ads}), energy gap (ΔE), surface area of molecule SA, and Log of the partition coefficient (AlogP). The findings demonstrated that there was a good connection between experimental inhibition efficiencies (IE_{exp}) and calculated inhibition efficiencies (IE_{calc}), with R2 and R values between IE_{exp} and IE_{calc} ranging from 0.857 to 0.967 and 0.926 and 0.974, respectively.

Density functional theory (DFT) was used (Lie et al., 2014) to examine the corrosion inhibitive performance of 4-chloro-acetophenone-O-1′-(1′,3′4′-triazolyl)-metheneoxime (CATM), 4-fluoro-acetophenone-O-1′-(1′,3′,4′-triazolyl)-metheneoxime (FATM), and 3,4-dichloro The energy gap (ΔE), Mulliken charges, hardness, dipole moment, and the fraction of electrons transferred (N) were calculated, among other quantum chemical parameters. The highest occupied molecular orbital energy (E_{HOMO}), the lowest unoccupied molecular orbital energy (E_{LUMO}), and the energy gap (E) were also calculated. To describe the inhibitory performance of the compounds under study, a composite index of the aforementioned descriptors known as a quantitative structure activity relationship (QSAR) approach was applied. In addition, research using Monte Carlo simulation (Figure 3) were used to find the ideal configurational space for the iron/triazole derivative system.

The corrosion protection of a triazole heterocyclic compound, 3-(4-ethyl-5-mercapto-1, 2, 4-triazol-3-yl)-1-phenylpropanone (EMTP), against 1M hydrochloric acid medium was experimentally determined (Israa Abd Alkadir Aziz et al., 2021). To calculate quantum chemical parameters and determine the relationship between molecular structure and protection performance, the DFT method was used. The absolute electronegativity value from inhibitor to unoccupied d-orbital of iron atoms was also determined and correlated with protection efficiency. Theoretical findings revealed that the protection performance of EMTP increased with increasing HOMO energy (Figure 4), and the nitrogen, oxygen, and sulphur atoms are the most likely positions for bonding via electron transfer to the d-orbital of iron atoms on the mild steel surface.

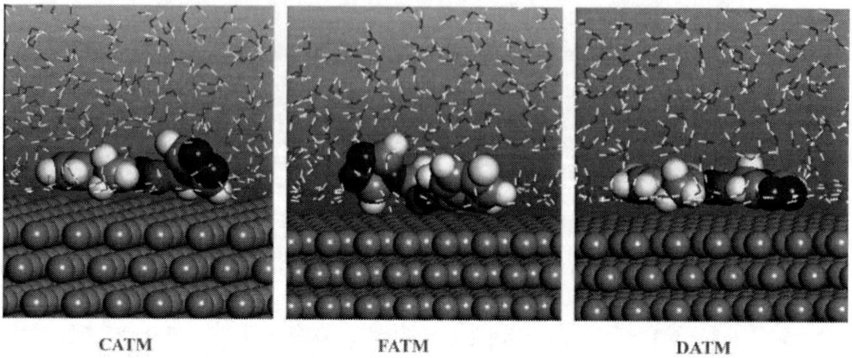

Figure 3. Equilibrium adsorption configurations of inhibitor molecules on Fe(11 0) surface (in aqueous phase).

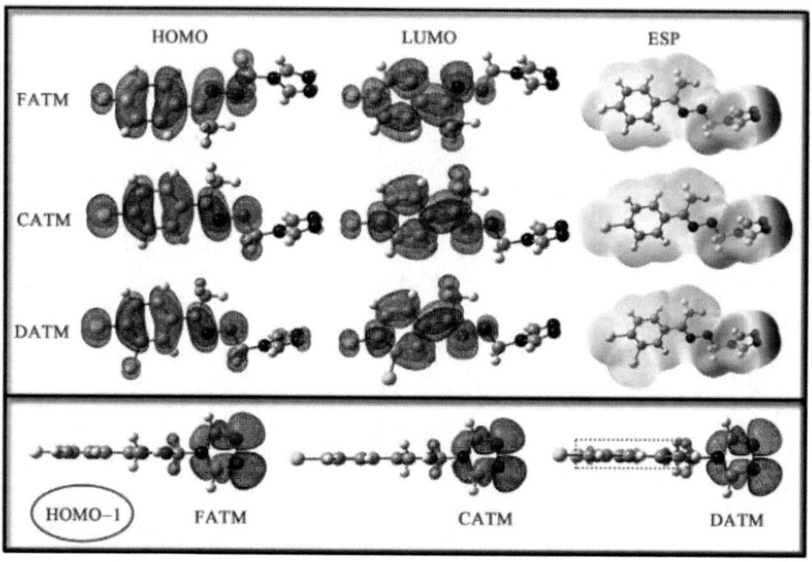

Figure 4. Electrostatic potential (ESP) and the frontier molecule orbital density distributions.

To investigate corrosion inhibition mechanisms of two 1,2,4-triazole derivatives with variable-length alkyl chains, 5-octylsulfanyl-1,2,4-triazole (TR8) and 5-decylsulfanyl-1,2,4-triazole (TR10), against mild steel (MS) corrosion in HCl, theoretical techniques like density functional theory (DFT) and molecular dynamics (MD) simulations were employed. Atomic force microscopy and scanning electron microscopy were used to examine the impact of molecules on the MS surface (AFM). It was shown that there is a relationship between corrosion inhibition abilities and physicochemical and electrical characteristics as measured by DFT and MD (Aoufir et al., 2020).

The corrosion inhibitory effect of two 3,5-bis(disubstituted)-4-amino-1,2,4-triazole derivatives, namely 3,5-bis(4-tolyl)-4-amino-1,2,4-triazole (K1) and 3,5-bis(3,4-dimethoxyphenyl)-4-amino-1,2,4-triazole (K2), for mild steel in 2M H_3PO_4 has been studied using various experimental measurements. It was found that K1 and K2 show corrosion inhibition efficiency of 80.00 and 86.35%, respectively at 10−3M and 308K. The correlation between inhibition efficiency and molecular structure of the inhibitors is investigated by determination of chemical indexes, which were performed using density functional theory (DFT) at (B3LYP/6-31G) (d,p) level. Monte Carlo simulations were further performed to simulate the adsorption of K1 and K2 on Fe (110) surface in the presence of water and the results show that K2 is more effective as corrosion inhibitor for mild steel in 2M H_3PO_4 medium than K1 (Belghiti et al., 2016).

Inhibitory characteristics of two 3,5-bis (disubstituted) derivatives of -4-amino-1, 2,4-triazoles, specifically 3,5-bis (4-methoxyphenyl) 3,5-bis and -4-amino-1,2,4-triazole (T1) and (4-chlorophenyl)-4-amino-1,2,4-triazole (T2) in a 2M H_3PO_4 solution at 308K was examined (Elbelghiti et al., 2016) using electrochemical impedance spectroscopy (EIS), Tafel polarisation curves, and weight loss measurements. Monte Carlo simulations were used to examine the adsorption behaviour of 4-amino-1, 2,4-triazole derivatives on Fe (110) surface in the presence of water in order to confirm the effectiveness of their corrosion inhibition properties. The findings showed that T1 had a higher adsorption energy than T2, which is consistent with the inhibition efficiency discovered through experimentation.

Mechanism of Adsorption and Inhibition

On the basis of adsorption phenomena, a feasible mechanism for the corrosion-inhibiting activity may be justified. 1,2,4-triazole can persist in acidic media as cationic entities. At the cathodic sites of MS, cationic substances adsorb, which slows down the events that lead to hydrogen evolution. Through specially adsorbed chloride ions, the protonated azole also adheres to the surface of MS (Figure 5) (Ahamad et al., 2010).

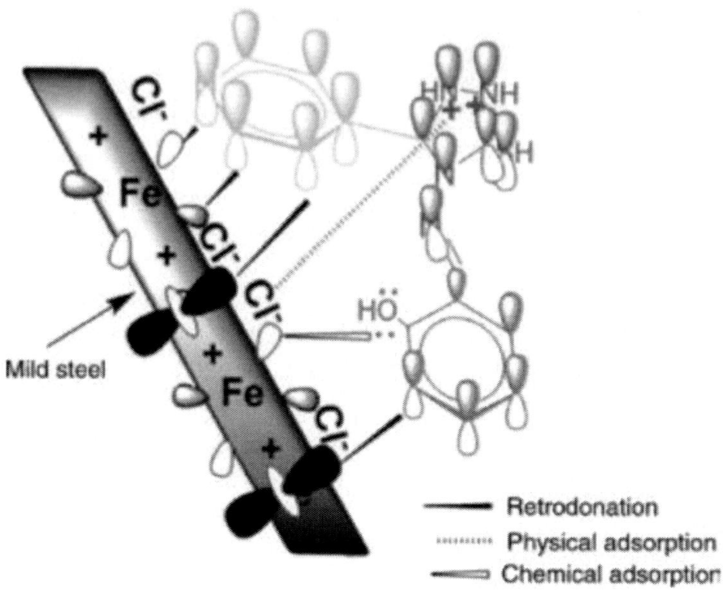

Figure 5. Adsorption and inhibition mechanism of triazoles on metal surface.

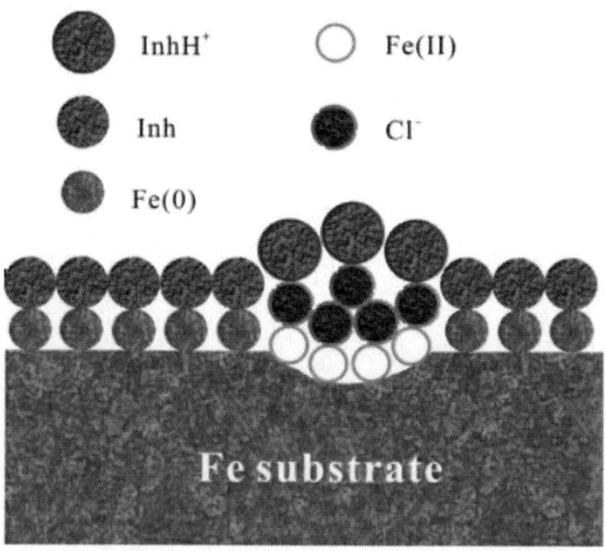

Figure 6. Adsorption model for inhibitors on metal surface in acid medium.

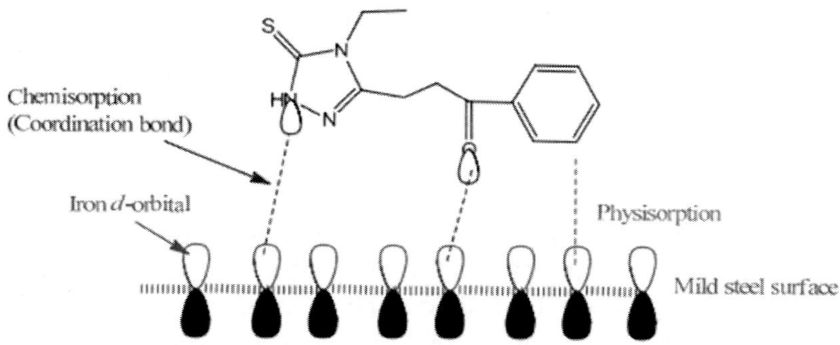

Figure 7. Suggested inhibition mechanism for inhibitors on metal surface in acid medium.

The metal-inhibitor complex is generated at the flat area for samples in inhibited acid solutions. Cl anions are first adsorbed onto the positively charged metal surface in the low-lying area. due to the possibility of protonated triazole compounds in acidic solutions. The inhibitor adsorption model on the iron surface is shown in Figure 6 & 7. Inhibitor interacts with Fe(0) fast, forming a thin, robust protective layer that is most likely a single monolayer. The protonated inhibitor molecule, however, interacts with Fe(II) and creates a substantial and protective complex. The protective coating ultimately prevented caustic ions from reaching the steel, successfully shielding it from corrosion (Guo et al., 2014; Israa Abd Alkadir Aziz et al., 2021).

Discussion

Summary of the current study suggests several triazole derivatives have been developed and studied for their corrosion inhibiting property for various metals in different environment. The nature and type of the substituents included in the core triazole molecule have a significant impact on the differences in the inhibitory efficacy of triazole derivatives. In general, the concentration of the triazole-based inhibitor rises together with the percentage inhibition, and the rate of corrosion rises along with the temperature. As a result, the triazole-based inhibitors work better at lower temperatures than they do at higher ones. This might be because the CI absorbs more readily on the surface of MS and ionises more quickly when exposed to stronger acid. In practically all the studies, there was good agreement between the electrochemical and gravimetric data. The findings also indicated that the majority of triazoles exhibit mixed-type inhibitory behaviour and that the Langmuir isotherm model is largely followed in their adsorption on metal surfaces. The impact of immersion time has also been researched in several papers in order to evaluate the stability of the triazole-based inhibitor behaviour over time. Most often, prolonging the immersion period up to 10–12 h increases the IE of the triazole derivatives, leading to the creation of a more protective layer. The inhibitory impact diminishes after 10 to 12 hours, and this may be caused by film flaws that allow aggressive ions to reach the acid-metal interface. According to the capacity of sulphate and chloride ions to adsorb on surfaces, the examined triazole derivatives are often more effective in HCl than in H_2SO_4. The negatively charged MS surface that results from this attracts the positive (cationic) species of the inhibitor, producing a barrier to the hostile acid solution, because Cl anions are likely more prominent than SO_4^{-2} anions. This review has summarized the anti-corrosion activity of triazole derivatives with respect to their IE, metal medium and method.

Disclaimer

None.

References

Ahmed Al-Amiery, Lina M Shaker, Abdul Amir H Kadhum, Mohd S Takriff, Synthesis, characterization and gravimetric studies of novel triazole-based compound, *International Journal of Low-Carbon Technologies*, (2020), 15 (2), 164–170.

Alkadir Aziz, I. A., Annon, I. A., Abdulkareem, M. H., Hanoon, M. M., Alkaabi, M. H., Shaker, L. M., Alamiery, A. A., Wan Isahak, W. N. R., Takriff, M. S. Insights into Corrosion Inhibition Behavior of a 5-Mercapto-1, 2, 4-triazole Derivative for Mild Steel in Hydrochloric Acid Solution: Experimental and DFT Studies. *Lubricants* (2021), 9, 122. https://doi.org/10.3390/lubricants9120122.

Al Maofari A, Ezznaydy G, Idouli Y, Gue´dira F, Zaydoun S, Labjar N, Hajjaji SE. Inhibitive action of 3,40 -bi-1,2,4- triazole on the corrosion of copper in NaCl 3% solution. *J Mater Environ Sci* (2014) 5:2081–2085.

Al Maofari A, Ezznaydy G, Idouli Y, Gue´dira F, Zaydoun S, Labjar N, Hajjaji SE, Inhibitive action of 3,40 -bi-1,2,4- triazole on the corrosion of copper in NaCl 3% solution. *J Mater Environ Sci*, (2014) 5:2081–2085.

Ansari KR, Quraishi MA, Singh A. Schiff's base of pyridyl substituted triazoles as new and effective corrosion inhibitors for mild steel in hydrochloric acid solution. *Corros Sci*, (2014), 79: 5–15.

Ansari KR, Sowmya Ramkumar, D. Nalini & M. A. Quraishi, Studies on adsorption and corrosion inhibitive properties of quinoline derivatives on N80 steel in 15% hydrochloric acid, *Cogent Chemistry*, (2016) 2:1, 1145032, DOI: 10.1080/23312009.2016.1145032.

Ansari KR, Sowmya Ramkumar, D.S. Chauhan, Md. Salman, D. Nalini, V. Srivastava and M.A. Quraishi, Macrocyclic compounds as green corrosion inhibitors for aluminium: electrochemical, surface and quantum chemical studies, *Int. J. Corros. Scale Inhib.*, (2018), 7(3), 443–459.

Arshad N, Akram AR, Akram M, Rasheed I. Triazolothiadiazine derivatives as corrosion inhibitors for copper, mild steel and aluminum surfaces: electrochemical and quantum investigations. *Protect Metals Phys Chem Surf*, (2017, 53: 343–358.

Awad MK, Mustafa MR, Elnga MMA, Computational simulation of the molecular structure of some triazoles as inhibitors for the corrosion of metal surface. *J Mol Struct* (Thoechem), (2010) 959:66–74.

Belghiti, Y. Karzazi, A. Dafali, I. B. Obot, E. E. Ebenso, K. M. Emran, I. Bahadur, B. Hammouti, F. Bentiss, Anti-corrosive properties of 4-amino-3,5-bis(disubstituted)-1,2,4-triazole derivatives on mild steel corrosion in 2M H3PO4 solution: Experimental and theoretical studies, *Journal of Molecular Liquids*, (2016), 216, 874-886.

Bentiss F, Traisnel M, Gengembre L, Lagrenée M. Inhibition of acidic corrosion of mild steel by 3, 5-diphenyl-4H-1,2,4-triazole. *App Surf Sci*, (2000) 161: 194–202.

Bentiss F, Bouanis M, Mernari B, Traisnel M, Lagrenee M. Effect of iodide ions on corrosion inhibition of mild steel by 3,5-bis(4-methylthiophenyl)-4H-1,2,4-triazole in sulfuric acid solution. *J App Electrochem*, (2002) 32: 671–678.

Bentiss F, Traisnel M, Vezin H, Lagrenée M. Linear resistance model of the inhibition mechanism of steel in HCl by triazole and oxadiazole derivatives: structure-activity correlations. *Corros Sci*, (2003) 45: 371–380.

Bentiss F, Bouanis M, Mernari B, Traisnel M, Vezin H, Lagrenee M. Understanding the adsorption of 4H-1,2,4-triazole derivatives on mild steel surface in molar hydrochloric acid. *Appl Surf Sci*, (2007) 253: 3696–3704.

Bentiss F, Jama C, Mernari B, El Attari H, El Kadi L, Lebrini M, Traisnel M, Lagrenée M. Corrosion control of mild steel using 3,5-bis(4-methoxyphenyl)-4-amino-1,2,4-triazole in normal hydrochloric acid medium. *Corros Sci*, (2009) 51: 1628–1635.

Bineet Kumar Singh, Munmun Basak, Martin West, Indranil Guha, "*Estimating the Cost of Corrosion in Indian industry*," Petrotech-2010, 31st October – 3 rd November 2010, New Delhi, India, Paper ID : 20100815.1.

Chaitra TK, Mohana KN, Tandon HC. Thermodynamic, electrochemical and quantum chemical evaluation of some triazole Schiff bases as mild steel corrosion inhibitors in acid media. *J Mol Liq* 2015, 211: 1026–1038.

Chebabe D, Dermaj A, Erramli H, Hajjaji N . Corrosion inhibition of bronze alloy B66 by 4-amino-3-methyl-1,2,4-triazole-5-thione in 3% NaCl solution. *Anti-Corros Methods Mater* (2014) 61:281–286.

Chen Y, Tang Y, Yang W, Liu Y, Wang J, Inhibition of 3,5-bis(2-thienyl)-4-amino-1,2,4-triazole on copper in HCl media. *Corros Sci Prot Technol*, (2009) 21:406–409.

Chieb T, Belmokre K, Benmessaoud M, Drissi SEH, Hajjaji N, Srhiri A, The inhibitive effect of 3-methyl 4-amino 1,2,4 triazole on the corrosion of copper-nickel 70–30 in NaCl 3% solution. *Mater Sci Appl*, (2011) 2:1260–1267.

Dhaif HK, Jasim EQ, Muhajjar ZA, Shanta AA. Corrosion Inhibition of Mild-Steel in 0.5 M HCl using some prepared 1,2,3-Triazoles Derivatives. *Mediterr J Chem* (2019) 9: 290–304.

Dheeraj Singh Chauhan, M. A. Quraishi, A. A. Sorour, Sourav Kr. Saha and Priyabrata Banerjee, Triazole-modified chitosan: a biomacromolecule as a new environmentally benign corrosion inhibitor for carbon steel in a hydrochloric acid solution, *RSC Adv.*, (2019a), 9, 14990.

Dheeraj Singh Chauhan, M.A. Quraishi, C. Carrière, A. Seyeux, P. Marcus, A. Singh, Electrochemical, ToF-SIMS and computational studies of 4-amino-5-methyl-4H-1,2,4-triazole-3-thiol as a novel corrosion inhibitor for copper in 3.5% NaCl, *Journal of Molecular Liquids*, (2019b), 289, 111113.

El Aoufir, R. Aslam, F. Lazrak, R. Marzouki, S. Kaya, S. Skal, A. Ghanimi, I. H. Ali, A. Guenbour, H. Lgaz, I. M. Chung, The effect of the alkyl chain length on corrosion inhibition performances of 1,2,4-triazole-based compounds for mild steel in 1.0 M HCl: Insights from experimental and theoretical studies, *Journal of Molecular Liquids*, (2020), 303, 112631.

El Attari H, El Bribri A, Mhaidra L, Bentiss F, Siniti M. Synthesis and anticorrosion for carbon steel of 4-amino-3,5 bis (4-hydroxy-3-methoxy)-1,2,4-triazole in hydrochloric acid solution. *Am J Eng Res* (2015) 4:44–51.

ElBelghiti, Y. Karzazi, A. Dafali, B. Hammouti, F. Bentiss, I.B. Obot, I. Bahadur, E.E. Ebenso, Experimental, quantum chemical and Monte Carlo simulation studies of 3,5-disubstituted-4-amino-1,2,4-triazoles as corrosion inhibitors on mild steel in acidic medium, *Journal of Molecular Liquids*, (2016), 218, 281-293.

El Mehdi B, Mernari B, Traisnel M, Bentiss F, Lagrenee M. Synthesis and comparative study of the inhibitive effect of some new triazole derivatives towards corrosion of mild steel in hydrochloric acid solution. *Mater Chem Phys*, (2003) 77: 489–496.

Elkhotfi Y, Forsal I, Rakib EM, Mernari B. Optimization of the inhibitor efficiency of a triazole on corrosion of mild steel in 1m HCl. *J Adv Electrochem* 2017, 27: 141–143.

El Sayed H. El Ashry, Ahmed El Nemr, Sami A. Esawy, Safaa Ragab, Corrosion inhibitors: Part II: Quantum chemical studies on the corrosion inhibitions of steel in acidic medium by some triazole, oxadiazole and thiadiazole derivatives, *Electrochimica Acta*, (2006), 51 (19), 3957-3968.

Ennajih, R. Al, G. Anton, E. Mokhtaressassi, L. De, U. Mohamed, V. A. Faculté, A. Ibn, B. Bp, R. Maroc, L. De Chimie, O. Hétérocyclique, S. Av, I. Battouta, P. O. Box, and E.- Valencia, —Inhibition of mild steel corrosion in 1M hydrochloric acid by Benzimidazolium bromide derivatives, Aust. J. Basic Appl. Sci., (2013), 7(6), 1-8.

Espinoza-Vázquez A, Negrón-Silva GE, González-Olvera R, Angeles-Beltrán D, Herrera-Hernández H, Romero-Romo M, Palomar-Pardavé M. Mild steel corrosion inhibition in HCl by di-alkyl and di-1,2,3-triazole derivatives of uracil and thymine. *Mater Chem Phys*, (2014), 145: 407–417.

Espinoza Vázquez, R. González-Olvera, D. Moreno Cerros, G. E. Negrón Silva, I. A. Figueroa, F. J. Rodríguez Gómez, M. Castro, A. Miralrio, L. Huerta, Inhibition of acid corrosion in API 5L X52 steel with 1,2,3-triazole derivatized from benzyl alcohol: Experimental and theoretical studies, *Journal of Molecular Structure,* (2021), 1242, 130731.

Fouad Bentiss, Michel Traisnel, Léon Gengembre, Michel Lagrenée, A new triazole derivative as inhibitor of the acid corrosion of mild steel: electrochemical studies, weight loss determination, SEM and XPS, *Applied Surface Science*, 152(3–4), 1999, 237-249.

Fu J, H. Zang, Y. Wang, S. Li, T. Chen, and X. Liu, —Experimental and theoretical study on the inhibition performances of Quinoxaline and its derivatives for the corrosion of mild steel in hydrochloric acid, *Ind. Eng. Chem. Res,* (2012), 51918),6377 -6386.

Gökhan Gece, Semra Bilgiç, Quantum chemical study of some cyclic nitrogen compounds as corrosion inhibitors of steel in NaCl media, *Corrosion Science*, (2009), 51(8), 876-1878.

González Olvera R, Román Rodríguez V, Negrón Silva G,Espinoza Vázquez A, Rodríguez Gómez F, Santillan R. Multicomponent Synthesis and Evaluation of New 1,2,3-Triazole Derivatives of Dihydropyrimidinones as Acidic Corrosion Inhibitors for Steel. *Molecules* (2016) 21: 250.

Guo L, Zhu S, Zhang S, He Q, Li W. Theoretical studies of three triazole derivatives as corrosion inhibitors for mild steel in acidic medium. *Corros Sci*, (2014) 87: 366–375.

Gurudatt DM, Mohana KN, Tandon HC. Synthesis of new 1,2,4-triazole derivatives and their anticorrosion properties on mild steel in hydrochloric acid medium. *J Mol Liq,* (2015) 211: 275–287.

Hackerman N, "A view of the History of Corrosion and its Control," In: Gundry, R. D. Ed. Corrosion 93, *Planery and Keynote Lectures*, Texas: NACE, Houston, 1– 5, 1993.

Hamdy H. Hassan, Essam Abdelghani, Mohammed A. Amin, Inhibition of mild steel corrosion in hydrochloric acid solution by triazole derivatives: Part I. Polarization and EIS studies, *Electrochimica Acta*, (2007), 52(22), 6359-6366.

Hassan HH, Abdelghani E, Amin MA. Inhibition of mild steel corrosion in hydrochloric acid solution by triazole derivatives: Part I. Polarization and EIS studies. *Electrochim Acta*, (2007) 52: 6359–6366.

Hazazi, A. Fawzy, M. Awad, Synergistic Effect of Halides on the Corrosion Inhibition of Mild Steel in H2SO4 by a Triazole Derivative: Kinetics and Thermodynamic Studies, *International Journal of Electrochemical Science*, (2014), 9(8), 4086-4103.

Hrimla M, Bahsis L, Boutouil A, Laamari MR, Julve M, Stiriba SE. Corrosion inhibition performance of a structurally well-defined 1, 2, 3-triazole derivative on mild steel-hydrochloric acid interface. *J Mol Struct* (2021) 1231: 129895.

Israa Abd Alkadir Aziz, Iman Adnan Annon, Makarim H. Abdulkareem, Mahdi M. Hanoon, Mohammed H. Alkaabi, Lina M. Shaker, Ahmed A. Alamiery, Wan Nor Roslam Wan Isahak, Mohd S. Takr, Insights into Corrosion Inhibition Behavior of a 5-Mercapto-1, 2, 4-triazole Derivative for Mild Steel in Hydrochloric Acid Solution: Experimental and DFT Studies, *Lubricants*, (2021), 9(12), 122, https://doi.org/10.3390/lubricants9120122.

Jiang L, Lan Y, He Y, Li Y, Li Y, Luo J. 1,2,4-Triazole as a corrosion inhibitor in copper chemical mechanical polishing. *Thin Solid Films* (2014) 556:395–404.

Khaled KF, Molecular simulation, quantum chemical calculations and electrochemical studies for inhibition of mild steel by triazoles, *Electrochimica Acta*, (2008), 53(9), 3484-3492.

Khiati Z, Othman AA, Sanchez-Moreno M, Bernard MC, Joiret S, Sutter EMM, Vivier V. Corrosion inhibition of copper in neutral chloride media by a novel derivative of 1,2,4-triazole. *Corros Sci* (2011) 53:3092–3099.

Kumar MS, Loganathan S, Kumar A, Sreekanth A. Anticorrosion potential of 4-amino-3-methyl-1,2,4-triazole-5-thione derivatives (SAMTT and DBAMTT) on mild Steel in hydrochloric acid solution. *Ind Eng Chem Res* (2012) 51:5408–5418.

Lebrini M, F. Robert, and C. Roos, —Adsorption properties and inhibition of C38 steel corrosion in hydrochloric solution by some Indole derivates: Temperature effect, activation energies, and thermodynamics of adsorption, *Int. J. Corros*., (2013), 2013, 1–13.

Lei Guo, Shanhong Zhu, Shengtao Zhang, Qiao He, Weihua Li, Theoretical studies of three triazole derivatives as corrosion inhibitors for mild steel in acidic medium, *Corrosion Science*, (2014), 87, 366-375.

Merimi I, A. Bitari, Y. Kaddouri, N. Rezki, M. Mohamed, R. Touzani and B. Hammouti, Metal corrosion inhibition by triazoles: A review, *Int. J. Corros. Scale Inhib*., 2022, 11(2), 524–540.

Merimi, Y. EL Ouadi, R. Benkaddour, H. Lgaz, M. Messali, F. Jeffali, B. Hammouti, Improving corrosion inhibition potentials using two triazole derivatives for mild steel in acidic medium: Experimental and theoretical studies, Materials Today: *Proceedings*, (2019), 13(3), 920-930.

Mert BD, Mert ME, Kardaş G. Experimental and theoretical investigation of 3-amino-1,2,4-triazole-5-thiol as a corrosion inhibitor for carbon steel in HCl medium. *Corros Sci*, (2011) 53: 4265–4272.

Meryem Hrimla, Lahoucine Bahsis, My Rachid Laamari, Miguel Julve and Salah-Eddine Stiriba, An Overview on the Performance of 1,2,3-Triazole Derivatives as Corrosion Inhibitors for Metal Surfaces, *Int J Mol Sci*, (2021), 23(1):16. doi: 10.3390/ijms23010016.

Mohamed K. Awad, Mohamed R. Mustafa, Mohamed M. Abo Elnga, Computational simulation of the molecular structure of some triazoles as inhibitors for the corrosion of metal surface, *Journal of Molecular Structure: THEOCHEM* (2010) 959 66–74.

Mounim Lebrini, Michel Traisnel, Michel Lagrenée, Bouchaib Mernari, Fouad Bentiss, Inhibitive properties, adsorption and a theoretical study of 3,5-bis(n-pyridyl)-4-amino-1,2,4-triazoles as corrosion inhibitors for mild steel in perchloric acid, *Corrosion Science*, (2008), 50(2), 473-479.

Musa AY, Kadhum AA, Mohamad AB, Takriff MS, Daud AR, Kamarudin SK. On the inhibition of mild steel corrosion by 4-amino-5-phenyl-4H-1,2,4-trizole-3-thiol. *Corros Sci*, (2010)52: 526–533.

Nalini D, Sudha D and Sowmya Ramkumar, Corrosion Inhibition of Indoloimidazole Derivative on Mild Steel in HCl, *Che Sci Rev Lett*, (2013), 2(5), 323-331.

Nalini D, K. S. Kohilah, Sowmya Ramkumar, A Combined Experimental and Theoretical Investigation on Pyrazolone Derivative as Corrosion Inhibitor for Mild Steel in 0.5 M Sulphuric Acid Media, *Portugaliae Electrochimica Acta*, (2014), 32(2), 109-123.

Natasa Kovacevic, Ingrid Milosev and Anton Kokalj, How relevant is the adsorption bonding of imidazoles and triazoles for their corrosion inhibition of copper?, *Corrosion Science,* (2017), 124, 25-34.

Negrón Silva G, Cruz Gonzalez D, González Olvera R, Lomas Romero L, Gutiérrez Carrillo A, Palomar Pardavé M, Romero Romo M A, Santillan R, Uruchurtu J. One-Pot Three-Component Synthesis of New Mono- and Bis-1,2,3-triazole Derivatives of 2-Benzimidazolethiol with a Promising Inhibitory Activity against Acidic Corrosion of Steel. *Synthesis* (2014) 46 : 1217–1223.

Nuha Wazzan, I. B. Obot, Taghreed M. Fagieh, The role of some triazoles on the corrosion inhibition of C1020 steel and copper in a desalination descaling solution, *Desalination*, (2022), 527, 115551.

Ouici H. B., O. Benali, Y. Harek, L. Larabi, and B. Hammouti, —Influence of the 2-Mercapto-1-methyl imidazole (mmi) on the corrosion inhibition of mild steel in 5% HCl, *Int. J. Electrochem. Sci*., (2012)7, 2304–2319.

Parameswari K, S. Rekha, S. Chitra, and E. Kayalvizhy, —Study on the inhibition of mild steel corrosion by Benzoisoxazole and Benzopyrazole derivatives in H2SO4 medium, *Port. Electrochim. Acta*, (2010), 28(3), 189–201.

Phadke Swathi, Seranthimata Samshuddin, Aeshah Hassan Alamri, Kedila Rasheeda, Vijaya D. P. Alva, Talal A. Aljohani, Experimental and theoretical investigation of a new triazole derivative for the corrosion inhibition of carbon steel in acid medium, *Egyptian Journal of Petroleum*, (2022), 31(2), 15-21.

Pour-Ali S, Kiani-Rashid A, Babakhani A. Improved corrosion inhibition of 3-amino-1,2,4-triazole on mild steel electrode in HCl solution using surface nanocrystallization. *Int J Mater Res* (2016) 107:1031–1040.

Qi Ma, Sijun Qi, Xiaohong He, Yongming Tang, Gang Lu, 1,2,3-Triazole derivatives as corrosion inhibitors for mild steel in acidic medium: Experimental and computational chemistry studies, *Corrosion Science*, (2017), 129, 91-101.

Quraishi, M. A., and R. Sardar. "Aromatic Triazoles as Corrosion Inhibitors for Mild Steel in Acidic Environments." *Corrosion* 58 (2002).

Quraishi MA and D. Jamal, Inhibition of mild steel corrosion in the presence of fatty acid triazoles, *Journal of Applied Electrochemistry*, (2002)32: 425–430.

Quraishi MA and Danish Jamal, Fatty acid triazoles: Novel corrosion inhibitors for oil well steel (N-80) and mild steel, *Journal of American Oil Chemists' Society*, (2000), 77 (10), 1107 – 1111.

Rahmani, H., K. I. Alaoui, M. EL Azzouzi, F. Benhiba, A. El Hallaoui, Z. Rais, M. Taleb, A. Saady, B. Labriti, A. Aouniti, A. Zarrouk, Corrosion assessement of mild steel in acid environment using novel triazole derivative as a anti-corrosion agent: A combined experimental and quantum chemical study, *Chemical Data Collections*, (2019), 24, 100302.

Rahmouni K, Hajjaji N, Keddam M, Srhiri A, Takenouti H, The inhibiting effect of 3-methyl 1,2,4-triazole 5-thione on corrosion of copper in 3% NaCl in presence of sulphide. *Electrochim Acta*, (2007) 52:7519–7528.

Ramesh S, Rajeswari S, Maruthamuthu S, Corrosion inhibition of copper by new triazole phosphonate derivatives. *Appl Surf Sci*, (2004) 229:214–225.

Raja PB, Ismail M, Ghoreishiamiri S, Mirza J, Ismail MC, Kakooei S, Rahim AA, Reviews on corrosion inhibitors: a short view. *Chem Eng Commun*, (2016) 203:1145–1156.

Rajkumar G, Sethuraman MG . Corrosion protection ability of self-assembled monolayer of 3-amino-5-mercapto-1,2,4-triazole on copper electrode. *Thin Solid Films* (2014) 562:32–36.

Ramzi. T. T. Jalgham, Theoretical, Monte Carlo Simulations and QSAR Studies on Some Triazole Derivatives as Corrosion Inhibitors for Mild Steel in 1 M HCl, *ES Energy & Environment, (2021), 13, 37-49.*

Rao BVA, Reddy MN, Sreedhar B. Self-assembled 1-octadecyl-1H-1,2,4-triazole films on copper for corrosion protection. *Prog Org Coat* (2014) 77:202–212.

Rouifi Z, F. Benhiba, M. El Faydy, T. Laabaissi, H. About, H. Oudda, I. Warad, A. Guenbour, B. Lakhrissi, A. Zarrouk, Performance and computational studies of new soluble triazole as corrosion inhibitor for carbon steel in HCl, *Chemical Data Collections*, (2019), 22100242, https://doi.org/10.1016/j.cdc.2019.100242.

Rugmini Ammal, M. Prajila, Abraham Joseph, Physicochemical studies on the inhibitive properties of a 1,2,4-triazole Schiff's base, HMATD on the corrosion of mild steel in hydrochloric acid, *Egyptian Journal of Petroleum*, (2018), 27(3), 307-317.

Sampath S, Vadivelu M, Ravindran R, Perumal P T, Velkannan V, Karthikeyan K. Synthesis of 1,2,3-Triazole Tethered 3-Hydroxy-2-oxindoles: Promising Corrosion Inhibitors for Steel in Acidic Medium and Their Anti-Microbial Evaluation. *ChemistrySelet* (2020) 5 : 2130–2134.

Sam John, K Mohammad Ali & Abraham Joseph, Electrochemical, surface analytical and quantum chemical studies on Schiff bases of 4-amino-4H-1, 2, 4-triazole-3,5-dimethanol (ATD) in corrosion protection of aluminium in 1N HNO_3, *Bull Mater Sci*, (2011), 34, 1245–1256. https://doi.org/10.1007/s12034-011-0234-x.

Sastri V. S., "*Corrosion Inhibitors: Principles and Applications*," Reprinted in 2001, copyrighted to John – Wiley & Sons Ltd, England.

Sherif ESM, El Shamy AM, Ramla MM, El Nazhawy AOH, 5-(Phenyl)-4H-1,2,4-triazole-3-thiol as a corrosion inhibitor for copper in 3.5% NaCl solutions. *Mater Chem Phys*, (2007) 102:231–239.

Sherif ESM, Erasmus RM, Comins JD, Corrosion of copper in aerated synthetic sea water solutions and its inhibition by 3-amino-1,2,4-triazole. *J Colloid Interface Sci*, (2007) 309:470–477.

Soumoue A, El Ibrahimi B, El Issami S, Bazzi L, Some triazolic compounds as corrosion inhibitors for copper in sulphuric acid. *Int J Sci Res*, (2014) 3:349–354.

Sounthari, A. Kiruthika, S. Chitra, and K. Parameswari, —Oxadiazole polyamides: new and effective corrosion inhibitors for mild steel in 1M H_2SO_4, *Int. J. Recent Sci. Res.*, (2013), 4, 1185–1194.

Sowmya Ramkumar, D. Nalini, Correlation between Inhibition Efficiency and Chemical Structure of New Indolo Imidazoline on the Corrosion of Mild Steel in Molar HCl with DFT Evidences, *Oriental Journal of Chemistry*, (2015), 31(2), 1057-1064.

Sowmya Ramkumar and D. Nalini, Enhanced corrosion resistance of mild steel in sulphuric acid medium by imidazole derivative: Experimental and computational studies, *Materials Today: Proceedings*, (2019a) 18, 1696–1708.

Sowmya Ramkumar, D. Nalini, M. A. Quraishi, Eno E. Ebenso, Chandrabhan Verma, Anti-corrosive property of bioinspired environmental benign imidazole and isoxazoline heterocyclics: A cumulative studies of experimental and DFT methods, *J Heterocyclic Chem*. (2019b), 1–17.

Stanley Udochukwu Ofoegbu, Tiago L. P. Galvão, José R. B. Gomes, João Tedim, Helena I. S. Nogueira, M. G. S. Ferreira and M. L. Zheludkevichac, Corrosion inhibition of copper in aqueous chloride solution by 1H-1,2,3-triazole and 1,2,4-triazole and their combinations: electrochemical, Raman and theoretical studies, *Phys Chem Chem Phys*, (2017), 19(8):6113-6129. doi: 10.1039/c7cp00241f.

Sudheer, M. A. Quraishi, Electrochemical and theoretical investigation of triazole derivatives on corrosion inhibition behavior of copper in hydrochloric acid medium, *Corrosion Science*, (2013), 70, 161-169.

Tang, YM., Chen, Y., Yang, WZ. et al., Electrochemical and theoretical studies of thienyl-substituted amino triazoles on corrosion inhibition of copper in 0.5 M H2SO4 . *J Appl Electrochem*, (2008), 38, 1553–1559 https://doi.org/10.1007/s10800-008-9603-6.

Turuvekere Krishnamurthy Chaitra, Kikkeri Narasimha Shetty Mohana, Harmesh Chander Tandon, Thermodynamic, electrochemical and quantum chemical evaluation of some triazole Schiff bases as mild steel corrosion inhibitors in acid media, *Journal of Molecular Liquids*, (2015), 211, 1026-1038.

Tourabi, K. Nohair, M. Traisnel, C. Jama, F. Bentiss, Electrochemical and XPS studies of the corrosion inhibition of carbon steel in hydrochloric acid pickling solutions by 3,5-bis(2-thienylmethyl)-4-amino-1,2,4-triazole, *Corrosion Science*, (2013), 75, 123-133.

Walsh F, "Faraday and his Laws of Electrolysis," *Bulletin of Electrochem*, (1991), 7(11), 481–489.

Wang, Y. Wan, Q. Wang, and Y. Ma, —Study of inhibitive effect of 1,4-Bis(benzimidazolyl) benzene on mild steel corrosion in 0.5 HCl and 0.25M H2SO4 solutions, *Int. J. Electrochem. Sci*., vol. 8, pp. 806–820, 2013.

Wang L, Zhu MJ, Yang FC, Gao CW. Study of a triazole derivative as corrosion inhibitor for mild steel in phosphoric acid solution. *Int J Corros*, (2012) 2012: 6.

Xu QJ, Li CX, Zhou GD, Zhu LJ, Lin CJ, Copper corrosion inhibition and adsorption behavior of 3-amino-1,2,4-triazole. *Acta Phys Chim Sin* (2009) 225:86–90.

Yu. I. Kuznetsov, Triazoles as a class of multifunctional corrosion inhibitors, *Int. J. Corros. Scale Inhib*., 2021, 10(1), 29-53. doi: 10.17675/2305-6894-2021-10-1-2.

Yu J, Jiang L, Gan F. Inhibition of copper corrosion in deionized water by 3-amino-5-mercapto-1,2,4-triazole. *J Chin Soc Corros Prot* (2010) 30:21–24.

Zarrouk A, Hammouti B, Al-Deyab SS, Salghi R, Zarrok H, Jama C, Bentiss F. Corrosion inhibition performance of 3,5- diamino-1,2,4-triazole for protection of copper in nitric acid solution. *Int J Electrochem Sci* (2012) 7:5997–6011.

Zhang, S, Tao, Z, Liao, S, Wu, F, Substitutional adsorption isotherms and corrosion inhibitive properties of some oxadiazoltriazole derivative in acidic solution. *Corros. Sci*, (2010), 52, 3126–3132.

Chapter 8

Bioactive 1,2,3-Triazoles: Applications in Agrochemistry

Poliana Aparecida Rodrigues Gazolla[1], PhD
Ângela Maria Almeida Lima[2], PhD
Alex Ramos de Aguiar[1], PhD
Samira Soares Santiago[1]
Danilo Aniceto da Silva[2]
Roberta Silva Evangelista[1]
Adilson Vidal Costa[2], PhD
and Róbson Ricardo Teixeira[1,*], PhD

[1]Grupo de Síntese e Pesquisa de Compostos Bioativos (GSPCB), Departamento de Química, Universidade Federal de Viçosa, Viçosa, Minas Gerais, Brazil
[2]Grupo de Estudo Aplicado em Produtos Naturais e Síntese Orgânica (GEAPS), Departamento de Química e Física, Universidade Federal do Espírito Santo, Alegre, Espírito Santo, Brazil

Abstract

In modern agriculture, farmers continuously face a battle to increase quality and productivity of crops. The optimization of agriculture efficiency demands, along with other requirements, the application of chemical crop protection agents (pesticides) to control a variety of pests, such weeds, insects, and fungi. Nowadays, there are several available pesticides that help farmers to keep under control a myriad of pests. However, the use of pesticides typically results in the development of

* Corresponding Author's Email: robsonr.teixeira@ufv.br.

In: Triazoles and Their Uses
Editor: Karl Roland
ISBN: 979-8-88697-800-1

species resistance to them. Moreover, increasing public concern with regard to environmental pollution deriving from agricultural practice strictly requires that pesticides would be endowed with low recalcitrance, and thus may be rapidly mineralized by the soil microflora. Low toxicity to mammals, high specificity, low application rates, and fast soilborne microorganism degradation are thus important characteristics sought in the development of new pesticides. In the search and development of new pesticides, researchers have prepared compounds containing the 1,2,3-triazole functionality and investigated their potential as agrochemicals. This book chapter will focus on the developments that have been reported concerning the bioactivities of 1,2,3-triazole in the agrochemistry field. Compounds presenting phytotoxic, insecticide, and fungicide activities will be covered. Results regarding the fungicide activity of 1,2,3-triazole based compounds from our research group will be described. In addition, the synthetic steps involved in the preparation of the bioactive 1,2,3-triazoles will be discussed.

Keywords: insecticides, fungicides, herbicides, click chemistry, CuAAC reaction, agrochemicals, 1,2,3-triazole

Introduction

The world population reached 8 billion people on 15th November 2022, having grown by 1 billion since 2010. It is projected that the population will surpass 9 billion around 2037 and 10 billion around 2058 (UN DESA, 2022).

The increase in world population poses several challenges to be faced by humanity, including scarcity of water, environmental degradation, adequate education, housing, and an increase in food production (United Nations, 2022a).

Food production needs to double by 2050 to feed an ever-increasing world population (United Nations, 2022b; Dijk et al., 2021). In this regard, productivity in agriculture must be increased which, among other necessities, requires the control of several pests. The most common strategy used in this regard corresponds to the employment of chemical agents collectively called pesticides. Herbicides (for the control of weeds) (Krämer and Shimer, 2007a), fungicides (that control fungi) (Krämer and Shimer, 2007b), and insecticides (that control insects) (Krämer and Shimer, 2007c) are the most consumed pesticides around the world (Phillips, 2020).

Nowadays, several available pesticides help farmers to keep control of a myriad of pests (Chio and Li, 2022). However, the use of pesticides typically results in the development of species resistant to them (HRAC, 2023; IRAC, 2023; FRAC, 2023). Moreover, increasing public concern about environmental pollution deriving from agricultural practice strictly requires that pesticides would be endowed with low recalcitrance, and thus may be rapidly mineralized by the soil microflora. Low toxicity to mammals, high specificity, low application rates, and fast soilborne microorganism degradation are thus important characteristics in the development of new pesticides. All these facts demonstrate that there is a demand for the development of new pesticides (Sparks and Lorsbach, 2017; Sparks et al., 2018).

In the search and development of new pesticides, researchers have prepared compounds containing the 1,2,3-triazole functionality and investigated their potential as agrochemicals. This book chapter will focus on the developments that have been reported concerning the bioactivities of 1,2,3-triazoles containing compounds in the agrochemistry field.

1,2,3-Triazoles

The five-membered ring heterocyclic compounds containing three nitrogen atoms are termed triazoles. There are two isomeric forms of triazoles: 1,2,3-triazole and 1,2,4-triazole (Figure 1).

Figure 1. Structures of 1,2,3- and 1,2,4-triazole rings.

Considering the 1,2,3-triazoles, they present aromatic character and remarkable stability under hydrolytic, oxidative, or reductive conditions. They are also stable towards enzymatic degradation. Besides, the 1,2,3-triazoles are capable of establishing hydrogen bonds and dipole-dipole interactions with biological receptors, mimics the amide group, and presents high dipole moment (Ali, 2021). All these features make this ring attractive to be employed in the research of new bioactive compounds. Numerous medicinal attributes have been described for compounds bearing this type of heterocyclic ring (Dheer, Singh, and Shankar, 2017; Bozorov, Zhao, and Aisa, 2019; Tron et al., 2008; Zhang, 2019; Agalave, Maujan, and Pore, 2011; Jiang et al., 2019;

Ali, 2021; da Silva et al., 2014; Alam, 2021; Xu, Zhao, and Liu, 2019; Forezzi et al., 2021; Razzaghi-Asl et al., 2020). Several drugs utilized nowadays in the clinical present the 1,2,3-triazole ring in their structures (Figure 2).

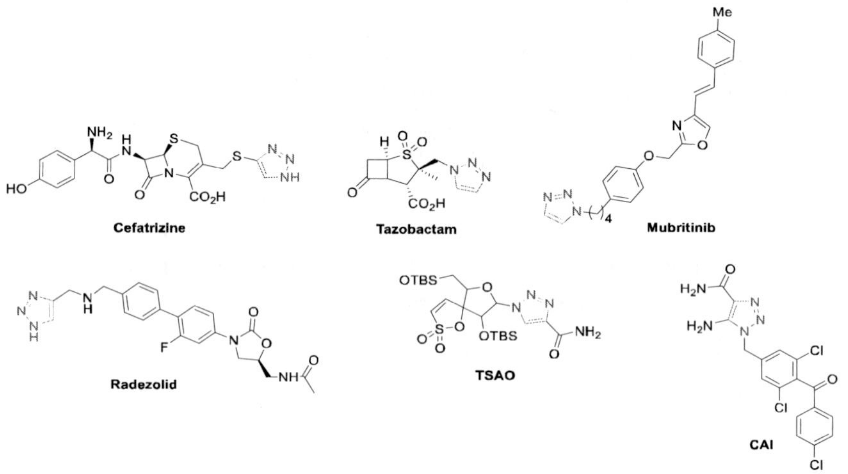

Figure 2. Structures of commercial drugs containing 1,2,3-triazole rings. The 1,2,3-triazole moiety is highlighted in red.

In addition to pharmaceuticals, the 1,2,3-triazoles have found applications in several fields such as organic synthesis, polymer, supramolecular, and carbohydrate chemistry, and material science (Liang and Astruc, 2011; Sykam, Donempudi, and Basak, 2022; Moses and Moorhouse, 2007; Kuswaha et al., 2013; Xi et al., 2014; Nandiva, Jiang, and Lahann, 2007).

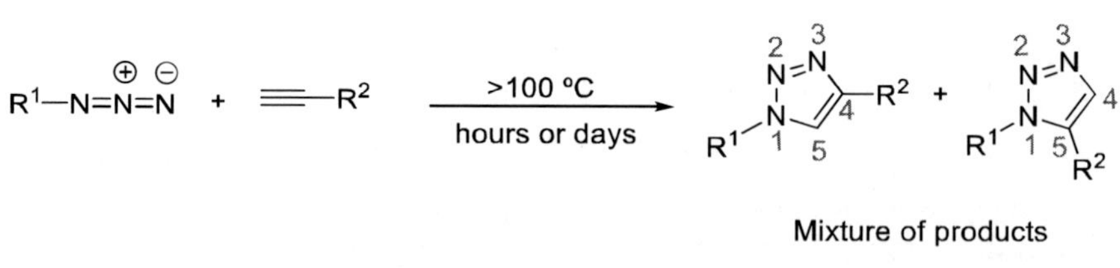

Figure 3. Husgein cycloaddition reaction.

From the preparation standpoint, there are several available methodologies to synthesize 1,2,3-triazoles (Aswar, 2014; Haldón, Nicasio, and Pérez, 2015; Ali, 2021; Dai et al., 2022). However, the most widely used methodology is the 1,3-dipolar cycloaddition reaction between alkynes and azides. This transformation was discovered by Husgein and collaborators in

1960 (Heravi et al., 2016; Breugst and Reissig, 2020). They reported that the thermal reaction between a terminal alkyne and an organic azide afforded 1,2,3-triazoles (Figure 3).

This reaction presented several drawbacks such as long reaction times, the necessity of high temperatures for its occurrence, and the formation of regioisomers (1,4- and 1,5-disubstituted 1,2,3-triazoles) as shown in Figure 3.

Independent investigations conducted by the research groups of Sharpless (Rostovtsev et al., 2002) and Meldal (Tornøe, Christensen, and Meldal, 2002) demonstrated that the presence of copper(I) species in the reaction media not only accelerate the cycloaddition but also afforded exclusively the 1,4-disubstituted-1,2,3-triazole (Figure 4). Since then, this reaction has been called the copper(I)-catalyzed alkyne-azide cycloaddition reaction (CuAAC reaction).

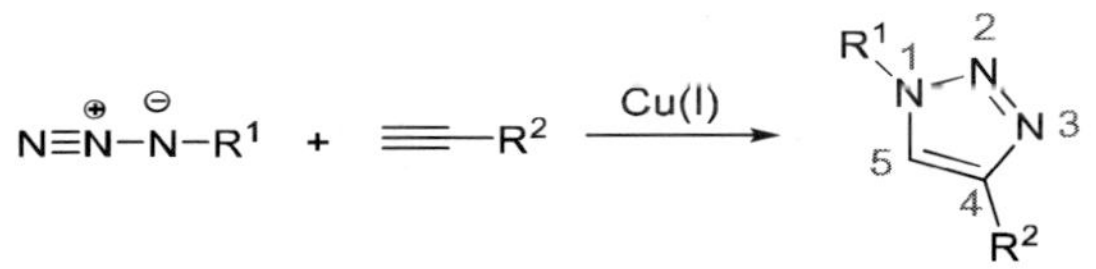

Figure 4. General scheme for the CuAAC reaction.

It should be mentioned that in 2001, Sharpless and co-workers published a paper introducing the concept of click chemistry reaction. They defined a series of criteria that a reaction has to meet to be considered a click process.

"The reaction must be modular, wide in scope, give very high yields, generate only inoffensive byproducts that can be removed by nonchromatographic methods, and be stereospecific (but not necessarily enantioselective). The required process characteristics include simple reaction conditions (ideally, the process should be insensitive to oxygen and water), readily available starting materials and reagents, the use of no solvent or a solvent that is benign (such as water) or easily removed, and simple product isolation. Purification – if required – must be by nonchromatographic methods, such as crystallization or distillation, and the product must be stable under physiological conditions." (Kolb, Finn, and Sharpless, 2001). The CuAAC reaction is one of the best examples of processes that meet the click reaction criteria. Therefore, the CuAAC reaction has also been called the click reaction.

In Figure 5, it is shown the catalytic cycle of the CuAAC reaction (Meldal and Tornøe, 2008).

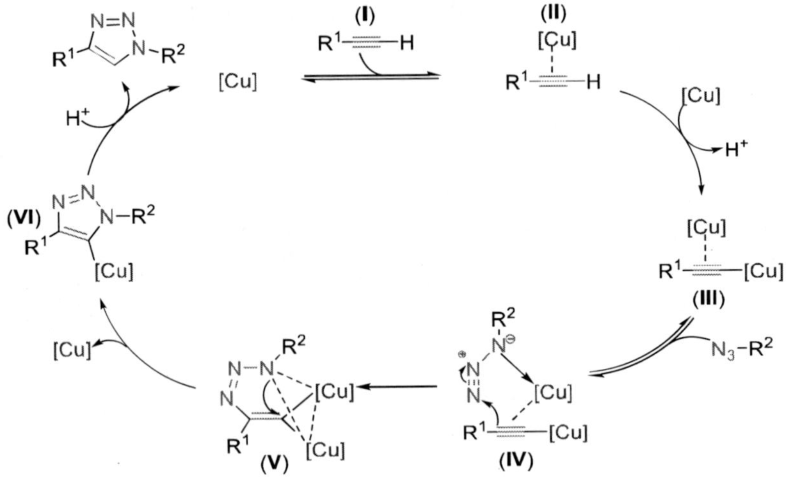

Figure 5. Catalytic cycle of CuAAC reaction.

The first step involves the coordination of terminal alkyne (**I**) with Cu(I) species to give the complex (**II**). This coordination decreases the pKa of the terminal alkyne from 26 to approximately 9, which facilitates the removal of a proton from species (**II**). Coordination of a second Cu(I) species gives rise to complex (**III**) which, in turn, coordinates with azide affording the complex (**IV**). From the dicopper species (**IV**) there is the first C-N bond formation in a six-membered metallacycle (**V**). Then, an intramolecular C-N bond formation yields the triazolyl-Cu(I) intermediate and dissociation of one of the two coppers lead to the formation of (**VI**). In the last step of the catalytic cycle, a proton release affords the 1,2,3-triazole-1,4-disubstituted and the regeneration of Cu(I) species.

The CuAAC reaction is affected by several factors including Cu(I) source, ligands on Cu(I) catalyst, and reactivity of alkyne and azide substrates (Haldón, Nicasio, and Pérez, 2015; Meldal and Tornøe, 2008, Singh, Chowdhury, and Koley, 2016; Freitas et al., 2011).

As quoted above, the 1,2,3-triazoles have found applications in several fields, including agrochemistry. In the following sections, it will be described the advances in this field regarding the use of 1,2,3-triazole functionality in the preparation of compounds endowed with phytotoxic, insecticide, and fungicide activities.

1,2,3-Triazoles with Phytotoxic Activity

The compound known as tyrosol (**1**) was isolated from cultures of the fungus *Alternaria euphorbiicola*. This natural product caused bleaching and necrotic lesions on leaves of the weed species *Euphorbia Heterophylla* at a concentration as low as 80 μmol L^{-1} (Varejão et al., 2013). The structure of (**1**) was modified to produce a series of tyrosol 1,2,3-triazole derivatives (compounds **3-25**), Figure 6). The preparation of compounds (**3-17**) involved the convertion of tyrosol (**1**) into 4-(2-bromoethyl)phenol via treatment of (**1**) with HBr. After that, the reaction of 4-(2-bromoethyl)phenol with sodium azide gave (**2**). A series of phenols, a 2-pyridone, and a coumarin were propargylated with propargyl bromide to afford the corresponding terminal alkynes. The CuAAC reaction between these alkynes and azide (**3**) afforded the compounds (**3-25**). On the other hand, tyrosol (**1**) was propargylated to furnish alkyne (18). The CuAAC reaction involving (**18**) and benzylic azides (obtained from benzyl bromides) gave 1,2,3-triazolic compounds (**19-25**). The triazolic derivatives were evaluated against the weed species *E. heterophylla*, *Conyza sumatrensis*, and *Cammelia benghalensis*. The tyrosol derivatives did not present herbicidal effect on *C. benghalensis* and *C. sumatrensis*. On the contrary, almost all the derivatives produced yellowing, bleaching, and necrosis on the leaves of *E. heterophylla* as determined by the perforated leaf assay. The compounds (**12**), (**18**), (**23**), and (**25**) had effects that were statistically equal or higher than diquat (herbicide used as positive control). These four compounds were also evaluated with respect to their capability of interfering with photosynthetic parameters. Among them, compounds (**12**) and (**25**) stood out interfering with stomatal conductance, net photosynthesis, internal carbon concentration, transpiration rate, water-use efficiency, and chlorophyll *a* and *b* contents. The most active compounds (**12**) and (**25**) were also tested on leaves of corn and soybean. These crops are among the ones that suffer from intense competition with *E. heterophylla*. None of the compounds produced lesions on the leaves of both crops, suggesting that the compounds are endowed with selectivity. No direct correlation between structure of the compounds and their activities were found (Franco et al., 2022).

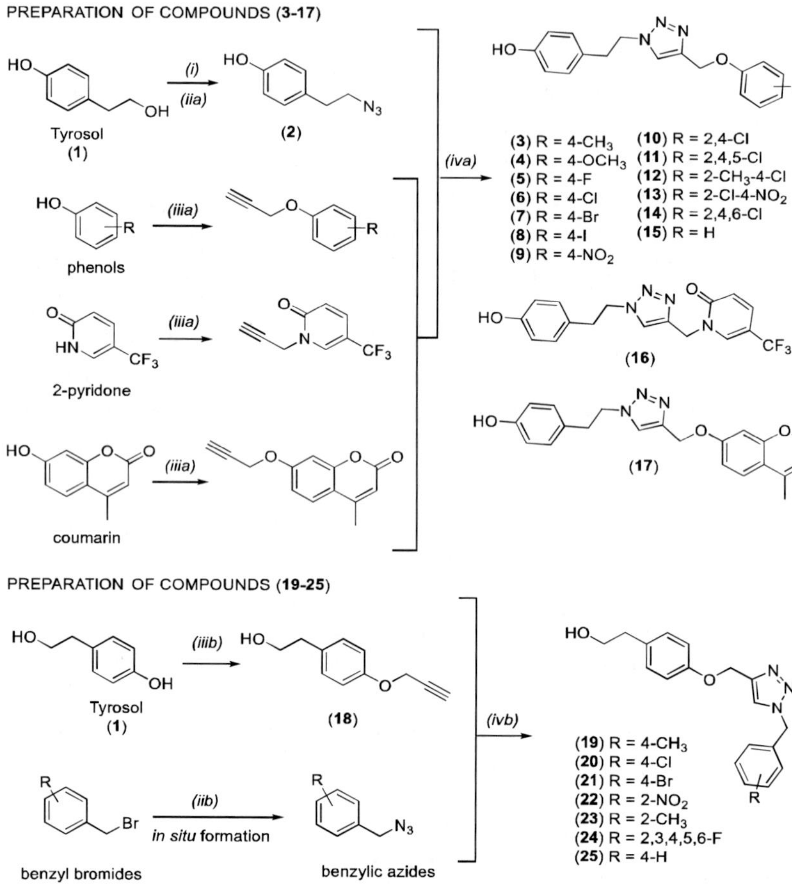

Figure 6. Synthesis of tyrosol 1,2,3-triazole derivatives. Reagents and conditions: (*i*) HBr 48% (w w^{-1}), 80 °C, 70% yield; (*iia*) NaN_3 (1.5 eq.), dimethylacetamide (DMAC), r.t., 4-5 h; (*iiia*) propargyl bromide solution in toluene (80% w w^{-1}, 1.5 eq.), anhydrous K_2CO_3 (2.0 eq.), anhydrous acetonitrile, nitrogen atmosphere, 50 °C, 24 h, 55-99% yield; (*iva*) $CuSO_4 \cdot 5H_2O$ (0.15 eq.), sodium ascorbate (0.3 eq.), DMAC/H_2O (4:1 v v^{-1}), r.t., 12 h, 27-98% yield (one-pot); (*iib*), sodium azide (1.5 eq.), DMSO (6 mL), r.t., 12 h; (*iiib*) propargyl bromide solution in toluene (80% w w^{-1}, 2.0 eq.) dissolved in anhydrous acetone (2.5 mL), anhydrous K_2CO_3 (3.0 eq.), anhydrous KI (10 mol%), anhydrous acetone (7.5 mL), nitrogen atmosphere, 50 °C, 24 h, 90% yield; (*ivb*) azide, alkyne (1.0 eq.), sodium ascorbate (0.4 eq.), $CuSO_4 \cdot 5H_2O$ (0.2 eq.), CH_2Cl_2/H_2O (4:1 v v^{-1}), r.t., 12 h, 70-83% yield.

One aspect deserves comment at this point. Looking at the conditions used to carry out the CuAAC reaction (Figure 6), someone realizes that the copper source corresponds to $CuSO_4 \cdot 5H_2O$, which is a Cu(II) containing species.

However, the presence of sodium ascorbate in the reaction media reduces Cu(II) to Cu(I), generating *in situ* the species that accelerates and makes the CuAAC reaction regioselective. In fact, $CuSO_4 \cdot 5H_2O$/sodium ascorbate is the most utilized combination to carry out cycloaddition reactions involving terminal alkynes and azides.

Borgatti and co-workers (2013) synthesized a series of thirteen 1,2,3-triazoles, most of them containing halogen atoms. Figure 7 depicts the steps involved in the preparation of ten of those compounds (triazoles **29a-j**).

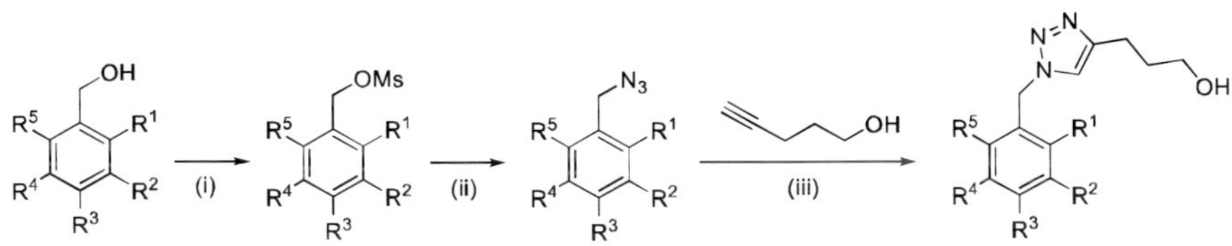

(26a-j) (27a-j) 77-100% (28a-j) 72-96% (29a-j) 44-90%

$R^1 = R^2 = R^3 = R^4 = R^5 = H$:	(**26-29a**)
$R^1 = R^2 = R^4 = R^5 = H$, $R^3 = F$:	(**26-29b**)
$R^1 = R^2 = R^4 = R^5 = H$, $R^3 = Cl$:	(**26-29c**)
$R^1 = R^2 = R^4 = R^5 = H$, $R^3 = Br$:	(**26-29d**)
$R^1 = R^2 = R^4 = R^5 = H$, $R^3 = I$:	(**26-29e**)
$R^1 = R^2 = R^4 = R^5 = H$, $R^3 = OCF_3$:	(**26-29f**)
$R^1 = R^2 = R^4 = R^5 = H$, $R^3 = CF_3$:	(**26-29g**)
$R^1 = R^4 = R^5 = H$, $R^2 = R^3 = Br$:	(**26-29h**)
$R^1 = Cl$, $R^2 = R^3 = R^5 = H$, $R^4 = Br$:	(**26-29i**)
$R^1 = R^3 = R^5 = Cl$, $R^2 = R^4 = H$:	(**26-29j**)

Figure 7. Synthetic steps involved in the preparation of compounds (**29a-j**). Reagents and conditions: (*i*) MsCl, Et_3N, CH_2Cl_2, -50 ºC, 30 min; (*ii*) NaN_3, DMSO, r.t., 15 h; (*iii*) pent-4-yn-1-ol, sodium ascorbate (40 mol%), $CuSO_4 \cdot 5H_2O$ (20 mol%), CH_2Cl_2/H_2O (50% v v^{-1}), r.t., 24 h.

Thus, commercially available benzylic alcohols (**26a-j**) were converted into the mesylates (**27a-j**) via treatment of the alcohols with mesyl chloride. Then, the S_N2 reaction between compounds (**27a-j**) and sodium azide gave the benzylic azides (**28a-j**). In the last step, the CuAAC reaction involving azides (**28a-j**) and pent-4-yn-1-ol resulted in the 1,2,3-triazoles (**29a-j**). The other three 1,2,3-triazoles (**30a**, **30b**, and **33**) were obtained according to the steps shown in Figure 8. The reactions utilized in the preparation of these three compounds were similar to that shown in Figure 7 (Borgatti et al., 2013).

The phytotoxicity of 1,2,3-triazoles were evaluated against the target species *Allium cepa* (onion, monocotyledonous species), *Lactuca sativus* (lettuce, dicotyledonous species), and *Cucumis sativus* (cucumber, dicotyledonous species). These species were chosen since they are sensitive and fast-responsive, even at low toxicity levels. It produces small seeds in large quantities, which present standardized germination and rapid initial development, in addition to having low cost and being easily purchased at most local markets. The effect of the compounds (inhibition or stimulation) on germination and radicle growth (shoot and root growth) were evaluated at 1 x 10^{-4} mol L^{-1}, 1 x 10^{-6} mol L^{-1}, and 1 x 10^{-8} mol L^{-1}. At the highest concentration, the evaluated compounds (**29a-j**), (**30a-b**), and (**33**) showed predominantly inhibitory effects against the evaluated species. The most active compound corresponded to the derivative (**32**), which inhibited almost completely the parameters evaluated. Compared to 2,4-D (a positive control), the derivative (**33**) presented similar activity on *L. sativus* and *C. sativus* and superior effect on *A. cepa* (Borgatti et al., 2013).

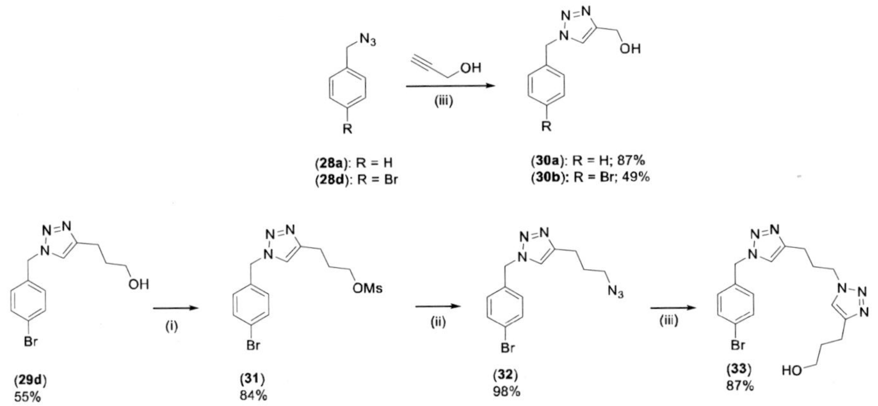

Figure 8. Steps involved in the preparation of 1,2,3-triazole (**30a**, **30b**, and **33**). Reagents and conditions: (*i*) MsCl, Et_3N, CH_2Cl_2, -50 °C, 30 min; (*ii*) NaN_3, DMSO, r.t., 15 h; (*iii*) pent-4-yn-1-ol, sodium ascorbate (40 mol%), $CuSO_4{\cdot}5H_2O$ (20 mol%), CH_2Cl_2/H_2O (50% v v^{-1}), r.t., 24 h.

Our research group has explored the glycerol containing 1,2,3-triazole scaffold. Within this scenario, the glycerol derivatives (**38a-i**) were prepared according to the steps shown in Figure 9. Thus, glycerol (**34**) was initially converted into the acetonide (**35**). Then, treatment of (**35**) with tosyl chloride gave the tosylate (**36**). The S_N2 reaction involving compound (**36**) and sodium azide resulted in the formation of organic azide (**37**). Finally, the CuAAC reaction between (**37**) and different terminal alkynes afforded the glycerol-containing 1,2,3-triazole derivatives (**38a-i**) (Costa et al., 2020).

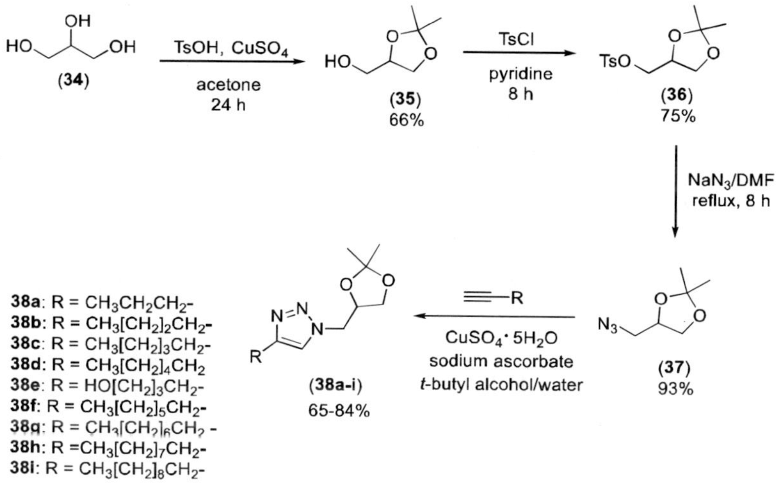

Figure 9. Preparation of glycerol containing 1,2,3-triazole derivatives (**38a-i**).

The phytotoxicity of compounds (**38a-i**) were evaluated against *L. sativa* at five different concentrations (1, 10, 100, 500, and 1000 μg mL^{-1}). The parameters percentage of germination and germination speed index were evaluated every 8 h for 48 h. After 48 h, the root length was measured for the indication of root growth. The germination was the less affected parameter. Except for compound (**38c**) at 1000 μg mL^{-1} (35% of germination inhibition), the other treatments were statistically identical to the negative control (dichloromethane + water). The derivatives (**38a**), (**38b**), (**38c**), (**38d**), (**38f**), and (**38h**) influenced the speed index and root length, particularly at 500 μg mL^{-1} and 1000 μg mL^{-1}. The most active compound corresponded to (**37c**), which at 1000 μg mL^{-1} reduced germination index and root growth by 68.53% and 85.78%, respectively. Those results were statistically similar to the herbicide pichloran used as a positive control (Costa et al., 2020).

Maslinic acid (**39**) is a triterpene found in *Olea europea* L. (Oleaceae). This natural compound was converted into 1,2,3-triazole hybrid derivatives (**46a-g**) as depicted in Figure 10. The hybrids present a phthalimide moiety tethered to maslinic acid by a 1,2,3-triazole functionality. Initially, maslinic acid (**39**) was converted into the propargylated derivative (**40**). The azides (**42-44** and **45a-d**) were obtained from phthalimide (**41**) according to the reactions shown in Figure 10. Then, the CuAAC reaction involving these azides and alkyne (**39**) derived from maslinic acid gave the hybrid compounds (**46a-g**) (Nejma et al., 2018).

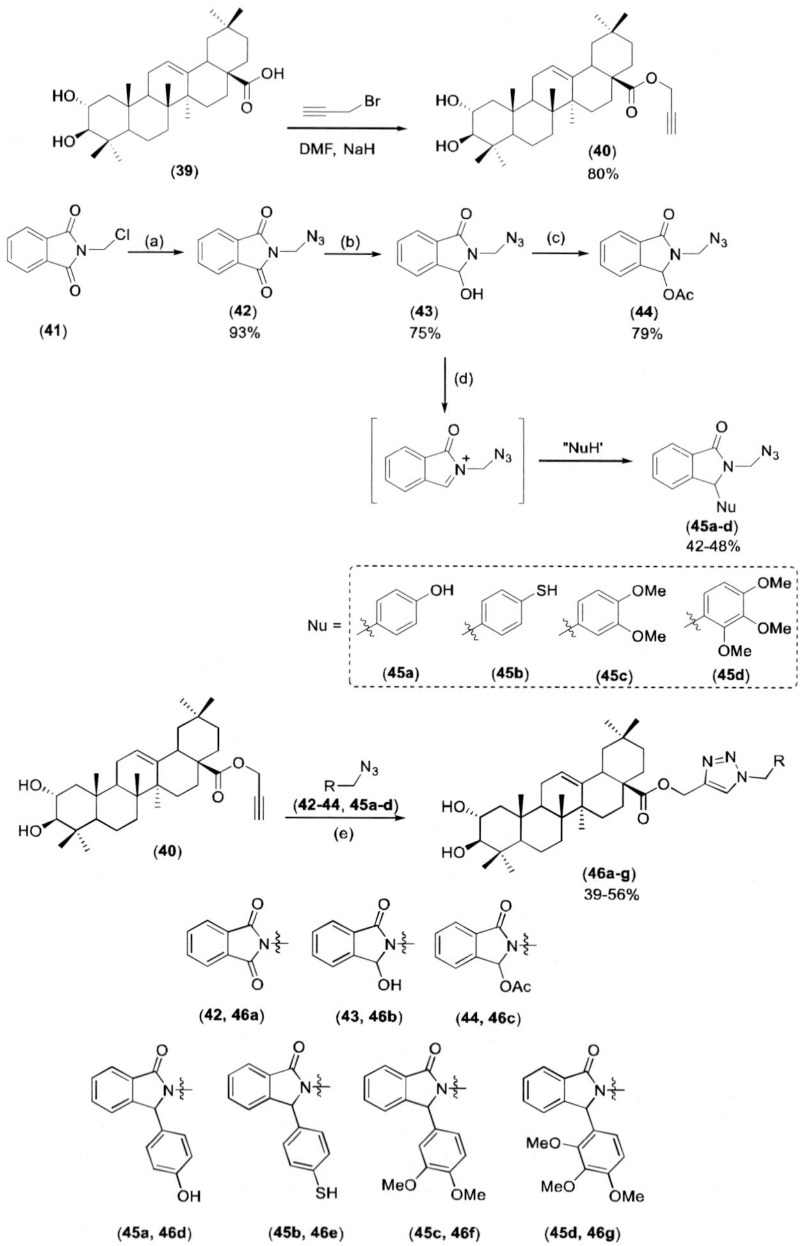

Figure 10. Conversion of maslinic acid (**39**) in 1,2,3-triazole derivatives (**46a-g**). (a) NaN_3, DMSO, 24 h, r.t.; (b) $LiEt_3BH$, CH_2Cl_2, 30 min, -80 °C; (c) Ac_2O, pyridine, 24 h, r.t.; (d) CH_2Cl_2, r.t., $Bi(OTf)_3$; (e) CuI, Et_3N, DMF, MW-250 W, 3-5 min.

The phytotoxicity of compounds (**46a-g**) was assessed against the model plant *L. sativa.* The authors also evaluated the effects of maslinic acid (**39**), the propargylated derivative (**40**), and the azides (**42-44** and **45a-d**). At 500 $\mu g\ mL^{-1}$, the hybrid derivatives presented superior effect on germination than the other evaluated compounds, showing percentages of germination inhibition varying from 91.79% to 100%. Seed germination was completely suppressed by compound (**46e**). The authors did not use a positive control in the experiments. In this sense, a comparison of the efficiency of the evaluated compounds and a commercial herbicide was not determined.

Insecticide Activity of 1,2,3-Triazoles

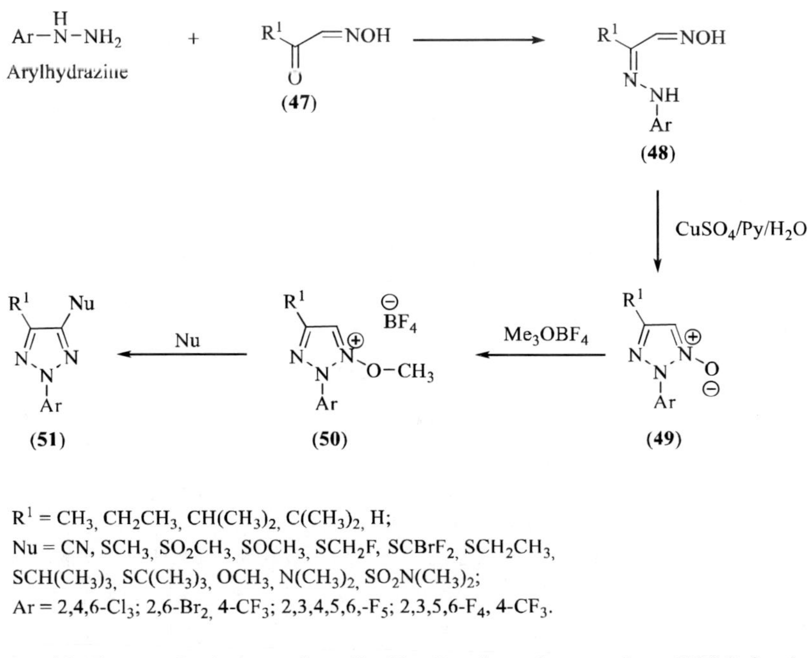

$R^1 = CH_3, CH_2CH_3, CH(CH_3)_2, C(CH_3)_2$, H;
Nu = CN, SCH_3, SO_2CH_3, $SOCH_3$, SCH_2F, $SCBrF_2$, SCH_2CH_3, $SCH(CH_3)_3$, $SC(CH_3)_3$, OCH_3, $N(CH_3)_2$, $SO_2N(CH_3)_2$;
Ar = 2,4,6-Cl_3; 2,6-Br_2, 4-CF_3; 2,3,4,5,6,-F_5; 2,3,5,6-F_4, 4-CF_3.

Figure 11. First synthetic route described by Boody and co-workers (1996) for the preparation of 1,2,3-triazoles (**51**).

It is outlined in Figures 11 and 12, two synthetic routes described by Boddy and co-workers for the preparation of 2-aryl-1,2,3-triazoles. In these cases, the preparation of the compounds did not involve the use of the CuAAC reaction. In the first synthetic route, arylhydrazine ($ArNHNH_2$) reacted with keto-oxime (**47**), affording oximinohydrazono (**48**). Compound (**48**) underwent copper-catalyzed cyclization to give cyclic *N*-oxide (**49**). Methylation of (**49**) gave rise to (**50**) that, in turn, reacted with suitable nucleophiles resulting in the formation of 1,2,3-triazoles (**51**) (Boddy et al. 1996).

The second route (Figure 12) involved the reaction of arylhydrazines ($ArNHNH_2$) with *bis*-oxime (**52**) to give hydrazone (**53**). Then, the monoacetylation of (**53**) followed by cyclization of (**54**) promoted by K_2CO_3 or Cs_2CO_3 afforded triazole-4-carbaldoxime (**55**). Hydrolyses of (**55**) gave the aldehyde (**56**), that was subsequently reacted with $NaBH_4$ to afford alcohol (**57**). Finally, conversion of the alcohol (**57**) into the corresponding mesylate followed by reaction of it with nucleophiles resulted in 1,2,3-triazoles (**58**) (Boddy et al. 1996).

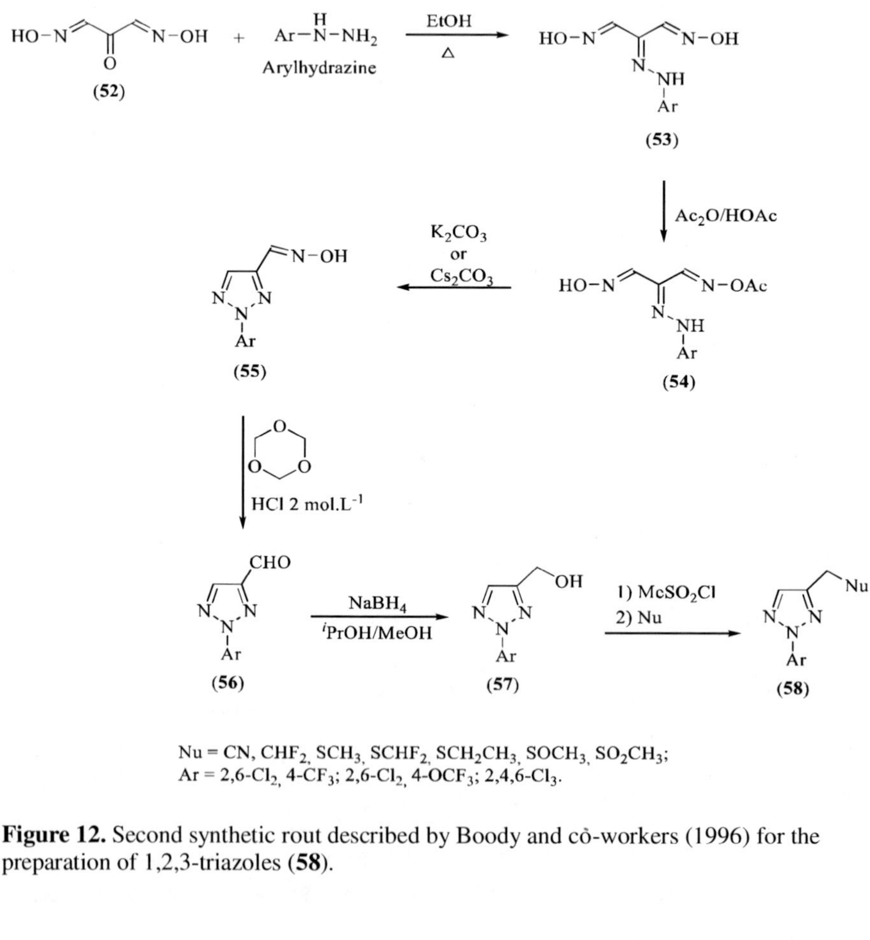

Figure 12. Second synthetic rout described by Boody and cô-workers (1996) for the preparation of 1,2,3-triazoles (**58**).

A total of thirty-eight 1,2,3-triazole compounds were synthesized using the two synthetic routes aforementioned. The compounds were biologically evaluated against ticks (*Boophilus microplus*), sheep blowfly (*Lucilia sericata*), housefly (*Musca domestica*), and cockroach (*Blatella germanica*). The activity of the compounds was particularly good against housefly and the structure of the most effective against this species is presented in Figure 13.

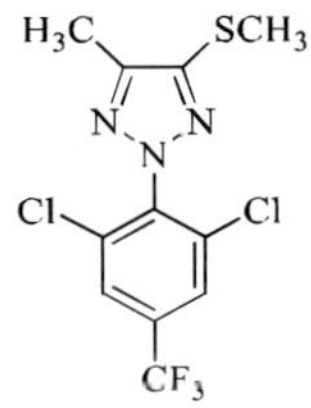

(59)

Figure 13. Structure of the most active 1,2,3-triazole (**59**) against housefly as described by Body and co-workers.

The triazole (**59**) presented three important characteristics, namely spectrum of activity (it was active against all the evaluated species, Table 1), potency, and residual activity (Boddy et al. 1996).

Table 1. Biological activity data concerning compound 59

Compound	Activity against				
	B. Microplus		*L. sericata*[a]	*M. domestica*[c]	*B. germanica*[c]
	Larval[a]	**Reproductive**[b]			
59	30	3	3	0.3-1	10-30

[a] Corresponds do lethal dose (LD_{50}) in mg L^{-1}.
[b] Corresponds do effective dose (ED_{50}) in μg per tick.
[c] Corresponds do lethal dose (LD_{50}) in mg m^{-2}.

According to the *International Union of Pure and Applied Chemistry* (IUPAC), bioisoster can be defined as "a compound resulting from the exchange of an atom or group of atoms with another, broadly similar, atom or group of atoms" (Wermuth et al., 1998). The amide group and the 1,2,3-triazole ring are bioisosteres which is associated with the similarities between them, namely *i*) their size (distance between substituents 3.8-3.9 Å in amides,5.0-5.1 Å in triazoles), *ii*) their dipolar character (amide *ca.* 4 debye, triazole *ca.* 5 debye), and *iii*) their H-bond acceptor capacity (the lone pairs of the amide O as well as of the triazole nitrogens as illustrated in (**A**) and (**B**) in

Figure 14). The triazole C(5)–H was considered to act as an H-bond donor similar to the amide NH.

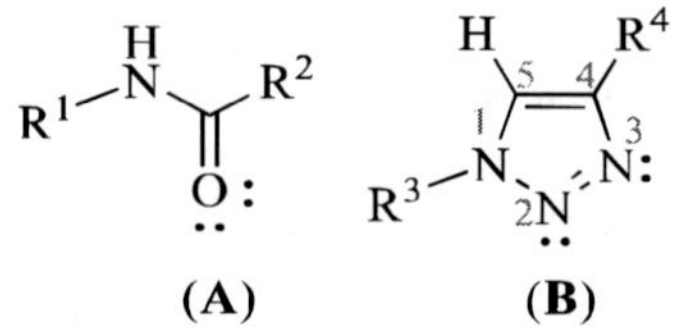

Figure 14. The bioisosters amide and 1,2,3-triazole ring.

Bioisosterism is an effective way to design bioactive compounds (Lima and Barreiro, 2005).

Taking the structure of chlorantranilipole as the source of inspiration and considering the bioisosterism concept, Mao and co-workers (2015) designed the synthesis of a series of 1,2,3-triazoles presenting the general structure (**60**), Figure 15.

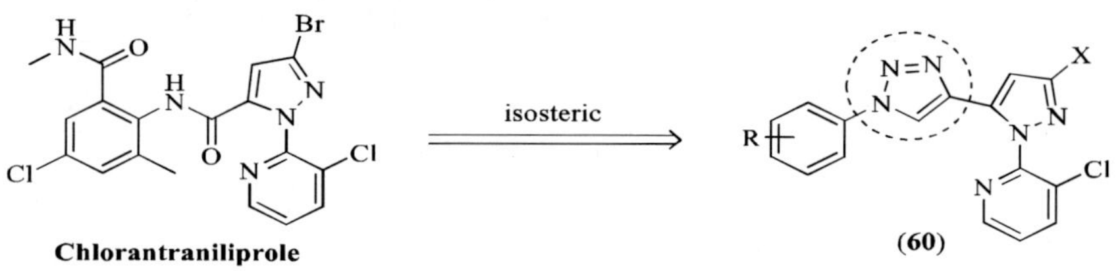

Figure 15. Chlorantraniliprole and 1,2,3-triazole of general structure (**60**).

The key step involved in the preparation of the compounds corresponded to the CuAAC reaction as depicted in Figure 16 (Mao et al., 2015).

When evaluated against the oriental army work (*Mythimna separata*), a severe pest in cereals, the compounds presented different degrees of efficiency. The triazoles were tested at 200, 100, 50, 25, and 5 mg mL^{-1} and the results are presented in Table 2.

The derivatives (**60c-f**, **60h-l**) were active only at the highest concentration. The most active was compound (**60b**), which at 200, 100, 50, 25, and 10 mg L^{-1} displayed 100% larvicidal activity, a result which is comparable to the positive control chlorantraniliprole. It also deserves comment the activity presented by compound (**60g**), albeit it is less active than (**60b**) (Mao et al., 2015).

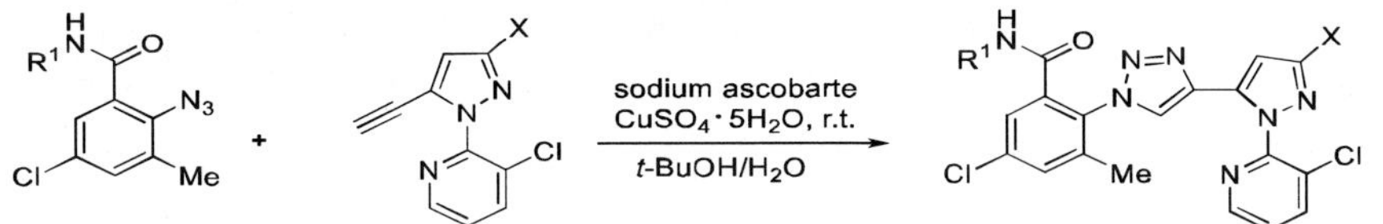

(60a-h)

(60a): R^1 = Me; X = Br
(60b): R^1 = ^{i}Pr; X = Br
(60c): R^1 = t-Bu; X = Br
(60d): R^1 = Bu; X = Br
(60e): R^1 = Pr; X = Br
(60f): R^1 = Me; X = Cl
(60g): R^1 = ^{i}Pr; X = Cl
(60h): R^1 = t-Bu; X=Cl

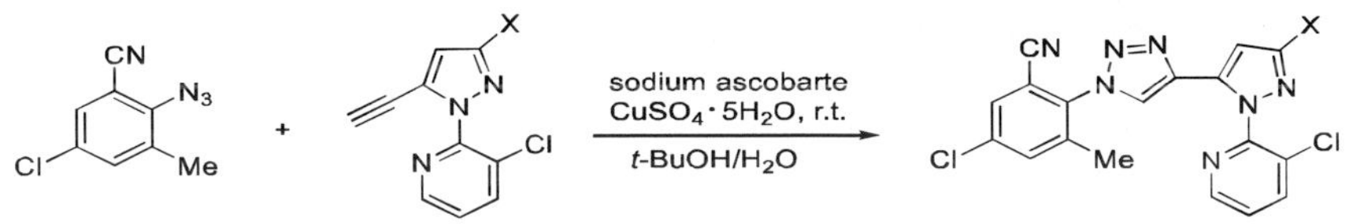

(60i-j)

(60i): X = Br
(60j): X = Cl

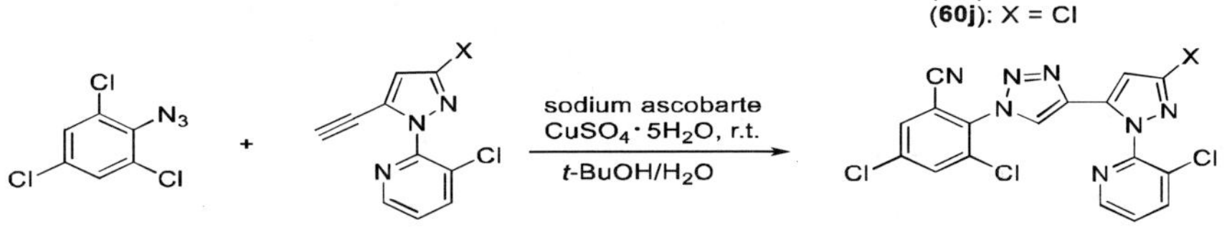

(60k-l)

(60k): X = Br
(60l): X = Cl

Figure 16. Preparation of 1,2,3-triazoles (**60a-l**).

Table 2. Biological activity data of compounds (60a-l) on *M. separata*

Compound	Larvicidal activity (%) at a concentration of (mg mL^{-1})					
	200	**100**	**50**	**25**	**10**	**5**
60a	100	13.3	-	-	-	-
60b	100	100	100	100	100	40
60c	60	-	-	-	-	-
60d	15	-	-	-	-	-
60e	56.7	-	-	-	-	-
60f	80	-	-	-	-	-
60g	100	100	100	100	60	-
60h	80	-	-	-	-	-
60i	20	-	-	-	-	-
60j	60	-	-	-	-	-
60k	40	-	-	-	-	-
60l	30	-	-	-	-	
Chlorantraniliprole	100	100	100	100	100	100

(-): no activity

Su and collaborators (2013) modified the commercial pesticide mandipropamid by replacing the 4-chlorophenyl group by a 1,2,3-triazole functionality (Figure 17).

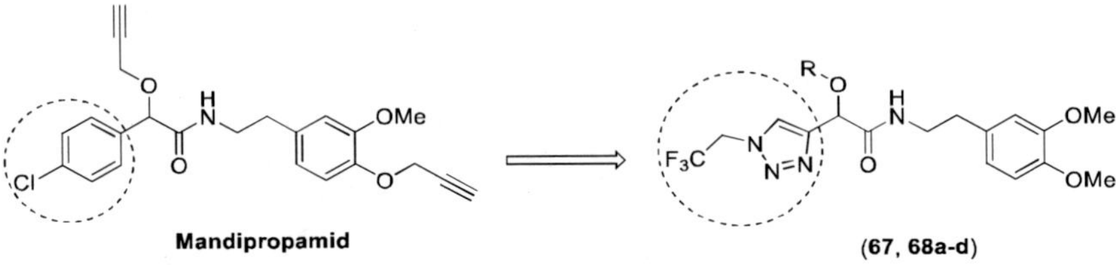

Figure 17. Mandipropamid and general structures of the analogues prepared by Su and collaborators (2013).

The synthetic steps involved in the preparation of triazolic derivatives (**67**) and (**68a-d**) are depicted in Figure 18. The CuAAC reaction was utilized to prepare the intermediate (**66**), which upon standard functional group interconversions gave rise to compounds (**67**) and (**68a-d**) (Su et al., 2013).

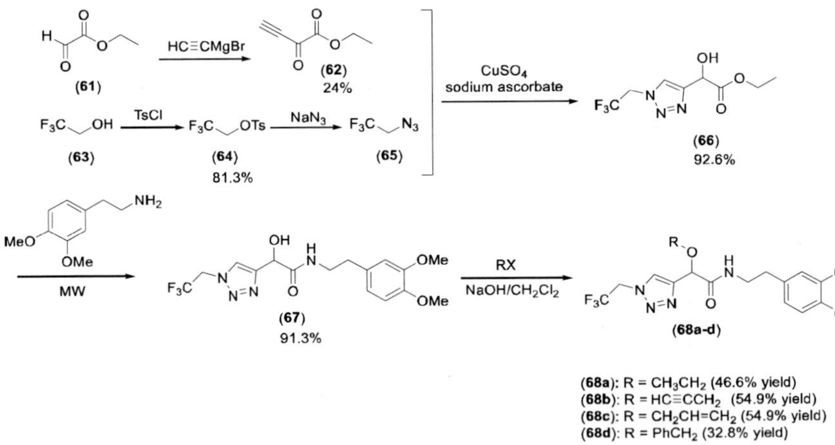

Figure 18. Steps involved in the preparation of mandipropamid analogues (**67**) and (**68a-d**).

The 1,2,3-triazoles (**67**) and (**68a-d**) presented moderate larvicidal activity (30-50%) against the fourth-instar larvae of *Culex pipiens pallens*, also known as common house mosquito (Su et al., 2013).

Fungicide Activity of 1,2,3-Triazoles

As quoted above, our research group has explored the bioactiviy profile of glycerol containin 1,2,3-triazole derivatives. Within this scenario and utilizing a synthetic route similar to that shown in Figure 9, we have investigated the fungicide activity of several 1,2,3-triazoles derived from glycerol. Table 3 shows the general structure of compounds belonging to this scaffold synthesized and biologically evaluated by us. The compounds (**37a**-**37i**), (**69a-h**), and (**70a-k**) were evaluated against *Colletotrichum* fungal species which cause papaya anthracnose. Regarding the compounds (**37a**-**37i**), they were assessed against *Colletotrichum gloesporioides* and presented lower efficiency in inhibiting the mycelial growth compared to the commercial fungicide tebuconazole. The compounds (**37d**, ED_{50} = 18.14 ppm) and (**37i**, ED_{50} = 10.14 ppm) were the most active, although less efficient than tebuconazole (ED_{50} = 0.26 ppm) used as a positive control. On the contrary, the derivatives (**37a**-**37i**) displayed inhibitory activity on sporulation compared to tebuconazole (ED_{50}< 1 ppm). It stood out compound (**37e**) which presented ED_{50} = 0.0002 ppm (Costa et al., 2020).

Table 3. Structures of 1,2,3-triazoles derived from glycerol

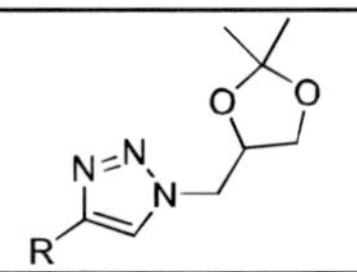

Compound	R group	Compound	R group
37a	$CH_3CH_2CH_2$-	**69f**	$CH_3CH(OH)CH_3$
37b	$CH_3[CH_2]_2CH_2$-	**69g**	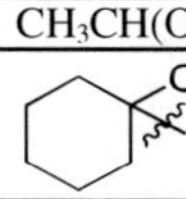
37c	$CH_3[CH_2]_3CH_2$-	**69h**	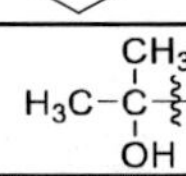
37d	$CH_3[CH_2]_4CH_2$-	**70a**	Phenyl
37e	$HO[CH_2]_3CH_2$-	**70b**	3-fluorophenyl
37f	$CH_3[CH_2]_5CH_2$-	**70c**	4-fluorophenyl
37g	$CH_3[CH_2]_6$ CH_2-	**70d**	2-fluorophenyl
37h	$CH_3[CH_2]_7CH_2$-	**70e**	3,4-difluorophenyl
37i	$CH_3[CH_2]_8CH_2$-	**70f**	2,4-difluorophenyl
69a	$HOCH_2$-	**70g**	3,5-difluorophenyl
69b	$HOCH_2CH_2$-	**70h**	4-trifluoromethylphenyl
69c	$HOCH_2CH_2CH_2$-	**70i**	3-trifluoromethylphenyl
69d	$CH_3CH(OH)CH_2$-	**70j**	2-trifluoromethhylphenyl
69e	$CH_3CH(OH)$-	**70k**	3,5-bis(trifluoromethyl)phenyl

The glycerol derivatives (**69a-69h**) were also evaluated against *C. gloeosporioides* and the compounds (**69f**) (ED_{50} = 394.80 μg mL^{-1}) and (**69g**) ED_{50} = 492.02 μg mL^{-1}) were the more efficient in controlling the mycelial growth of the fungal species, albeit their activity was much lower than tebuconazole (ED_{50} = 0.26 μg mL^{-1}). In terms of sporulation, compounds (**69g**) (ED_{50} = 0.44 μg mL^{-1}) and (**69h**) (ED_{50} = 0.83 μg mL^{-1}) inhibited it with efficiency compared to tebuconazole (ED_{50}< 1 μg mL^{-1}) (Costa et al., 2017). More recently, we prepared a series of glycerol-fluorinated triazole derivatives (**70a-k**) (Barcelos et al., 2022). Evaluation of them against *C. gloesporioides* showed that (**70d**) (ED_{50} = 59.14 μg mL^{-1}) was slightly more active than commercial fungicide tebuconazole (ED_{50} = 61.35 μg mL^{-1}). Compound (**70d**) presented attractive physicochemical features for agrochemical purposes as revealed by the calculated physicochemical parameters.

Considering fluorinated-triazole derivatives, our research group prepared several compounds derived from eugenol (**71**) (Figure 19).

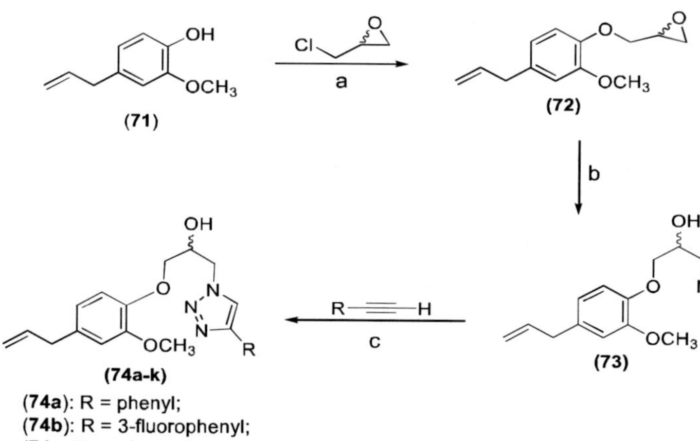

(**74a**): R = phenyl;
(**74b**): R = 3-fluorophenyl;
(**74c**: R = 4-fluorophenyl;
(**74d**): R = 2-fluorophenyl;
(**74e**): R = 3,4-difluorophenyl;
(**74f**): R = 2,4-difluorophenyl;
(**74g**): R = 3,5-difluorophenyl;
(**74h**): R = 4-(trifluoromethyl)phenyl;
(**74i**): R = 3-(trifluoromethyl)phenyl;
(**74j**): R = 2-(trifluoromethyl)phenyl;
(**74k**): R = 3,5-*bis*(trifluoromethyl)phenyl

Figure 19. Synthesis of 1,2,3-triazole compounds (**74a-k**) derived from eugenol. (a) KOH, $(Bu)_4NBr$, 0 °C → r.t. (yield 88%); (b) ammonium chloride, NaN_3, CH_3OH, H_2O, reflux, 60 °C (yield 94%); (c) sodium ascorbate, $CuSO_4 \cdot 5H_2O$, CH_2Cl_2, H_2O, r.t. (yield 48-80%).

In the first step, epoxide (**72**) was prepared via S_N2 reaction between eugenol (**71**) and epichlorohydrin, in the presence of potassium hydroxide and tetrabutylammonium bromide. Subsequently, epoxide was treated with sodium azide and ammonium chloride, resulting in the formation of azide (**73**). In the third and last step, aiming to evaluate the effect of the R-group linked to the triazole portion on the biological activity of the compounds, eleven eugenol derivatives (**74a-k**) were synthesized via CuAAC reaction between azide (**73**) and different fluorinated aromatic alkynes (Lima et al., 2022). The evaluation of eugenol (**71**) and derivatives (**74a-k**) against *Colletotrichum* sp. (CM01) utilizing the well- diffusion method showed that derivatives exhibited a moderate inhibitory effect (inhibition zone 2.3–5.1 mm) on the mycelial growth of the phytopathogen, when compared with the effect of the fungicide tebuconazole (inhibition zone 19.0 mm) and eugenol itself (10.7 mm). The most active eugenol derivative corresponded to (**74d**). When the derivatives were evaluated in terms of their inhibitory effect on conidial germination and fungal growth, the minimum inhibitory concentration (MIC) values were within 1.00 and 5.00 mg mL^{-1}, being compounds (**74d**) (MIC = 1.50 mg mL^{-1}) and (**77f**) (MIC = 1.02 mg mL^{-1}) the most effective derivatives. Eugenol (MIC < 0.500 mg mL^{-1}) and tebuconazole (MIC < 0.500 mg mL^{-1}) were again more active than derivatives (**74a-k**). Eugenol (**71**) presents recognized fungicide activity and the results described above demonstrated that it is more active on *Colletotrichum* sp. than compounds (**74a-k**). However, it should be mentioned that the use of eugenol (**71**) for agrochemical purposes has limitations due to its low stability since it is easily oxidized or decomposed in the presence of oxygen, light, or heat (Lima et al., 2022).

By using the synthetic sequence shown in Figure 20, Dai and collaborators obtained twenty-six 1,2,3-triazole phenylhydrazone derivatives.

In the first step, aromatic anilines were converted into aromatic azides via the diazotization process. Then, the CuAAC reaction between azides and propargyl alcohol gave 1,2,3-triazoles (**75**). After oxidation with MnO_2, the triazoles were converted to aldehydes (**76**) that, in turn, were condensed with different substituted phenylhydrazines to afford derivatives (**77**). The R^1 and R^2 in the structures of compounds (**77**) corresponded to halogens (F, Cl, Br), OMe, Me, and CF_3 at different positions of the aromatic rings. The fungicide activity of 1,2,3-triazoles (**77**) was assessed on *Rhizoctonia solani*, *Sclerotinia sclerotiorum*, *Fusarium graminearum*, and *Phytophthora capsici*. The most active compound (**78**) (Figure 20) exhibited significant *in vitro* anti-phytopathogenic activity, with EC_{50} values of 0.18 (*R. solani*), 2.28 (*S. sclreotiorum*), 1.01 (*F. graminearum*), and 1.85 (*P. capsici*) μg mL^{-1}.

Considering the species *R. solani*, the mentioned compound was more effective than the commercial fungicide carbendazim (1.42 μg mL^{-1}) and validamycin A (5.07 μg mL^{-1}). *In vivo* assessment demonstrated that (**78**) was effective in the control of rice sheath blight, rape sclerotinia rot, and fusarium head blight. This investigation also described the built of tridimensional Quantitative-Structure Activity Relationship (3D-QSAR), so that it allows the exploitation of this scaffold to obtain more active 1,2,3-triazole phenylhydrazones derivatives (Dai et al., 2015).

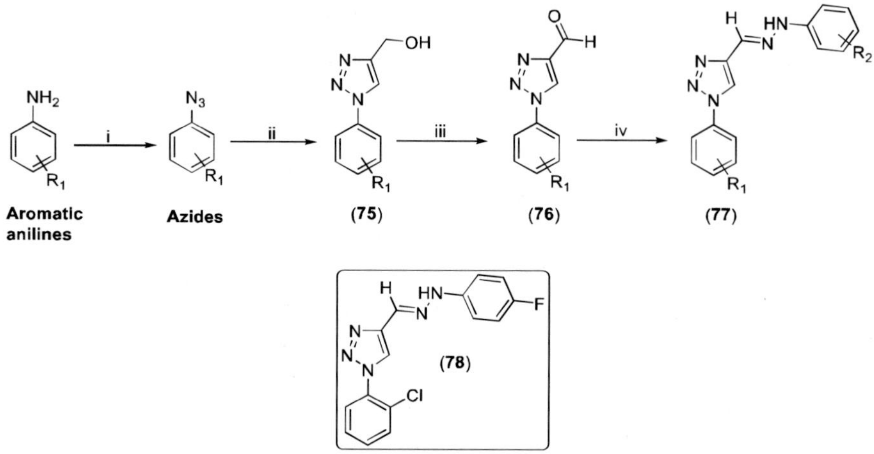

Figure 20. Preparation of 1,2,3-triazole phenylhydrazone derivatives (**77**) and structure of the more active compound (**78**) described by Dai and collaborators (2015). Reagents and conditions: (i) $NaNO_2$, HCl (10%), 0-5 °C; NaN_3, 24 h, r.t. (60-100% yield); (ii) propargyl alcohol, $CuSO_4$, sodium ascorbate, H_2O/*t*-butanol = 1:1 v v^{-1}, 24 h, r.t. (40-80% yield); (iii) MnO_2/EtOAc, 1 h, r.t. (85-100% yield); (iv) MeOH, substituted phenylhydrazine, 0.5 h (68-85% yield).

In a subsequent investigation, Chen and co-workers continued to explore the 1,2,3-triazole phenylhydrazone scaffold. After a bioactive-guided investigation, it was found that compound (**79**) was a promising fungicide candidate for further development. This compound displayed an EC_{50} value of 0.28 μg mL^{-1} against *F. graminearum*, which is 3.6 times lower than the aforementioned reported EC_{50} (1.01 μg mL^{-1}). In addition, derivative (**79**) also exhibited good inhibitory activities against *R. solani* and *S. sclerotiorum* with EC_{50} values of 0.86 and 1.66 μg mL^{-1}, respectively. *In vivo* evaluation of (**79**) at 200 μg mL^{-1} demonstrated that it effectively suppresses the disease development of fusarium head blight, one of the most serious diseases in cereals, with a protective efficacy of 80.6% (Chen et al., 2020).

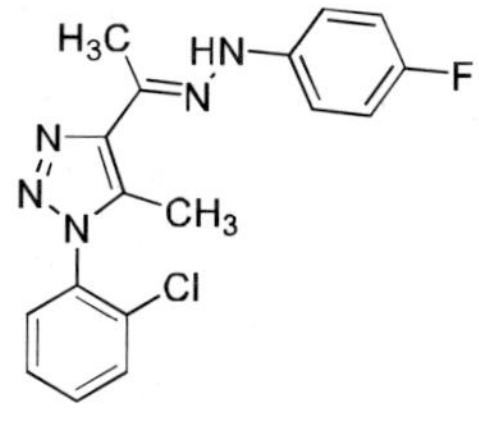

(**79**)

Figure 21. Structure of 1,2,3-triazole phenylhydrazone (**79**).

Oleanolic acid (**80**) was used as starting material for the preparation of eighteen 1,2,3-triazoles, according to the transformations outilined in Figure 22 (Chen et al., 2022).

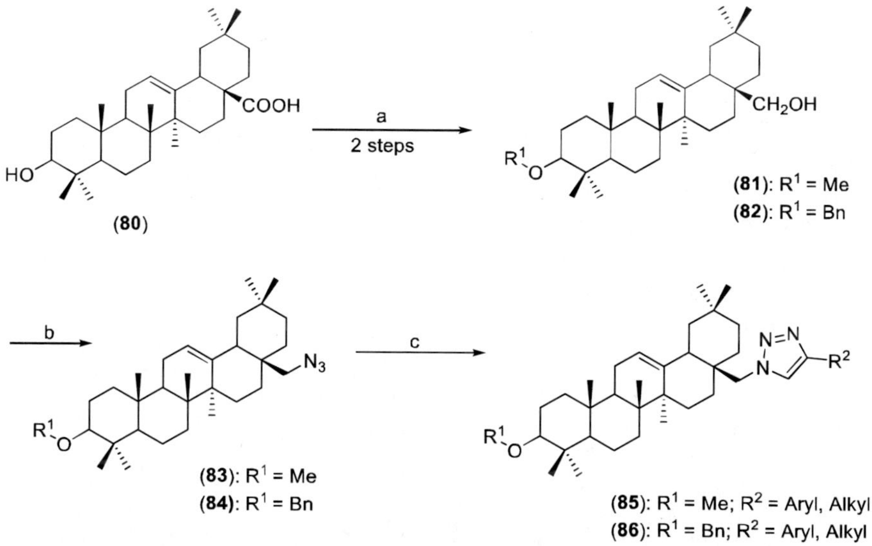

Figure 22. Synthetic steps involved in the preparation of 1,2,3-triazoles derived from oleanolic acid (**80**). Reagents and conditions: (step a): (1) MeI/BnBr, NaH, THF, 0 °C to 40 °C, 12 h; (2) LAH, THF, 0 °C to r.t., 4 h; 75% for (**81**), 72% for (**82**); (step b): Under N_2 protection, Tf_2O, pyridine, 0 °C to r.t., 15 min, then NaN_3, DMF, 60 °C; 87% to (**83**), 89% to (**84**); (step c): $CuSO_4{\cdot}5H_2O$, sodium ascorbate, then substituted phenylacetylene or alkyl alkyne, CH_3OH/H_2O (1:1 v v^{-1}), r.t., 1 h, 77-86% for (**85**), 74-89% for (**86**).

Compound (**80**) was converted into (**81**) and (**82**) after alkylation and reduction reactions. Conversion of (**81**) and (**82**) into the corresponding triflates followed by reaction with sodium azide resulted in the preparation of (**83**) and (**84**). The CuAAC reaction between azides and different terminal alkynes resulted in triazolic oleanolic acid derivatives (**85**) and (**86**). The antifungal activity of compounds (**85**) and (**86**) was tested on six target species, *S. sclerotiorum*, *Phytophtora boehmeriae Saw*, *Botrytis cinerea Pers*, *R. solani Kuhn*, *Pyricularia oryzae Cav.*, and *Fusarium oxysporum Schl. F.sp. vasinfectum (Atk.) Snyd. & Hans*, at 50 μg mL^{-1}, using the mycelial growth inhibition bioassay. The best inhibitory effects were observed on *R. solani Khun* for compounds (**85a-c**) and (**86a-c**) as presented in Table 4.

Table 4. Results of the antifungal activity on *R. solani Khun* of the most active 1,2,3-triazole oleanolic acid derivatives

Compound	**R^1**	**R^2**	**Growth inhibition (%)**
85a	CH_3	4-Cl-C_6H_4	85.6
85b	CH_3	4-NO_2-C_6H_4	83.1
85c	CH_3	4-F-C_6H_4	87.6
86a	Bn	4-Cl-C_6H_4	86.8
86b	Bn	4-NO_2-C_6H_4	87.7
86c	Bn	4-F-C_6H_4	89.6
Oleanolic acid (**80**)	-	-	20.1
Chlorothalonil	-	-	92.7

R^1 and R^{2}correspond to the groups shown in structures (**88**) and (**89**) in Figure 22.

It should be noted that the most active compounds on *R. solani* present as a common feature electronegative atoms (Cl or F) or group (NO_2) attached to the *para*-position of the aromatic ring in the structures of these compounds. The fungicide activity of triazolic derivatives (**85a-c**) and (**86a-c**) are superior than the parent compound oleanolic acid (**80**). Besides, the derivatives presented lower efficiency than the commercial fungicide chlorothalonil used as a positive control (Chen et al., 2022).

Two synthetic routes were designed to prepare a series of 1,2,3-carboxamides. In the synthetic route show in Figure 23, aromatic azides (prepared from aromatic anilines) were converted into 1,2,3-triazolic amines (**87a-h**). The amidation reactions of compounds (**87a-h**) with acyl chlorides gave the carboxamides (**88Aa-h, 88Ba-h,** and **88Ca-h**) (Yan et al., 2019).

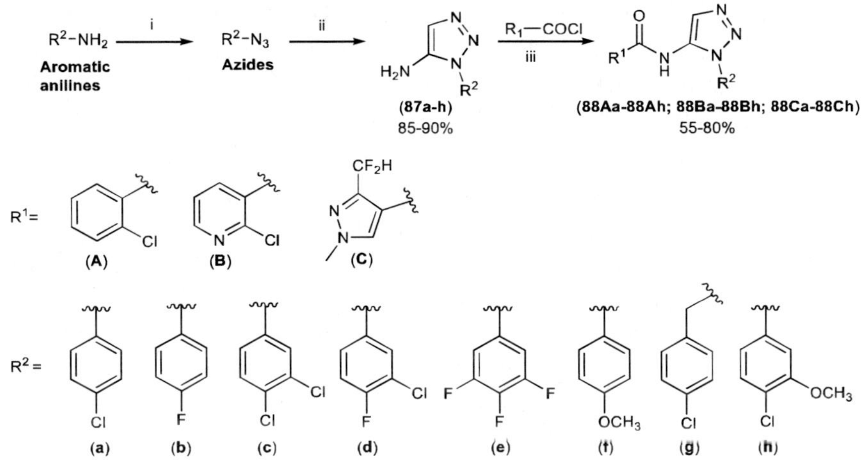

Figure 23. Synthetic steps involved in the preparation of carboxamides (**88Aa-88Ah, 88Ba-88Bh, and 88Ca-88Ch**). Reagents and conditions: (i) $NaNO_2$, HCl (3 mol L^{-1}), 0 °C; NaN_3, 2-3 h; (ii) acetonitrile, BuLi, THF, -78 °C; (iii) THF, NaH, acyl chloride, r.t.

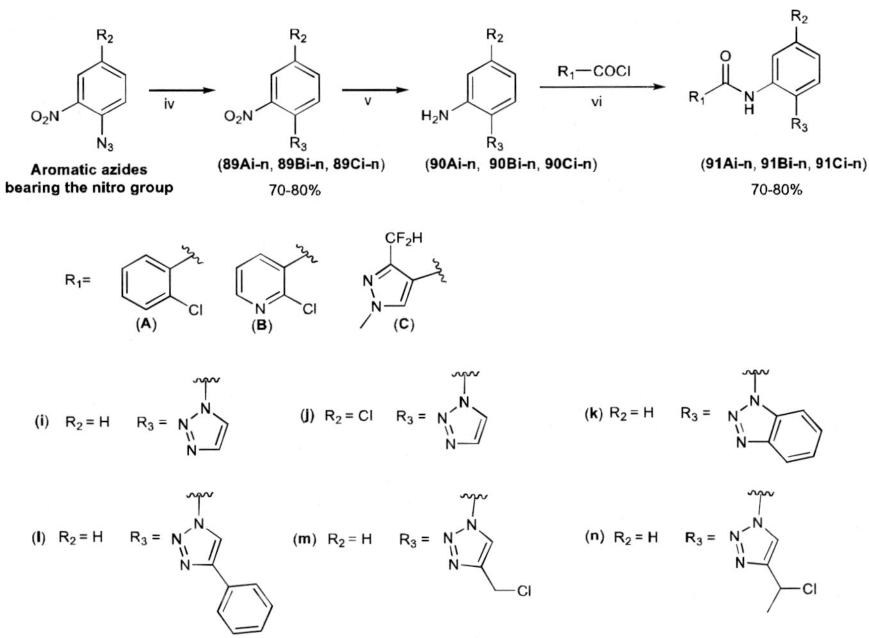

Figure 24. Synthetic steps involved in the preparation of carboxamides (**91Ai-n, 91Bi-n,** and **91Ci-n**). iv) different methodologies; (v) NH_4Cl, Fe, CH_3OH/H_2O (3:1 v v^{-1}); (vi) acyl chlorides, triethyl amine, CH_2Cl_2, r.t.

In the second synthetic approach (Figure 24), aromatic azides bearing the nitro group were converted into triazolic derivatives (**89Ai-n**, **89Bi-n,** and **89Ci-n**) by means of the CuAAC reaction. Reduction of nitro groups present in the aforementioned compounds provided the compounds (**90Ai-n**, **90Bi-n**, and **90Ci-n**), followed by amidation reactions with different carboxylic acids resulted in carboxamides (**91Ai-n**, **91Bi-n**, and **91Ci-n**) (Yan et al., 2019).

The obtained carboxamides were screened *in vitro* via mycelial growth inhibition assay against *S. sclerotiorum, B. cinerea, Fusarium solani, Rhizoctonia cerealis, Gaeumannomyces graminsis, Alternaria solani, Colletotrichum orbiculare, Fusarium moniliforme,* and *P. capsi* at 50 μg mL^{-1}. This initial screening selected compound (**88Ac**) as the most active carboxamide, which displayed relevant fungicidal activity with EC_{50} values of 1.08 (*S. scleoreotiorum*), 8.75 (*B. cinerea*), 1.67 (*R. cerealis*), and 5.30 (*G. graminsis*) μg mL^{-1}. Boscalid, an antifungal commercial carboxamide, was used as a positive control and the IC_{50} against *S. scleoreotiorum*, *B. cinerea*, *R. cerealis*, and *G. graminsis* were, respectively, 0.21, 1.04, 0.20, and 22.03 μg mL^{-1}. At 200 μg mL^{-1}, carboxamide (**88Ac**) was effective in suppressing rape sclerotinia rot, cucumber grey mold, and wheat powdery mildew, that are caused, respectively, by *S. scleoreotiorum*, *B. cinerea*, and *B. gramnis*. Besides, it was demonstrated the efficiency of (**89Ac**) against the succinate dehydrogenase (SDH, IC_{50} = 41.9 μmol L^{-1}), which is the target of fungicides belonging to the carboxamide class. The authors also reported a docking study that revealed the interaction of (**88Ac**) with ARG43, TYR58, and TRP173 aminoacid residues of SDH through hydrogen bond and pi-pi interaction, which could explain the probable mechanism of action between the inhibitor and the target enzyme (Yan et al., 2019).

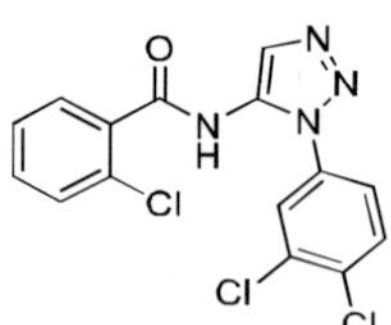

(88Ac)

Figure 25. Structure of carboxamide (**88Ac**).

Benzamidines displaying 1,2,3-triazole moieties were prepared according to the steps outilined in Figures 26-28 (Chen et al., 2014). Conversion of 4-bromomethyl benzonitrile (**92**) in triazol (**94**) was accomplished after reaction of (**92**) with sodium azide followed by CuAAC reaction of (**93**) with propargyl alcohol. Then, the Pinner reaction of (**94**) followed by treatment of Pinner salt with different amines produced the benzamidines (**95a-e**) (Figure 25).

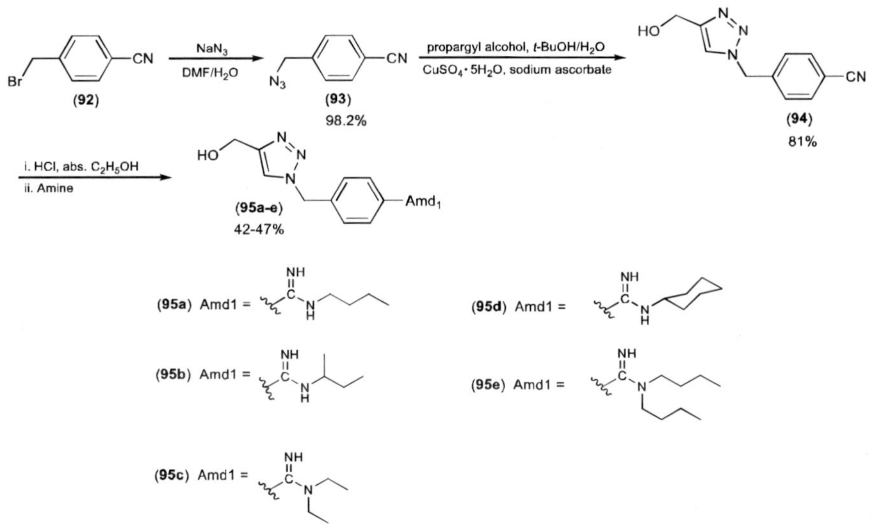

Figure 26. Preparation of benzamidines (**95a-e**).

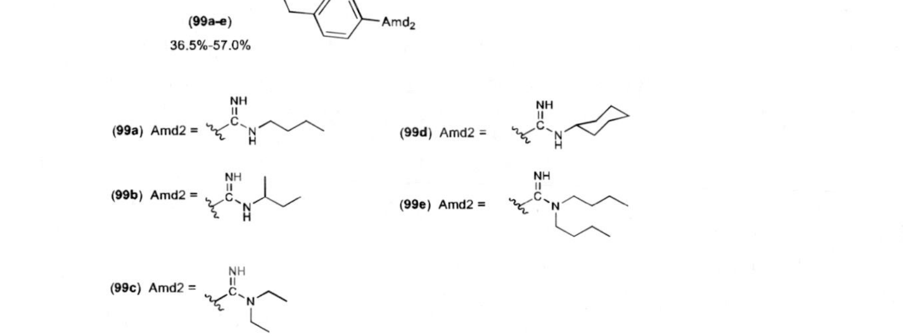

Figure 27. Synthesis of benzamidines (**99a-e**) from phenol (**96**).

Similar procedures converted phenol (**96**) to benzamidines (**99a-e**) (Figure 27), and phenol (**100**) and nitrile (**102**) in benzamidines (**105a-d**), and (**106a-d**) (Figure 28).

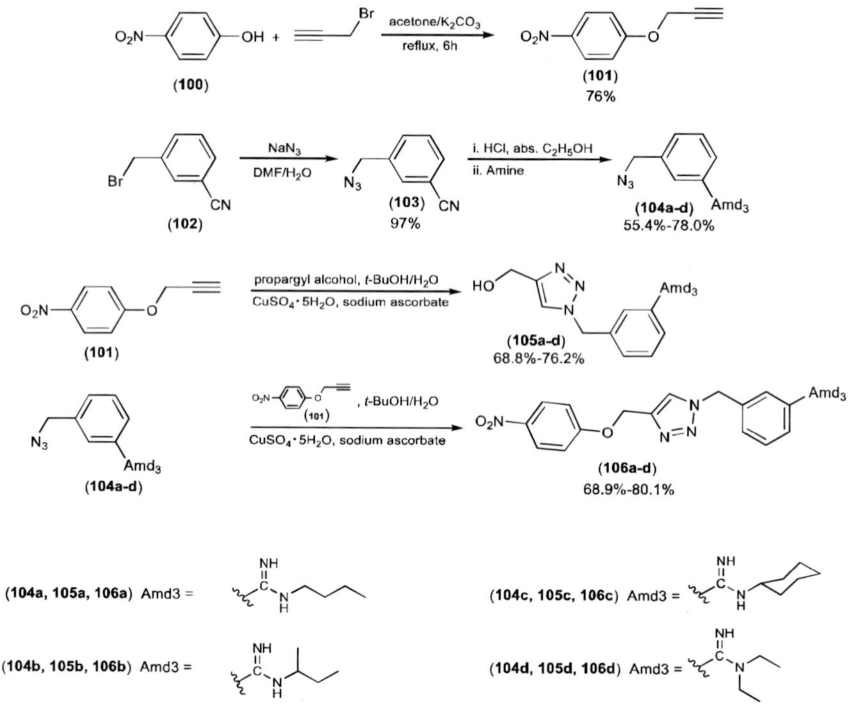

Figure 28. Synthesis of benzamidines (**105a-d**) and (**106a-d**).

When assessed *in vitro* on *Colletotrichum lagenarium* and *B. cinerea*, via the mycelial growth inhibition assay, the synthesized benzamidines presented weak antifungal activity (8-35% on *C. lagenarium* and 0-44% on *B. cinerea* at 200 μg mL^{-1}). On the contrary, when evaluated *in vivo* at 300 μg mL^{-1}, the antifungal activity of benzamidines was, in general, higher compared to the *in vitro* results. The compound (**99b**) against *C. lagenarium* (79% inhibition) and (**106b**) on *B. cinerea* (61%) were the most effective *in vivo* (Chen et al., 2014).

Indole (**107**) corresponded to the starting material used to obtain the triazolic derivatives (**111, 112, 115a-k**). The propargylation of indole (**107**) with propargyl bromide, under the specified conditions shown in Figure 29, afforded the key terminal alkyne (**108**). The CuAAC reactions conducted with $CuCl_2$/Zn as the catalytic system, involving the alkyne (**108**) and azides (**110**,

1-(azidomethyl)-4-nitrobenzene, and **114a-k**) resulted in triazolic indole derivatives within 80–93% yield (Xu et al., 2016).

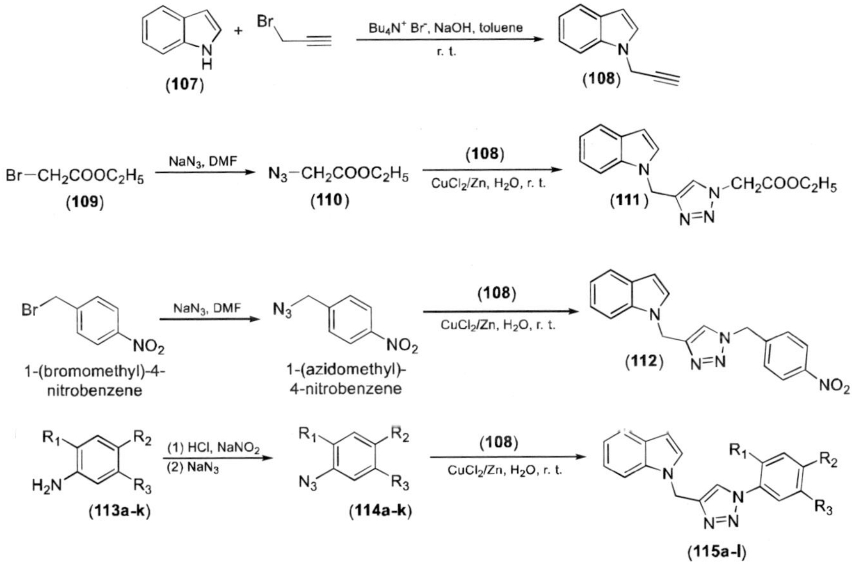

(**113a**, **114a**, **115a**): $R^1 = R^2 = R^3 = H$
(**113b**, **114b**, **115b**): $R^1 = NO_2$, $R^2 = H$, $R^3 = H$
(**113c**, **114c**, **115c**): $R^1 = H$, $R^2 = H$, $R^3 = NO_2$
(**113d**, **114d**, **115d**): $R^1 = H$, $R^2 = NO_2$, $R^3 = H$
(**113e**, **114e**, **115e**): $R^1 = CH_3$, $R^2 = H$, $R^3 = NO_2$
(**113f**, **114f**, **115f**): $R^1 = CH_3$, $R^2 = H$, $R^3 = H$
(**113g**, **114g**, **115g**): $R^1 = H$, $R^2 = CH_3$, $R^3 = H$
(**113h**, **114h**, **115h**): $R^1 = H$, $R^2 = OH$, $R^3 = H$
(**113i**, **114i**, **115i**): $R^1 = CF_3$, $R^2 = H$, $R^3 = H$
(**113j**, **114j**, **115j**): $R^1 = F$, $R^2 = H$, $R^3 = H$
(**113k**, **114k**, **115k**): $R^1 = Cl$, $R^2 = H$, $R^3 = H$
(**113l**, **114l**, **115l**): $R^1 = H$, $R^2 = H$, $R^3 = Cl$

Figure 29. Preparation of indole derivatives with 1,2,3-triazole fragments (**111**, **112**, and **115a-l**).

When evaluated on *Colletotrichum capsica* at 20 μg mL^{-1} via the mycelial growth assay, the indole derivatives (**112**), (**115a**), and (**115d**) were inactive and the most active derivative (83% inhibition) corresponded to compound (**111**). The indole derivatives were also assessed, at the same concentration, against cotton *Physalospora*. In this case, while compounds (**115c**) and (**115h**) did not inhibit the mycelial growth, the derivative (**115k**) was the most effective (65% inhibition) (Xu et al., 2016). None of the compounds presented superior activity than the commercial fungicides hexaconazole and flutriafol used as positive controls.

Huo and co-workers (2018) developed a methodology to synthesize *β*-carboniles bearing aryl-1,2,3-triazoles fragments. After reaction optimization, the investigators found that the combination of the propargylated *β*-carbonile (**116**) with sodium azide and boronic acids afford compounds (**117a-k**) (Figure 30).

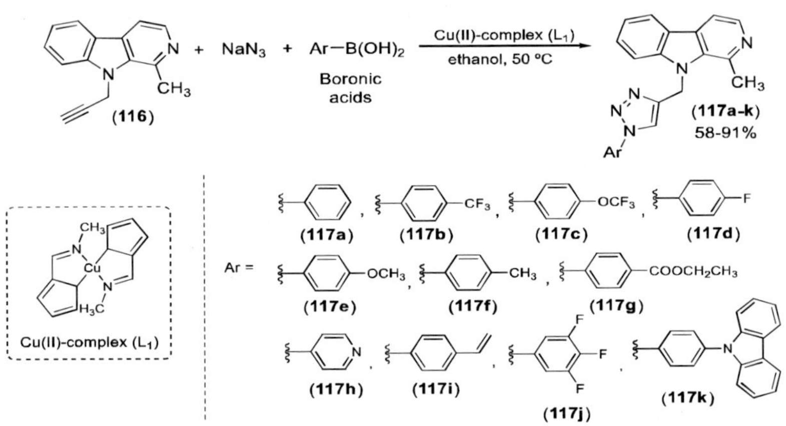

Figure 30. *β*-carbonile derivatives with 1,2,3-triazole portions (**117a-k**).

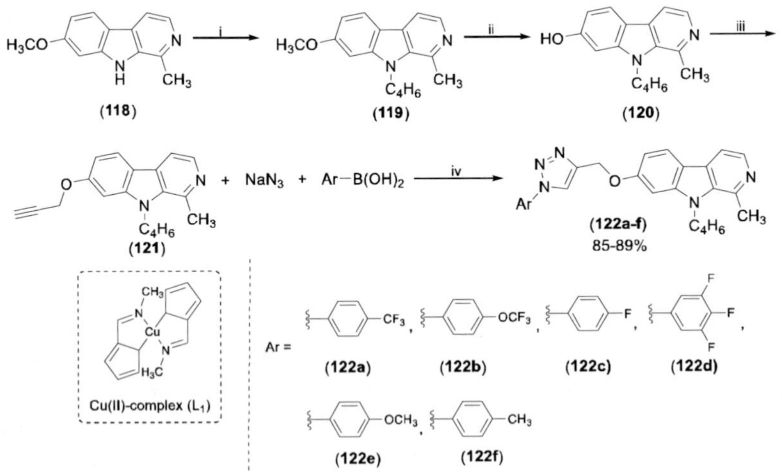

Figure 31. Harmine derivatives with 1,2,3-triazole fragments (**122a-f**). Reagents and conditions: (i) DMF, NaH, iodobutane, stirred at r.t.; (ii) HBr, HOAc, reflux (iii) DMF, NaH, 3-bromopropyne, stirred at r.t.; (iv) Cu(II)-complex (L_1), ethanol, 50 °C.

It is important to mention that in these methodologies the catalytic species is the Cu(II)-complex (L_I) (Figure 30) (Huo et al., 2018). The authors also demonstrated the possibility of preparing triazolic β-carboniles (**122a-f**) from the natural alkaloid harmine (**118**) as shown in Figure 31.

The evaluation of the fungicide activity on *R. solani*, *F. oxysporum*, *B. cinerea*, sunflower sclerotinia rot, and rape sclerotinia rot revealed that the compounds inhibited the mycelial growth rate with different degrees of efficiency. It stood out the inhibition growth effects of the compounds (**117a**) (85%), (**117b**) (87%), (**117c**) (86%), and (**122b**) (85%) which were comparable to carbendazim (~ 90%) used as positive control (Huo et al., 2018).

The amides (**126a-p**) and esters (**127a-e**) were obtained from the β-keto ester (**123**) according to the steps shown in Figure 32. The CuAAC methodology was not utilized to access the amides and esters. In the first step, organic azides were condensed with (**123**) to provide the 1,2,3-triazoles (**124a-e**). The basic ester hydrolysis of (**124a-e**) gave rise to carboxylic acids (**125a-e**). The acids were converted to the corresponding acyl chlorides by treatment of the acids with thionyl chloride. The reactions of acyl chlorides with amines or phenols produced the amides (**126a-p**) and esters (**127a-e**) (Wang et al., 2014).

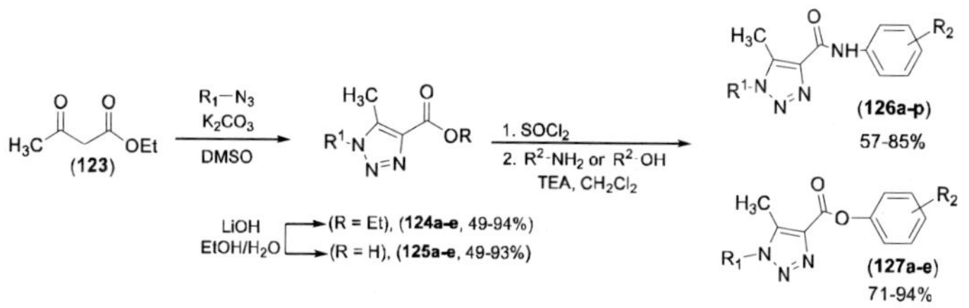

R^1 = $PhCH_2$ (**124a, 125a**)
R^1 = $(CH_2)_2CH_3$ (**124b, 125b**)
R^1 = 4-Cl phenyl (**124c, 125c**)
R^1 = CH_2CH_3 (**124d, 125d**)
R^1 = 2-Cl phenyl (**124e, 125e**)

(**126a**): R^1 = $(CH_2)_2CH_3$; R^2 = 2-Cl; (**126b**): R^1 = 4-Cl phenyl; R^2 = 2-CH_3, 6-CH_3; (**126c**): R^1 = $(CH_2)_2CH_3$; R^2 = 3-Cl, 4-CH_3;
(126d): R^1 = 4-Cl phenyl; R^2 = 4-Cl; (**126e**): R^1 = CH_2CH_3; R^2 = 4-Cl; (**126f**): R^1 = CH_2CH_3; R^2 = 2-NO_2;
(**126g**): R^1 = CH_2CH_3; R^2 = 3-Cl; (**126h**): R^1 = 4-Cl phenyl; R^2 = 2-CH_3; (**126i**): R^1 = CH_2CH_3; R^2 = 2-CH_3;
(**126j**): R^1 = $PhCH_2$; R^2 = 2-Cl; (**126k**): R^1 = $PhCH_2$; R^2 = 2-CH_3, 6-CH_3; (**126l**): R^1 = 2-Cl phenyl; R^2 = 2-CH_3, 6-CH_3;
(**126m**): R^1 = CH_2CH_3; R^2 = 2-Cl, 4-Cl; (**126n**): R^1 = CH_2CH_3; R^2 = 2-Br; (**126o**): R^1 = CH_2CH_3; R^2 = 2-Cl, 6-Cl;
(**126p**): R^1 = CH_2CH_3; R^2 = 2-Cl; (**127a**): R^1 = CH_2CH_3; R^2 = 2-Cl; (**127b**): R^1 = 2-Cl phenyl; R^2 = 2-Cl;
(**127c**): R^1 = 2-Cl phenyl; R_2 = 2-NO_2; (**127d**): R^1 = 2-Cl phenyl; R^2 = 2-CH_3, 6-CH_3; (**127e**): R^1 = $PhCH_2$; R^2 = 2-CH_3, 6-CH_3;

Figure 32. 1,2,3-triazolic amides (**126a-p**) and esters (**127a-e**).

In terms of fungicide activity, these compounds were evaluated against *Physalospora piricola*, *A. solani*, *Cercospora arachidicola*, *Gibberella zeae*, *P. capsici*, *Phythophthora infestans*, *B. cinerea*, *S. sclerotiorum*, and *Thanatephorus cucumeris* using the mycelial growth inhibition assay and the best performance was noticed on *S. sclerotiorum*, being the results outilined in Table 5.

Table 5. Best results of the fungicide activity evaluation of amides (126a-p) and esters (127a-e) on *S. sclerotiorum*

Compound	**EC_{50} (μg mL^{-1})**
126k	23.3
126l	1.3
127a	15.4
127d	8.9
127e	7.8
Chlorothalonil	7.0

It deserves comment the fact that the compound (**126l**) was approximately five times more potent than chlorothalonil (positive control) and compound (**127e**) was equipotent to it (Wang et al., 2014).

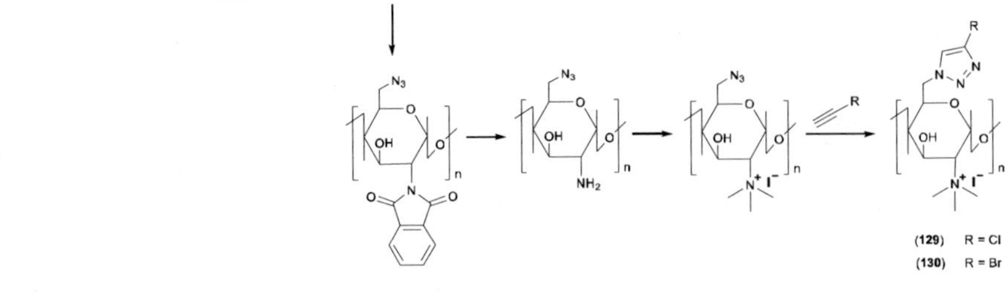

Figure 33. Preparation of water-soluble chitosan derivatives (**128-130**). Reaction conditions were not described by the authors of the investigation.

Three water soluble 1,2,3-triazolic chitosan derivatives (**128-130**) were synthesized according to the steps shown in Figure 33 and the CuAAC reaction corresponded to the key step to afford the derivatives (Li et al., 2016).

Table 6 depicts the results of the fungicide evaluation activity against three fungi species.

Table 6. Results of the fungicide activity evaluation of compounds 128-130

Compound	% of inhibitory growth activity		
	Fusarium oxysporumf.sp. niveum	*Colletotrichum lagenarium*	*Fusarium oxysporum.sp.cucume-brium Owen*
Chitosan	24.2	30	14.7
128	58.5	85.1	86.1
129	84.0	93.2	91.7
130	81.8	88.1	91.7

Compared to chitosan, the fungicide activity of the triazolic derivatives was significantly higher, which was ascribed to the introduction of the 1,2,3-triazole group in the framework of chitosan.

Conclusion

In this book chapter, we gathered information concerning the application of 1,2,3-triazoles in the agrochemistry field. Results regarding phytoxic, insecticide, and fungicide activities of this class of heterocycles were described. In general, the reports utilized the CuAAC reaction for the introduction of 1,2,3-triazole ring into several compounds' framework. This is certainly because of the facility in carrying out this transformation as well as the good yields obtained with the use of the CuAAC reactions. The results of the performance of these compounds 1,2,3-triazole fragments in terms of phytotoxicity, insecticide, and fungicide activities are encouraging and it is possible to predict that new studies will be published in the future expanding the agrochemistry applications of 1,2,3-triazoles.

Disclaimer

The authors declare no conflict of interest.

Acknowledgments

The authors are grateful to Fundação de Amparo à Pesquisa do Estado de Minas Gerais (FAPEMIG), Fundação de Amparo à Pesquisa e Inovação do Espírito Santo (FAPES), Coordenação de Aperfeiçoamento de Pessoal de Níel Superior (CAPES), and Conselho Nacional de Desenvolvimento Científico e Tecnológico (CNP_q) for financial support.

References

Agalave SG, Maujan SR, Pore VS. Click chemistry: 1,2,3-Triazoles as pharmacophores. *Chemistry an Asian Journal* (2011) 6:2696-2718.

Alam, MM. 1,2,3-Triazole hybrids as anticancer agents: A review. *Arch der Pharmazie* (2022) 355:e2100158.

Ali, AA. *1,2,3-Triazoles: Synthesis and biological application*. In: Kuznetsov A (ed). Azoles – Synthesis, properties, applications and perspectives, Chapter 4, 18 pages, London, UK: IntechOpen, 2021.

Aswar NA. Synthesis of triazoles: an overview. *International Journal of Current Research* (2014) 6:5248-5251.

Barcelos FF, da Silva LF, Gazolla PAR, Teixeira RR, de Queiroz VT, Morais PAB, Fonseca VR, Romão W, Lacerda Júnior V, Scherer R, de Oliveira FM, Fontes MMP, Costa AV. Synthesis of glicerol-fluorinated triazole derivatives and evaluation of their fungicidal activity. *Química Nova* (2022) 45:788-796.

Boddy IK, Briggs GG, Harrison RP, Jones TH, O'Mahony MJ, Marlow ID, Roberts BG, Willis RJ, Bardsley R, Reid J. The synthesis and insecticidal activity of a series of 2-aryl- 1,2,3-triazoles. *Pesticide Science* (1996) 48:189-196.

Borgati TF, Alves RB, Teixeira RR, de Freitas RP, Perdigão TG, da Silva SF, dos Santos AA, Bastidas AJO. Synthesis and phytotoxic activity of 1,2,3-triazole derivatives. *Journal of the Brazilian Chemical Society* (2013) 24:953-961.

Bozorov K, Zhaoa J, Aisa HA. 1,2,3-Triazole-containing hybrids as leads in medicinal chemistry: A recent overview. *Bioorganic & Medicinal Chemistry* (2019) 27:3511-3531.

Breugst M, Reissig HU. The Huisgen reaction: Milestones of the 1,3-dipolar cycloaddition. *Angewandte Chemie Int*ernational *Editions* (2020) 59:12293-12307.

Chen G, Zhou Y, Cai C, Lu J, Zhang X. Synthesis and antifungal activity of benzamidine derivatives carrying 1,2,3-triazole moieties. *Molecules* (2014) 19:5674-5691.

Chen Y, Yao K, Wang K, Xiao C, Li K, Khan B, Zhao S, Yan W, Yonghao Y. Bioactive-guided structural optimization of 1,2,3-triazole phenylhydrazones as potential fungicides against *Fusarium graminearum*. *Pesticide Biochemistry and Physiology* (2020) 164:26-32.

Chen Z, Jiang Y, Xu C, Sun X, Ma C, Xia Z, Zhao H. Oleanane-type triterpene conjugates with 1*H*-1,2,3-triazole possessing of fungicidal activity. *Molecules* (2022) 27:4928.

Chio ED, Li QX. Pesticide research and development: General discussion and Spinosad case. *Journal of Agriculture and Food Chemistry* (2022) 70:8913-8919.

Costa AV, Moreira LC, Pinto RT, Alves TA, Schwan VV, de Queiroz VT, Fontes MMP, Teixeira RR, Morais PAB, Jesus Júnior WC. Synthesis of glycerol-derived 4-alkyl-substituted 1,2,3-triazoles and evaluation of their fungicidal, phytotoxic, and antiproliferative activities. *Journal of the Brazilian Chemical Society* (2020) 31:821-832.

Costa AV, de Oliveira MVL, Pinto RT, Moreira LC, Gomes EMC, Alves TA, Pinheiro PF, de Queiroz VT, Vieira LFA, Teixeira RR, Jesus Júnior WC. Synthesis of novel glycerol-derived 1,2,3-triazoles and evaluation of their Fungicide, phytotoxic and cytotoxic activities. *Molecules* (2017) 22: 1666.

Dai J, Tian S, Yang X, Liu Z. Synthesis methods of 1,2,3-/1,2,4-triazoles: A review. *Frontiers in Chemistry* (2022) 10:891484.

Daí ZC, Chen YF, Zhang M, Li SK, Yang TT, Shen L, Wang JX, Qian SS, Zhu HL, Ye YH. Synthesis and Antifungal Activity of 1,2,3-Triazole Phenylhydrazone Derivatives. *Organic and Biomolecular Chemistry* (2015) 13:477-486.

Dheer D, Singh V, Shankar R. Medicinal attributes of 1,2,3-triazoles: Current developments. *Bioorganic Chemistry* (2017) 71:30-54.

Forezi LSM, Lima CGS, Amaral AAP, Ferreira PG, de Souza MCBV, Cunha AC, da Silva FC, Ferreira VF. Bioactive 1,2,3-triazoles: An account on their synthesis, structural diversity and biological applications. *The Chemical Record* (2021) 21:2782-2807.

Franco CA, da Silva TI, Dias MG, Ferreira BW, de Sousa BL, Bousada GB, Barreto RW, Vaz BG, Lima GS, dos Santos MH, Grossi JAS, Varejão EVV. Synthesis of Tyrosol 1,2,3-Triazole derivatives and their phytotoxic activity against *Euphorbia heterophylla*. *Journal of Agriculture and Food Chem*istry (2022) 70:2806−2816.

Freitas LBO, Ruela FA, Pereira GR, Alves RB, de Freitas RP. A reação “click” na síntese de 1,2,3-triazóis: Aspectos químicos e aplicações. *Química Nova* (2011) 34:1791-1804.

Fungicide Resistance Action Committee (FRAC). Avaialable at https://www.frac.info/fungicide-resistance-management/by-frac-mode-ofaction-group. Accessed 7 January 2023).

Haldón E, Nicasio MC, Pérez PJ. Copper-catalysed azide-alkyne cycloadditions (CuAAC): an update. *Organic and Biomolecular Chemistry* (2015) 13:9528-9550.

Heravi MM, Tamimi M, Yahyavi H, Hosseinnejad T. Huisgen’s cycloaddition reactions: A full perspective. *Current Organic Chemistry* (2015) 20:1591-1647.

Herbicide Resistance Action Committee (HRAC). Avaialable at https://hracglobal.com/tools/hrac-mode-of-action-classification-2022-map. Accessed 07 January 2023.

Huo XY, Guo L, Chen XF, Zhou YT, Zhang J, Han XQ, Dai B. Design, synthesis, and antifungal activity of novel aryl-1,2,3-triazole-β-carboline hybrids. *Molecules* (2018) 23:1344.

Insecticide Resistance Action Committee (IRAC). Avaialable https://irac-online.org/mode-of-action. Accessed 07 January 2023).

Jiang X, Hao X, Jing L, Wu G, Kang D, Liu X, Zhan P. Recent applications of click chemistry in drug discovery. *Expert Opinion on Drug Discovery* (2019) 14:779-789.

Kolb HC, Finn MG, Sharpless KB. Click chemistry: Diverse chemical function from a few good reactions. *Angewandtie Chemie International Edition* (2001) 40:2004-2021.

Krämer W, Schirmer U (eds). *Modern Crop Protection Compounds,* Volume 1, Weinheim, Germany: WILEY-VCH Verlag, 2007a.

Krämer W, Schirmer U (eds). *Modern Crop Protection Compounds,* Volume 2, Weinheim, Germany: WILEY-VCH Verlag, 2007b.

Krämer W, Schirmer U (eds). *Modern Crop Protection Compounds,* Volume 3, Weinheim, Germany: WILEY-VCH Verlag, 2007c.

Kushwaha D, Dwivedi P, Kuanar SK, Tiwari VK. Click Reaction in carbohydrate chemistry: Recent developments and future perspective. *Current Organic Synthesis* (2013) 10:90-135.

Li Q, Tan W, Zhang C, Gud G, Guo Z. Synthesis of water soluble chitosan derivatives with halogeno-1,2,3-triazole and their antifungal activity. *International Journal of Biological Macromolecules* (2016) 91:623-629.

Liang L, Astruc D. The copper(I)-catalyzed alkyne-azide cycloaddition (CuAAC) "click" reaction and its applications. An overview. *Coordination Chemistry Reviews* (2011) 255:2933-2945.

Lima AMA, de Paula WT, Leite ICHL, Gazolla PAR, de Abreu LM, Fonseca VR, Romão W, Lacerda Júnior V, de Queiroz VT, Teixeira RR, Costa AV. Synthesis of eugenol-fluorinated triazole derivatives and evaluation of their fungicidal activity. *Journal of the Brazilizam Chemical Society* (2022) 33:1200-1210.

Lima LM, Barreiro EJ. Bioisosterism: A useful strategy for molecular modification and drug design. *Current Medicinal Chemistr*y (2005) 12:23-49.

Mao M, Li Y, Liu Q, Xiong L, Zhang X, Li Z. Synthesis and biological evaluation of novel *N*-pyridylpyrazole derivatives containing 1,2,3-triazole moieties. *Journal of Pesticide Science* (2015) 40:138-142.

Meldal M, Tornøe CW. Cu-catalyzed azide-alkyne cycloaddition. *Chemical Reviews* (2008) 108:2952-3015.

Moses JE, Moorhouse AD. The growing applications of click chemistry. *Chemical Society Rev*iew (2007) 36:1249-1262.

Nandivada H, Jiang X, Lahann J. Click chemistry: Versatility and control in the hands of materials scientists. *Advanced Materials* (2007) 19:2197-2208.

Nejmaa AB, Znati M, Daich A, Othman M, Lawson AM, Jannet HB. Design and semisynthesis of new herbicide as 1,2,3-triazole derivatives of the natural maslinic acid. *Steroids* (2018) 138:102-107.

Phillips MWA, Agrochemical industry development, trends in R&D and the impact of regulation. *Pest Management Science* (2020) 76:3348-3356.

Razzaghi-Asl N, Sepehri S, Ebadi A, Karami P, Nejatkhah N, Johari-Ahar M. Insights into the current status of privileged *N*-heterocycles as antileishmanial agents. *Molecular Diversity* (2020) 24:525-569.

Rostovtsev VV, Green LG, Fokin VV, Sharpless KB. A stepwise Huisgen cycloaddition process: Copper(I)-catalyzed regioselective "ligation" of azides and terminal alkynes. *Angewandtie Chemie International Edition* (2002), 41:2596-2599.

Silva FC, Cardoso MFC, Ferreira PG, Ferreira VF. *Biological Properties of 1H-1,2,3- and 2H-1,2,3-Triazoles*. In: Dehaen W, Bakulev V (eds). Chemistry of 1,2,3-triazoles.

Topics in heterocyclic chemistry, Volume 40, 117-165, Berlin, Germany: Springer-Verlag 2014.

Singh MS, Chowdhury S, Koley S. Advances of azide-alkyne cycloaddition-click chemistry over the recent decade. *Tetrahedron* (2016) 72:5257-5283.

Sparks TC, Hunter JE, Lorsbach BT, Hanger G, Gast, RE, Kemmitt G., Bryant RJ. Crop protection discovery: Is being the first best? *Journal of Agriculture and Food Chemistry* (2018) 66:10337−10346.

Sparks TC, Lorsbach B. Perspectives on the agrochemical industry and agrochemical discovery. *Pest Management Science (*2017) 73:672-677.

Su NN, Li Y, Yu SJ, Zhang X, Liu XH, Zhao WG. Microwave-assisted synthesis of some novel 1,2,3-triazoles by click chemistry, and their biological activity. *Research on Chemical Intermediates* (2013) 39:759-766.

Sykam K, Donempudi S, Basak P. 1,2,3-Triazole rich polymers for flame retardant application: A review. *Journal of Applied Polymer Science* (2022) 139:e52771.

Tornøe CW, Christensen C, Meldal M. Peptidotriazoles on solid phase: [1,2,3]-Triazoles by regiospecific copper(I)-catalyzed 1,3-dipolar cycloadditions of terminal alkynes to azides. *Journal of Organic Chemistry* (2002) 67:3057-3064.

Tron GC, Pirali T, Billington RA, Canonico PL, Sorba G, Genazzani AA. Click chemistry reactions in medicinal chemistry: Applications of the 1,3-dipolar cycloaddition between azides and alkynes. *Medicinal Research Reviews* (2008) 28:278-308.

van Dijk M, Morley T, Rau ML, Saghai, Y. A meta-analysis of projected global food demand and population at risk of hungerfor the period 2010-2050. *Nature Food* (2021) 2:494-501.

Varejão EVV, Demuner AJ, Barbosa LCA, Barreto RW. Phytotoxic effects of metabolites from *Alternaria euphorbiicola* against its host plant *Euphorbia heterophylla*. *Química Nova,* (2013) 36:1004-1007.

United Nations, 2022a. Avaialalbe at https://www.un.org/en/global-issues/population#:~:text=*Our growing population&text=The world's population is expected, billion in the mid-2080s*. Acessed 01 december 2022.

United Nations, 2022b. Avaialable at https://press.un.org/en/2009/gaef3242.doc.htm. Acessed 01 december 2022.

Xi W, Scott TF, Kloxin CJ, Bowman CN. Click chemistry in materials science. *Advance Functional Materials* (2014) 24:2572-2590.

Xu G, Zhao J, Jiang Y, Zhang P, Li W. Design, synthesis and antifungal activity of novel indole derivatives linked with the 1,2,3-triazole moiety *via* the CuAAC click reaction. *Journal of Chemical Research* (2016) 40:269-272.

Xu Z, Zhao SJ, Liu Y. 1,2,3-Triazole-containing hybrids as potential anticancer agents: Current developments, action mechanisms and structure-activity relationships. *European Journal of Medicinal Chemistry* (2019) 183:111700.

Yan W, Wang X, Li K, Li TX, Wang JJ, Yao KC, Cao LL, Zhao SS, Ye YH. Design, synthesis, and antifungal activity of carboxamide derivatives possessing 1,2,3-triazole as potential succinate dehydrogenase inhibitors. *Pesticide Biochemistry and Physiology* (2019) 156:160-169.

Wang ZJ, Gao Y, Hou YL, Zhang C, Yu SJ, Bian Q, Li ZM, Zhao WG. Design, synthesis, and fungicidal evaluation of a series of novel 5-methyl-1*H*-1,2,3-trizole-4-carboxyl amide and ester analogues. *European Journal of Medicinal Chemistry* (2014) 86:87-94.

Wermuth CG, Ganellin CR, Lindberg P, Mitscher LA. Glossary of terms used in medicinal chemistry (IUPAC Recommendations 1998). *Pure and Applied Chemistry*, (1998) 70:1129-1143.

Zeifman L, Hertog S, Kantorova V, Wilmonth J. A world of 8 billion. *UN DESA* (2022) 140:1-10. https://www.un.org/development/desa/dpad/publication/un-desa-policy-brief-no-140-a-world-of-8 billion/. Acessed 01 December 2022.

Zhang B. Comprehensive review on the anti-bacterial activity of 1,2,3-triazole hybrids. *European Journal of Medicinal Chemistry* (2019) 168:357-372.

Index

#

1,2,3-triazole scaffold, 5, 214
5a, 53
5b, 53
5c, 53
5d, 53
6a, 53
6b, 53
6c, 53
6d, 53

A

acylation reactions, 136
adsorption, 171, 174, 175, 178, 179, 180, 181, 182, 183, 186, 189, 191, 192, 193, 194, 195, 196, 197, 200, 201, 203
agrochemicals, x, xi, 84, 170, 206, 207
Alkyl Triazole Glycosides (ATGs), 157, 158, 159, 160, 161, 162, 168
alkylation reactions, 111, 112, 115
alkynes, vii, 1, 2, 4, 5, 8, 9, 11, 12, 13, 14, 15, 16, 17, 19, 20, 21, 22, 26, 27, 28, 29, 34, 35, 36, 43, 44, 56, 149, 208, 211, 213, 214, 225, 228, 240, 241
amination reactions, 134, 136
antimicrobial activity, 48, 52, 53, 57, 60, 61
antitubercular activity, viii, 47, 48, 49, 50, 51, 52
arylation reactions, 129, 134
azide, vii, 2, 4, 5, 6, 8, 9, 14, 18, 20, 22, 23, 26, 27, 29, 31, 32, 33, 34, 35, 39, 40, 41, 42, 43, 44, 45, 46, 145, 157, 161, 163, 165, 209, 210, 211, 212, 213, 214, 225, 227, 230, 234, 239, 240, 241

B

Bizerte Lagoon, v, viii, ix, 63, 64, 65, 66, 68, 72, 74, 76, 77, 78, 79, 80, 81, 83, 84, 85, 86, 87, 90, 92, 94, 95, 96, 97, 98, 99, 100

C

cascade reaction, v, 1, 2, 4, 6, 8, 29, 30, 31, 40, 42, 43, 45
click chemistry, ix, 41, 43, 44, 48, 49, 50, 53, 56, 144, 145, 153, 168, 206, 209, 239, 240, 241
conjugation of surfactant assemblies, ix, 153, 154
copper catalyzed azide alkyne cycloaddition (CuAAC), 4, 44, 164, 165, 206, 209, 210, 211, 212, 213, 214, 215, 217, 220, 222, 225, 227, 230, 232, 235, 236, 237, 239, 240, 241
corrosion inhibitors, vi, x, 169, 170, 171, 172, 173, 174, 175, 178, 179, 181, 182, 185, 188, 189, 191, 197, 198, 199, 200, 201, 202, 203
critical assembly concentration (CAC), 164
CuAAC reaction, 4, 206, 209, 210, 211, 212, 213, 214, 215, 217, 220, 222, 225, 227, 230, 232, 236, 237
cycloaddition, vii, 1, 2, 4, 6, 8, 9, 16, 18, 19, 23, 25, 26, 27, 29, 34, 35, 39, 41, 42, 43, 44, 45, 56, 145, 146, 208, 209, 213, 238, 239, 240, 241

D

distribution, v, viii, ix, 63, 65, 72, 78, 79, 81, 83, 84, 85, 90, 95, 100, 110, 154, 165, 181, 186

E

extraction, 67, 68, 69, 79, 80, 88, 89, 91, 98, 99, 100

F

fungicides, 64, 79, 80, 81, 84, 99, 100, 101, 206, 230, 234, 238

G

gas chromatography (GC), 69, 81, 89, 100, 241
gram-negative, 146

H

herbicides, 84, 206
heterocycles, viii, ix, 7, 43, 47, 48, 103, 104, 105, 127, 144, 147, 149, 150, 151, 152, 170, 173, 237, 240

I

inhibition, 30, 51, 61, 144, 146, 171, 174, 175, 177, 178, 179, 180, 181, 182, 183, 185, 188, 189, 190, 191, 192, 193, 194, 195, 196, 197, 198, 199, 200, 201, 202, 203, 214, 215, 217, 225, 228, 230, 232, 233, 235, 236
insecticides, 84, 206

K

Knoevenagel condensation, 41

L

limits of detection (LOD), 67, 68, 72, 87, 91
lowest unoccupied molecular orbital (LUMO), 41, 189, 190, 191

M

marine environment, viii, 64, 83, 84
Michael addition, 113, 114, 145, 149
molecular docking study, viii, 48, 55, 57, 60, 61
molecular dynamic (MD) method, 189
multicomponent reaction, vii, 2, 4, 22, 38, 43, 45, 46
Mycobacterium tuberculosis, viii, 47, 49
Mycobacterium tuberculosis H37Rv (MTB), viii, 47, 49

N

non-alkynes, vii, 2, 37
non-ionic surfactants, 155
non-regioselective *N*-alkylations, 126

O

oxo-triazole derivative (DTP), 188
oxylation, 140

P

packing theory, 154, 158, 159
pesticides, viii, x, 63, 64, 65, 68, 70, 72, 73, 74, 75, 76, 77, 78, 79, 80, 81, 84, 88, 90, 93, 94, 95, 97, 98, 99, 100, 205, 206, 207
phytosanitary products, v, 63, 83
prototropic tautomerism, ix, 103, 106, 107, 108, 109, 110, 134, 140, 147

Q

QuECHERS, 67, 69, 88, 91

R

regioselectivity, 5, 9, 22, 35, 41, 110, 111, 113, 114, 115, 116, 117, 119, 122, 123, 129, 130, 131, 132, 138
renewable resources, ix, 153, 154, 155, 167

S

sediment cores (SCs), v, viii, 63, 64, 65, 66, 68, 70, 72, 74, 75, 78, 79, 81, 100
sediments, v, ix, 65, 66, 69, 70, 72, 74, 78, 79, 81, 83, 84, 85, 88, 96, 100
silylation, 110, 140, 141, 149
structure-activity relationship (SAR), viii, 48, 241
sugar-based surfactants, ix, 153, 154, 155, 165, 166, 168
sulfonylation, 110, 140, 142, 143, 151, 152
surfactant(s), ix, 153, 154, 155, 156, 157, 158, 159, 160, 161, 162, 163, 164, 165, 166, 167, 168
synthesis, v, vii, viii, ix, 1, 2, 4, 5, 6, 7, 8, 9, 10, 11, 12, 13, 14, 15, 16, 20, 23, 24, 26, 27, 28, 29, 30, 31, 33, 34, 35, 36, 37, 38, 39, 40, 41, 42, 43, 44, 45, 46, 47, 49, 56, 57, 58, 60, 61, 62, 104, 113, 119, 120, 122, 128, 131, 132, 134, 135, 137, 141, 144, 145, 146, 148, 149, 150, 151, 152, 153, 157, 162, 163, 164, 167, 168, 196, 198, 199, 201, 202, 208, 212, 220, 224, 231, 232, 238, 239, 240, 241

T

tautomerism, v, ix, 103, 104, 106, 107, 108, 109, 110, 112, 126, 134, 147, 148
tebuconazole, 67, 71, 72, 73, 74, 76, 78, 79, 81, 88, 91, 92, 93, 95, 96, 98, 223, 224, 225
total organic carbon (TOC), ix, 66, 68, 70, 76, 77, 78, 79, 83, 87, 95, 96
triazole pesticides, ix, 64, 65, 68, 69, 70, 71, 72, 74, 78, 79, 81, 83, 84, 87, 88, 90, 92, 93, 94, 95, 96, 97, 98, 100

V

vesicular drug delivery, 162, 163, 166
vinylation, 110, 140, 141, 152

Y

ynamides, 6, 7, 30, 46